C0-ALO-954

Playback and Studies of Animal Communication

NATO ASI Series

Advanced Science Institutes Series

A series presenting the results of activities sponsored by the NATO Science Committee, which aims at the dissemination of advanced scientific and technological knowledge, with a view to strengthening links between scientific communities.

The series is published by an international board of publishers in conjunction with the NATO Scientific Affairs Division

A	**Life Sciences**	Plenum Publishing Corporation
B	**Physics**	New York and London
C	**Mathematical and Physical Sciences**	Kluwer Academic Publishers
D	**Behavioral and Social Sciences**	Dordrecht, Boston, and London
E	**Applied Sciences**	
F	**Computer and Systems Sciences**	Springer-Verlag
G	**Ecological Sciences**	Berlin, Heidelberg, New York, London,
H	**Cell Biology**	Paris, Tokyo, Hong Kong, and Barcelona
I	**Global Environmental Change**	

Recent Volumes in this Series

Volume 222—The Changing Visual System: Maturation and Aging in the Central Nervous System
edited by P. Bagnoli and W. Hodos

Volume 223—Mechanisms in Fibre Carcinogenesis
edited by Robert C. Brown, John A. Hoskins, and Neil F. Johnson

Volume 224—Drug Epidemiology and Post-Marketing Surveillance
edited by Brian L. Strom and Giampaolo Velo

Volume 225—Computational Aspects of the Study of Biological Macromolecules by Nuclear Magnetic Resonance Spectroscopy
edited by Jeffrey C. Hoch, Flemming M. Poulsen, and Christina Redfield

Volume 226—Regulation of Chloroplast Biogenesis
edited by Joan H. Argyroudi-Akoyunoglou

Volume 227—Angiogenesis in Health and Disease
edited by Michael E. Maragoudakis, Pietro Gullino, and Peter I. Lelkes

Volume 228—Playback and Studies of Animal Communication
edited by Peter K. McGregor

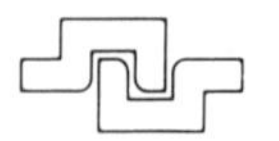

Series A: Life Sciences

Playback and Studies of Animal Communication

Edited by

Peter K. McGregor

The University of Nottingham
Nottingham, United Kingdom

Plenum Press
New York and London
Published in cooperation with NATO Scientific Affairs Division

Proceedings of a NATO Advanced Research Workshop on
Playback and Studies of Animal Communication: Problems and Prospects,
held August 5-9, 1991,
at Thornbridge Hall, near Chesterfield, United Kingdom

NATO-PCO-DATA BASE

The electronic index to the NATO ASI Series provides full bibliographical references (with key-words and/or abstracts) to more than 30,000 contributions from international scientists published in all sections of the NATO ASI Series. Access to the NATO-PCO-DATA BASE is possible in two ways:

—via online FILE 128 (NATO-PCO-DATA BASE) hosted by ESRIN, Via Galileo Galilei, I-00044 Frascati, Italy.

—via CD-ROM "NATO-PCO-DATA BASE" with user-friendly retrieval software in English, French, and German (© WTV GmbH and DATAWARE Technologies, Inc. 1989)

The CD-ROM can be ordered through any member of the Board of Publishers or through NATO-PCO, Overijse, Belgium.

Library of Congress Cataloging-in-Publication Data

Playback and studies of animal communication / edited by Peter K. McGregor.
p. cm. -- (NATO ASI series. Series A, Life sciences ; v. 228.)
"Proceedings of a NATO Advanced Research Workshop on Playback and Studies of Animal Communication: Problems and Prospects, held August 5-9, 1991, at Thornbridge Hall, near Chesterfield, United Kingdom"--T.p. verso.
"Published in cooperation with NATO Scientific Affairs Division."
Includes bibliographical references and index.
ISBN 0-306-44205-1
1. Animal communication--Congresses. 2. Animal behavior--Research--Methodology--Congresses. 3. Animal sounds--Recording and reproducing--Congresses. I. McGregor, Peter K. II. NATO Advanced Research Workshop on Playback and Studies of Animal Communication: Problems and Prospects (1991 : Chesterfield, England) III. North Atlantic Treaty Organization. Scientific Affairs Division. IV. Series.
QL776.P57 1992
591.59--dc20 92-11598
CIP

ISBN 0-306-44205-1

A Division of Plenum Publishing Corporation
233 Spring Street, New York, N.Y. 10013

Printed in the United States of America

Preface

Playback is the technique of rebroadcasting natural or synthetic signals to animals and observing their response. The ability to present a putative signal in isolation, without the potential confounding effects of other activities of the signaller, is the main reason for the depth and range of our knowledge of communication systems. To date, playback of sound signals has predominated, but playback of electric signals and even video playback of visual signals suggests that playback will become just as prevalent in studies of communication in other sensory modalities.

This book is one of the outcomes of a workshop on playback held at Thornbridge Hall in the Peak District National Park, England during August 1991. There were two reasons for organising the workshop. First, the considerable and lively debate in the literature about the design and analysis of playback experiments - the *pseudoreplication debate* - was in danger of generating more heat than light. A workshop forum seemed the obvious place to clarify and, if possible, resolve the debate. Second, with the number of new playback and analysis techniques increasing rapidly, it seemed an opportune moment to discuss these techniques and to review some rapidly developing areas of interest in sound communication.

Debate and disagreement are essential parts of the fabric of science, but they also have negative aspects. The pseudoreplication debate in the literature illustrated disagreement over some important features of playback design and analysis. However, it became clear from the comments of colleagues in other areas of research that the debate was being interpreted as a message that the technique of playback was critically flawed, that previous playback results were worthless, or, in some cases, both. It seemed a good time to sort out what the pseudoreplication issue was and, if possible, to come up with a generally agreed resolution of the issue. My initial tentative suggestions to other playback practitioners produced such an overwhelmingly positive response to the idea of a workshop that there was no going back - the workshop had to be organised. The result is the first chapter of this book, which is a true consensus reached during the workshop. It would be wrong to give the impression that such a consensus was arrived at easily, but it is a tribute to the professionalism and open-mindedness of the participants that a consensus was reached at all. The chapter aims to lay the spectre of pseudoreplication in two senses. First, that pseudoreplication will cease to be a knee-jerk criticism of any and all playback studies by those who do not perhaps fully appreciate the arguments and the fact that the pseudoreplication issue is inherent in all scientific studies, not just playback. Second, that the chapter will be a useful starting point for future studies, ensuring that the pitfalls are avoided. Unfortunately the chapter cannot convey the positive atmosphere prevalent at the workshop. We are convinced that playback will go from strength to strength and the chapters making up the rest of this book give some indication of its excellent state of health.

The remaining chapters are not organised into subsections because, to a greater or lesser extent, they all have features in common, indeed, all of the chapters incorporate changes arising from discussions during the workshop. However, the chapters are not a random assemblage, there are two groups with broad similarities. The first grouping (Falls to McComb) deals with playback techniques, while the second (Nelson to Ratcliffe & Weisman) focuses on the role of playback in investigating specific topics, all of which are an aspect of variability in sound signals or signalling behaviour.

The first of the playback technique chapters is a personal historical perspective of playback studies by Bruce Falls, an early playback pioneer. As the first talk of the conference it was something of a presidential address and therefore the spoken English has been retained. In the next chapter, Clive Catchpole discusses how playback integrates into studies of broad questions. Irene Pepperberg then draws important parallels between social learning theory and playback, emphasising practical aspects that may have been overlooked. Carl Gerhardt uses examples from studies of anuran female preference to illustrate important methodological considerations and also espouses an alternative (Bayesian) statistical rationale. A multivariate technique for handling the interrelationship of measures of response to playback is the subject of Peter McGregor's chapter. Torben Dabelsteen explains how advances in digital technology have allowed flexible, rapidly changing, interactions between experimenter and subject, and also discusses the potential and limitations of such interactive playback. The chapter by Karen McComb acts as link between the two groups of chapters; she discusses how playback can be used to investigate contests between groups, using social carnivores as examples.

The chapters in the second grouping all deal with one of the most striking aspects of acoustic communication, the variability of signals and signalling behaviour. In the first chapter, Doug Nelson looks at how variability is acquired. Marcel Lambrechts then discusses the relationship between variability and signals of male quality. Anuran vocalisations have relatively little variation in the structure of the signal, but as Georg Klump and Carl Gerhardt discuss, there is plenty of variability in the timing of vocal interactions. The first three chapters in this group have looked at variation from the male viewpoint. Bill Searcy's chapter redresses the balance somewhat by reviewing techniques for assessing the significance to females of variation in male signals. The final three chapters deal with how birds perceive variation. Andy Horn considers the evidence from field experiments for categorical perception. An alternative approach is to employ laboratory operant conditioning techniques, Danny Weary presents the first results and discusses the advantages and limitations of the approach. Laurene Ratcliffe and Ron Weisman consider a different perceptual question, how pitch is perceived, and demonstrate that laboratory and field experiments may give rather different results.

In addition to the participants, many people contributed to the success of this workshop. The Behaviour and Ecology Research Group, Department of Life Science, University of Nottingham allowed the use of various pieces of group equipment from computers to vehicles, and provided a conducive and supportive atmosphere during the planning and execution of the workshop and book. Xanthe Whittaker cheerfully and efficiently assumed the role of workshop *gopher*. Francis Gilbert was an invaluable sounding board and also provided E-mail connections. Peter Davies and Susan Scarborough of IBiS helped me to use various electronic translation and scanning devices so that most of the participants' computers eventually spoke the same language. Rachel Scudamore proof-read many of the chapters. Greg Safford of Plenum, New York advised on page layout and style. Above all, Leonie McGregor provided support at all stages of the exercise, as well as proof-reading every chapter. It is not an exaggeration to say that the workshop would not have been attempted, nor this book written, without her understanding, patience and encouragement.

Peter McGregor

Nottingham November 1991

Contents

Design of playback experiments: the Thornbridge Hall NATO ARW Consensus
Peter K. McGregor et al. 1

Playback: a historical perspective
J.Bruce Falls 11

Integrating playback: a wider context
Clive K. Catchpole 35

What studies on learning can teach us about playback design
Irene M. Pepperberg 47

Conducting playback experiments and interpreting their results
H. Carl Gerhardt 59

Quantifying responses to playback: one, many, or composite multivariate measures?
Peter K. McGregor 79

Interactive playback: a finely tuned response
Torben Dabelsteen 97

Playback as a tool for studying contests between social groups
Karen McComb 111

Song overproduction, song matching and selective attrition during development
Douglas A. Nelson 121

Male quality and playback in the great tit
Marcel M. Lambrechts 135

Mechanisms and function of call-timing in male-male interactions in frogs
Georg M. Klump and H.Carl Gerhardt 153

Measuring responses of female birds to male song
William A. Searcy 175

Field experiments on the perception of song types by birds
Andrew G. Horn 191

Bird song and operant experiments: a new tool to investigate song perception
Daniel M. Weary 201

Pitch processing strategies in birds: a comparison of laboratory and field studies
Laurene Ratcliffe and Ron Weisman 211

Participants 225

Index 227

DESIGN OF PLAYBACK EXPERIMENTS:

THE THORNBRIDGE HALL NATO ARW CONSENSUS

Peter K. McGregor, Clive K. Catchpole, Torben Dabelsteen, J. Bruce Falls, Leonida Fusani, H. Carl Gerhardt, Francis Gilbert, Andrew G. Horn, Georg M. Klump, Donald E. Kroodsma, Marcel M. Lambrechts, Karen E. McComb, Douglas A. Nelson, Irene M. Pepperberg, Laurene Ratcliffe, William A. Searcy, Daniel M. Weary [1]

Introduction to the Issues

Playback is an experimental technique commonly used to investigate the significance of signals in animal communication systems. It involves replaying recordings of naturally occurring or synthesised signals to animals and noting their response. Playback has made a major contribution to our understanding of animal communication, but like any other technique, it has its limitations and constraints.

This section of the workshop was intended to address two different issues. The first concerned the design of playback experiments and the analysis of the subsequent responses. The second issue was the range and type of practical pitfalls involved in actually carrying out playback experiments.

The First Issue

A paper by Hurlbert (1984) on the design of ecological field experiments stimulated an examination of the design of playback experiments (Kroodsma 1986; 1989a). Kroodsma suggested that the design (and analysis) of many playback experiments was inappropriate for the questions being investigated. The suggestion triggered a lively debate about such issues in the literature (Searcy 1989; Catchpole 1989; Kroodsma 1989b, 1990a, 1990b, *in press*; Weary and Mountjoy in press). One of the purposes of this workshop was to bring together practitioners of playback with interests in diverse topics and animal groups in an effort to reach a consensus on this controversial area. The first section of this chapter attempts to identify clearly the nature of the problems of playback design and analysis that are at the root of the controversy, and then to assess the implications for playback experiments and make recommendations for future work.

---

1. The details of authors' affiliations are given in the list of participants.

The Second Issue

Most experimenters with experience of playback have a list of factors that they consider to be important in a well-executed playback study. By incorporating these factors into their design, experimenters try to ensure that the experiment presents the animals with stimuli that differ only (or at least mainly) in the signal feature of interest. Although journals are often reluctant to print such details in the methods section of papers, the information is needed by experimenters trying to replicate studies. Our workshop, which hosted researchers with expertise on different taxonomic groups and areas of interest, presented an obvious opportunity to collate a list of factors considered to be important in running playback experiments. Although specific questions and specific animal groups will require additional factors to be considered, the list presented at the end of the chapter (Table I) is a starting point of general features to which more specific factors can be added.

Appropriate Design and Analysis - the Pseudoreplication Debate

The issues of appropriate design and analysis have come to be referred to as *the pseudoreplication problem* in the literature, following Hurlbert's (1984) title. This section aims to explain what is meant by this expression, and how to avoid the problem, and how it relates to external validity and the limits on generalisation.

What is Pseudoreplication?

Hurlbert (1984) defines *pseudoreplication* as "the use of inferential statistics to test for treatment effects with data from experiments where either treatments are not replicated (though samples may be) or replicates are not statistically independent." Hurlbert was mainly concerned with cases from field ecology, in which, for example, only one control field and only one experimental field would be compared statistically by using sub-samples drawn from each field, or in which experimental plots were spatially segregated from control plots. Pseudoreplication, however, is a problem in a great many areas of science.

In bioacoustics, the term has been applied most frequently to cases in which some general hypothesis is stated about response to general classes of stimuli, and the hypothesis is tested using insufficient numbers of exemplars from each class (Kroodsma 1989). The problem with the latter test is that the stimuli almost certainly vary within each class as well as between classes, so that any difference in response cannot necessarily be ascribed to the between-class difference in stimuli. An example of such a problem would be playing a number of birds one song from their own dialect and one song from a distant dialect. The two test tapes will vary in a number of features, only one of which is the feature of interest, that is, the signal structure that distinguishes own from distant dialect.

As there has been some debate over exactly how Hurlbert's (1984) definition applies to playback experiments, we wish to present our own definition, which we believe is clearer. We define pseudoreplication as the use of an n (sample size) in a statistical test that is not appropriate to the hypothesis being tested. Thus whether pseudoreplication can be said to occur in a given experiment depends on the hypothesis that is stated as being tested. Some hypotheses will dictate that we sample a sufficient number of stimuli from a particular class of stimuli, some that we sample a sufficient number of animals from a population of animals, some that we sample a sufficient number of groups of animals, and so on.

The application of this definition can be made clearer with a specific example. We will first state this example in as simple a manner as possible, shorn of statistical terminology. Next, we restate the example using the language of analysis of variance (ANOVA), which we have found makes the example more understandable for some and less understandable for others. The example is based on the phenomenon of bird song dialects because here little specific background is needed to grasp the questions addressed by playback.

A Specific Example The question of interest is the difference in response shown by birds to playback of different dialects. If the hypothesis is framed very narrowly, for example that birds of dialect X respond differently to song X_1 of their own dialect than to song Z_1 of a specified foreign dialect (Z), then it can be admissible to use only single exemplars of the two dialects, in this case X_1 and Z_1. If the hypothesis is stated more broadly, i.e. that birds of dialect X respond differently to own dialect (X) than to a specified foreign dialect (Z), then using only two exemplars, one from X (X_1) and one from Z (Z_1), and using the number of subjects as the *n* in a statistical test, would be to pseudoreplicate. To avoid pseudoreplication in this case, one would have to use a sample of songs from each dialect (X_1 X_2 X_3 X_4 X_5 etc. and Z_1 Z_2 Z_3 Z_4 Z_5 etc.), using enough songs to be sure that a statistical test could be done with the number of songs as the sample size. If the hypothesis is stated even more broadly, for example that birds of dialect X respond differently to own dialect than to foreign dialects in general, then songs from several foreign dialects (U, V, W Z etc.) must be played to avoid pseudoreplication.

The Example Restated in ANOVA Terminology Suppose that the hypothesis is that response to songs of their own dialect (X) is different from response to songs of a specified foreign dialect (Z). The easiest way to visualise the design is as a diagram (Fig. 1):

	OWN X					FOREIGN Z				
Fixed effect treatment (dialects)										
Random effect (songs within dialects)	1	2	3	4	5	1	2	3	4	5
Response of birds (i.e. actual data)	a	d	g	j	m	p	s	v	y	ß
	b	e	h	k	n	q	t	w	z	δ
	c	f	i	l	o	r	u	x	α	ϵ

Figure 1. A diagrammatic representation of an ANOVA design to test the hypothesis that response to own song dialect (X) is different from a specified foreign dialect (Z). The figure shows five male birds' renditions of own dialect (X_1 to X_5) and five of foreign dialect (Z_1 to Z_5). Each letter (a to ϵ) indicates the data (such as approach, amount of song, etc.) collected from a single male subject.

The design is a two-level mixed-model nested ANOVA. The treatments (own *v.* foreign dialect, i.e X *v.* Z) represent fixed effects since they are determined by the experimenter and are repeatable. However, while dialects are fixed, songs within dialects (X_1 to X_5 and Z_1 to Z_5) vary unpredictably between male birds (for example there may be individual differences in rendition of the dialect) and possibly within males also (for example there may be song by song differences in the male's renditions, i.e. male 1's version of X (X_1) may vary X_{1i} X_{1ii} X_{1iii} etc.). Thus a mixed-model nested ANOVA is appropriate, since this design allows for variation in response to randomly different songs within fixed treatments. If a single exemplar for each dialect were to be used, we could never be certain that the observed differences in response were really due to the different dialects: it is possible that uncontrolled factors might cause differences in response to songs even if they had come from the same dialect. The only way to separate uncontrolled from dialect effects is to replicate songs within dialects. Using *responses* (a to c, and p to r if the single exemplars were X_1 and Z_1 respectively) as replicates for *dialect treatments* (X *v.* Z), rather than *songs within dialects* (X_1 to X_5), is to pseudoreplicate.

In this experimental design, nothing is really gained by having replicate responses to any given song (X_3, say), except that it provides a better estimate of the average response to that song. The vital component is the replication of songs within dialects, which allows a test of the hypothesis that the average response to own dialect is different from the average response to a foreign dialect.

Sokal and Rohlf (1981, Table 10.2) run through the calculation of an exactly similar design: the only decision to make is whether to pool the within-group and songs-within-dialect mean squares before testing the significance of the between-dialect mean square. Sokal and Rohlf give the criteria for making this decision.

If own and foreign dialects are both played to an individual bird (i.e. response of birds is a-o for X and a-o for Z in Fig. 1), the design is different. If the hypothesis is that birds respond differently to own dialect than to a foreign dialect, then formally the test is that the average of (own *minus* foreign) response $< >$ 0, and can be tested with a t-test.

While there may be only one level for any given explicitly stated question or hypothesis where replication is mandatory, it may be desirable to estimate variance in lower level variables using a model II nested ANOVA. These estimates may be interesting in their own right and could help to explain a lack of a treatment effect. For example, no difference in response to dialects may be due to variation among exemplars within a dialect and/or variation among responses of individual subjects.

Avoiding Pseudoreplication

The two key features in avoiding problems of pseudoreplication are being explicit about the question being addressed by the playback experiment and deciding on the number of exemplars.

How Many Exemplars are Adequate? There is no simple answer to this question. Indeed, statistics texts will say that no answer is possible at all unless there is an estimate of the variability of the items of interest (Sokal and Rohlf 1981, p.262).

In the past it has been argued that the variability of the signal gives an indication of the extent of replication necessary. For example, if the signal appears to an observer to be stereotyped then fewer replicates would be required to represent the variation adequately than if a signal were more variable. There are two problems with this approach. First, there is no good *a priori* reason for the variation that is apparent to the human visual system when inspecting sound spectrograms or oscillograms (the same is true for measures taken from spectrograms) to be the same as the variation perceived by the study animal. Second, the feature of interest is the variation in *response* to the signal (playback); such variation may not be directly related to the variation perceived in the signal by both humans and the study animal (see chapters by Weary and by Ratcliffe and Weisman in this volume). There are three possible reasons for this difference: first, the animal may not perceive the variation (e.g. lack of perceptual ability, artifact of presentation); second, the animal may perceive the variation but it elicits no difference in response because this would be biologically inappropriate (e.g. there are competing behaviours such as mate guarding and feeding); finally, the measures of response taken by the experimenter may simply be too crude to show a difference between stimuli.

Workshop participants suggested a solution to the problem of determining the appropriate number of replicates; that is, for a two-stage approach. The first stage is to use previous experience with, and other work on, the study species to make an informed guess at the level of replication necessary and to use this level to examine the question. The second stage is to use the information from the first stage to refine the number of replicates needed. This procedure may involve formal measurement of the variation of response, possibly with principal components analysis to reduce complex features of the original signals to a manageable number, and the application of the standard formulae available to estimate the sample size necessary to show a difference of the required magnitude and probability level (e.g. Sokal and Rohlf 1981, Box 9.13; see also the section on Bayesian statistics in the chapter by Gerhardt in this volume). Once again the discussion during the workshop emphasised the importance of being clear about the question that playback was being designed to answer, as this will obviously affect the level of replication necessary.

Can Use of Synthetic Calls Help to Avoid Pseudoreplication?

Pseudoreplication typically arises in a playback experiment because of lack of control over the differences in our treatments. For example, in the dialect case, if we take song A_1 from one dialect and song B_1 from another, they must differ in dialect features, but they may also differ in all sorts of other features, e.g. motivation of the male when singing, quality of recording, etc. Our only hope of controlling these other differences is to use several examples of each dialect, so that those differences will average out. In contrast, if we use artificially modified stimuli we have better control over the differences between our stimuli. For example, if the experiment tests response to one natural song versus exactly the same song in manipulated form, then a statistical test can be done with n as the number of subjects to test the hypothesis that this manipulation of this particular song affects response. It may not be clear to what aspect of the changed stimulus the subjects are attending - for example if the manipulation is to halve a song, the subjects could be attending to the changed duration or the missing acoustic elements etc. - but still it is clear that the manipulation is responsible for the difference in response.

The proper use of synthetic sounds avoids many of the design problems arising from the multidimensional nature of natural signals and variations in recording quality. In principle, an investigator can explore the behavioural relevance of the entire perceptual or preference space that is delimited by variation in a set or sets of acoustic signals of particular interest: within-population, between-population (dialects), between different signals in the repertoire, and between species. There will be considerable practical difficulties of studying systems in which there are very many acoustic properties of potential significance, but we are hopeful that in many of these the set of relevant properties will be some relatively small subset of the possible properties. In fact, a few studies have begun to tackle the problem of varying two properties of known pertinence at the same time (Nelson 1988; Date et al. 1991; Gerhardt and Doherty 1988; Dooling 1982). As the number of simultaneously varying properties being examined increases, the design, execution and interpretation of such experiments will probably warrant another workshop like this one.

A first step is to generate a synthetic signal that is comparable in its behavioural effectiveness to a typical natural signal. The signal can be synthesised *de novo* or produced by modifying a naturally occurring sound. Normally, such a signal would have properties with values equal to the estimated mean values in the set of signals of interest. Ideally, in comparative tests, the synthetic standard call is neither more, nor less, attractive than a series of natural exemplars.

The second step is to develop criteria for choosing the amount of change in the value of a given parameter. In our view any of the three following criteria are appropriate, depending on the question being asked:

1) Change the value of a parameter by units equal to the standard deviation (or some other measure of variance) of the property in the natural set of interest. This procedure would be appropriate for estimating the proportion of signallers in a population that might be favoured by mate choice based on the property in question;

2) Vary the value of the parameter by some constant percentage, guided perhaps by any existing psychoacoustical data;

3) Third, if the distributions of a parameter in two classes of natural signals do not overlap, choose a difference between the property that corresponds to the minimum observed difference between the two sets (species, populations, dialects, signals within a repertoire). If the animals discriminate, then the difference in that property at least is adequate for recognition of the two natural classes of signals. If the animals do not respond selectively, then the difference can be increased systematically until there is a differential response. This information could then be related to the natural variation in the two sets of signals to provide an estimate of the proportion of individuals that would be likely to be distinguished in natural situations.

External Validity

External validity refers to the degree to which we can generalise from the results of a specific experiment. Limited external validity is a problem for all fields of science. An experiment can avoid pseudoreplication and still have limited external validity. In other words, no matter how good the design and execution of our experiments, there is always a logical danger in generalising from our results.

Although there is no theoretical reason for internal and external validity to be linked, practical field constraints may mean that adequate controls for the many features of the stimulus signal and its presentation will limit the range of the question posed and therefore the extent of external validity. For example, the logistics of carrying out experiments at two widely separated field sites and the constraints to carry out the playback experiments at the same time of day, breeding season or year may restrict the experiments to one field site, which will in turn limit the external validity.

Generalising

If all practical playback experiments are limited in their external validity, to what extent is it possible and desirable to generalise from these experiments? An extreme stance is that comments in the discussion section of a paper or manuscript should be restricted to the specific effect found, for example, "neighbour/stranger discrimination was shown for 16 first year males in the northeastern corner of the study site." It seems more reasonable to view each specific experiment as a step on the road to more general explanations and unless there is something unusual about the data set that could affect neighbour/stranger discrimination, such as that particular corner of the study population is the only one where males are territorial as first year animals, then the discussion would seem the obvious place for a consideration of the advantages of neighbour/stranger discrimination in general. Implicit in this approach is that results from the first experiment form the basis for a general hypothesis that will be tested in subsequent experiments. The hypothesis is only modified when the predictions are not supported by subsequent experiments; then a new hypothesis consistent with the literature is proposed for subsequent testing. Provided that the language in the discussion makes it quite clear what is being proposed as an hypothesis derived from the experiment as opposed to a result, then the distinction between test and idea will be maintained.

A variety of simple and complex problems can be examined using playback. Testing broad hypotheses and replicating treatments at a correspondingly broad level is certainly one valuable method for making progress in science. However, we also recognise the value of testing narrower hypotheses, with replication at a lower level (e.g. at the level of subjects rather than stimuli). Through a series of simple experiments, one can then approach the larger question. Both approaches are valuable and both demand explicit statement of hypotheses and appropriate replication.

Conducting Playback - Important Features to Consider

The recognition that experimental execution is critical to any investigation is the second issue addressed by this chapter. This point is also stressed by the paper that triggered the pseudoreplication debate (Hurlbert 1984). To quote Hurlbert (1984, p.189): "Yet in a practical sense, execution is a more critical aspect of experimentation than is design." The reasoning underlying his statement is that errors of execution are more common, more variable, more subtle and more difficult to detect at all stages of an experiment than design errors.

We have compiled a list of features (Table I) that should be considered by any playback experimenter in order that execution errors may be minimised or at least recognised. This is not a list of recommended procedures; neither is it an exhaustive list. Rather,it is a list of factors that can be important in the execution of playback experiments and whose importance must be judged by the experimenter in the context of his particular experimental situation. The methods section of any publication resulting from a playback experiment ought to state which of these features were judged to be important by the experimenter and how their effects were controlled.

Summary

Pseudoreplication as it applies to playback studies is a consequence of a lack of rigour in specifying the question being addressed by the study. Commonly a lack of replicates at the level of interest results in an inability to answer the question as posed. However, pseudoreplication is not ubiquitous in playback studies, nor does it impose a constraint on their usefulness. Pseudoreplication is a potential problem not only in playback, but in all areas of research. Remedying the problem is straightforward. Care has to be taken when specifying the hypothesis to be tested and an appropriate number of replicates must be used when conducting the experiment.

Equal care must be taken to appreciate the factors that can influence an animal's response to playback. Some important sources of execution error are identified in Table I, but we urge playback experimenters to include information on how execution errors were minimised in the methods section of resulting publications.

There is no reason to doubt that playback, with the continuing development of new techniques and experimental designs, will remain one of the most powerful tools available for the investigation of animal communication.

Table I. A list of some of the features affecting execution errors in playback experiments. SPL = sound pressure level, s/n = signal to noise.

Test tapes and test sounds
Sound per unit time, total amount of sound.
Degradation (distortion), SPL, s/n ratio of source sounds.
Encoded information on status, motivation, identity etc.
Level and type of background noise.
Filtering and editing to remove background noise.
Environmental conditions
Time of year (influences background noise, vegetation, activity of subjects and other species).
Weather conditions (same influences as time of year).
Time of day (degradation effects, see also time of year).
Test Animals
Subject location in relation to territory boundaries.
Effects of stage of breeding cycle.
Time of day effects.
Proximity of resources (mates, food etc.).
Activity of conspecifics (neighbours, intruders etc.).
Predator activity.
Playback equipment
Speaker directionality.
Fidelity of equipment (s/n ratio, frequency range, etc.).
Procedure
Position and behaviour of observer.
Use of blind experiments (observer bias likely?).
Loudspeaker position.
Information on failed tests.
Response measures (single, multiple).

References

Catchpole, C.K. 1989. Pseudoreplication and external validity: playback experiments in avian bioacoustics. *Trends in Ecology & Evolution*, **4**, 286-287.

Date, E.M., Lemon, E.R., Weary, D.M. and Richter, A.K. 1991. Species identity by birdsong: discrete or additive information. *Anim. Behav.*, **41**, 111-120.

Dooling, R.J. 1982. Auditory perception in birds. In: *Evolution and Ecology of Acoustic Communication in Birds. Vol.II.* (Ed. by D.E. Kroodsma, E.H. Miller & H. Ouellet), pp. 95-130. Academic Press, New York.

Kroodsma, D.E. 1986. Design of song playback experiments. *Auk*, **103**, 640-642.

Kroodsma, D.E. 1989a. Suggested experimental designs for song playbacks. *Anim. Behav.*, **37**, 600-609.

Kroodsma, D.E. 1989b. Inappropriate experimental designs impede progress in bioacoustic research: a reply. *Anim. Behav.*, **38**, 717-719.

Kroodsma, D.E. 1990a. Using appropriate experimental designs for intended hypotheses in 'song' playbacks, with examples for testing effects of song repertoire size. *Anim. Behav.*, **40**, 1138-1150.

Kroodsma, D.E. 1990b. How the mismatch between the experimental design and the intended hypothesis limits confidence in knowledge, as illustrated by an example from bird-song dialects. In: *Interpretation and Explanation in the Study of Animal Behaviour. Vol. II.* (Ed. by M. Bekoff & D. Jamieson), pp. 226-245. Westview Press, Boulder, Colorado.

Kroodsma, D.E. *in press*. Much ado creates flaws. *Anim. Behav.*

Gerhardt, H.C. and Doherty, J.A. 1988. Acoustic communication in the gray treefrog, *Hyla versicolor*: evolutionary and neurobiological implications. *J. comp. Physiol. A.*, **162**, 261-278.

Hurlbert, S.H. 1984. Pseudoreplication and the design of ecological field experiments. *Ecological Monographs*, **54**, 187-211.

Nelson, D.A. 1988. Feature weighting in species song recognition by the field sparrow (*Spizella pusilla*). *Behaviour*, **106**, 158-181.

Searcy, W.A. 1989. Pseudoreplication, external validity and the design of playback experiments. *Anim. Behav.*, **38**, 715-717.

Sokal, R.R. and Rohlf, F.J. 1981. *Biometry*. 2nd Edition. W.H. Freeman & Co., New York.

Weary, D.M. and Mountjoy, *in press*. On designs for testing the effect of song repertoire size. *Anim. Behav.*,

PLAYBACK: A HISTORICAL PERSPECTIVE

J.Bruce Falls

Department of Zoology
University of Toronto
Toronto, Ontario M5S 1A1
CANADA

Introduction

We are here to re-evaluate a technique that has been very productive in the study of animal communication. So in a sense this is a celebration of success. I have been asked to provide a historical overview and, as I look around the room, I see that my grey hair gives me at least one qualification for the task. Playback and I arrived on the scene at about the same time.

In the time at my disposal, I cannot provide anything approaching a thorough review of the enormous literature involving playback. As most of you know, I have worked with bird vocalisations and I hope you will forgive me if I draw heavily on my own experience and that of my students in what follows. I should like to begin by attempting to define what we mean by playback, then describe some of its beginnings, consider some basic features of methodology, and outline some of the uses to which it has been put in various groups of animals.

Definitions and Implications

Our conference title speaks of animal communication and I expect most of us would agree that communication occurs when a signal from one individual, the sender, alters the behaviour of another individual, the receiver. We would probably also agree that the signal resulted from some structure or act specially adapted for the purpose, rather than as a by-product of some other behaviour. The effect of the signal benefits the sender and may benefit the receiver. I like John Smith's (1965) representation of the process with a few additions.

Figure 1 shows a sender producing a signal, the structure of which is correlated with some aspect of the sender's internal state, behaviour or environment. Information about the sender's condition is encoded in the signal and constitutes the message or broadcast information. The signal involves some sensory modality. It must be detected and decoded by the receiver. Depending on the context, the receiver interprets its meaning - the transmitted information - and acts accordingly.

Playback and Studies of Animal Communication
Edited by P.K. McGregor, Plenum Press, New York, 1992

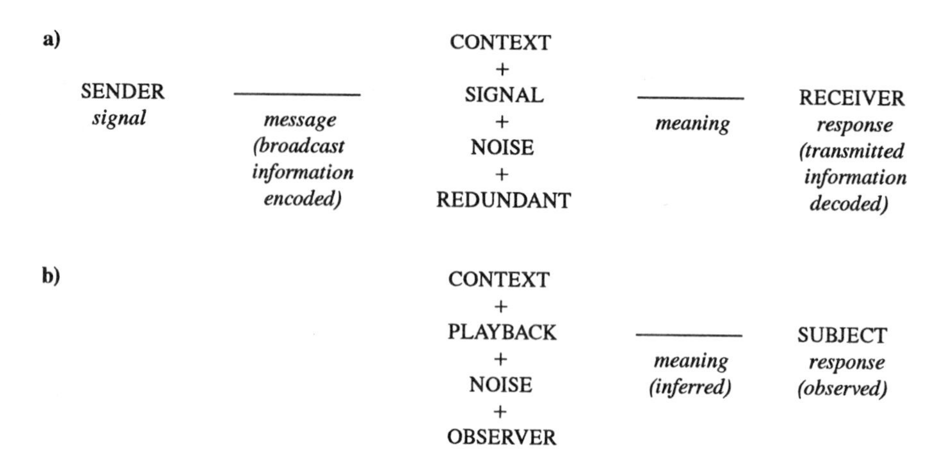

Figure 1. The communication process: a) natural, b) experimental.

When we use the term playback, we usually refer to signals in the sound channel using the auditory modality. In principle, the use of visual models, or smells, or even electrical discharges (see Kramer 1990) are essentially similar to playback and these can be combined in various ways. Indeed, video playback is already upon us. I shall confine my remarks to the use of sound signals. The term playback implies that an animal's acoustic signal has been recorded and is being reproduced in the presence of that individual or others. In practice, we usually include modifications of the signal or artificial copies in our use of the term. Essentially, the method consists of introducing a simulated signal and following the process (Fig. 1) from that point on.

This is an experimental method which means that we have some control over the situation. For example, we can broadcast a sound without any accompanying visual stimulus (except a loudspeaker) and hope to determine the effect of the sound signal alone. We can make rare events common and have them occur at our convenience. We can modify these signals and even substitute artificial copies with defined properties. With proper controls, we can hope to discover causal rather than correlational relationships between a signal and its effect. Essentially, we can study the completion of the process of communication.

To quote Colin Beer (1982), "the animal holds the answer to the question of what constitutes a signal". Thus, to take an example, we can measure individual variation in signals, but individual recognition is a matter for the animal to decide. Playback experiments often seem to be establishing the obvious but they are putting questions to the animal.

If we accept the concepts illustrated in Fig. 1, several things follow. We can observe the effect of a signal only by the response of a receiver, from which we can try to infer the meaning. However, if the response is not obvious or is delayed, we may not be able to observe it. In the laboratory we can monitor physiological functions and detect subtle change (e.g. Diehl and Helb 1986). I see no reason why this cannot be extended to the field using telemetry - in effect a sort of lie detector test.

Playback is essentially an analytical tool for investigating the sufficiency of sound. But the responses we study are holistic, influenced even by other sensory modalities. We are on relatively safe ground in limiting our investigation to sound with long distance signals or in situations where communication by other means is unlikely. However, even in these cases, our model (Fig. 1) tells us that the meaning taken and the response given depend not only on the signal itself but also on the context, which includes both social and ecological factors. Environmental noise, that may interfere with detection, is also part of the picture. All these components are subject to study and can be manipulated to some extent but the point I want to make is that we should be careful about defining the circumstances of our experiments. Since we want to understand the normal behaviour of animals, the more natural situation the better. I believe this point will be emphasised by Irene Pepperberg.

A related consideration is that playback assists us in studying only certain aspects of the communication process. We can manipulate the structure of the signal and observe the response but we are dependent on observation, including signal analysis, for the rest. Indeed, as some of you will stress, we need a background of observation to design our experiments and interpret their results.

Playback can be used either in the laboratory or in the field. There are trade-offs whichever way we go. There is more precise control of the conditions and more precise measurement in the laboratory but precision doesn't necessarily imply accuracy. Especially for large and mobile animals, the laboratory is not a natural context. The field, with all its unknowns and uncontrollable variables is where behaviour may be most uninhibited, natural and complete. Ideally, we can have the best of both venues by linking laboratory and field studies. Playback helps us to do that.

Beginnings

What led up to the use of this technique? Playback is a special case of methods used for a long time by physiologists to study stimulus-response systems. What makes it special? Mainly, that until the advent of portable tape recorders, it was very difficult to use experimental methods for the study of animal sounds, especially in the field. Caged animals and imitations had been used but left a lot to be desired. All at once, we had the means to record accurately, store and, with the near-simultaneous appearance of audiospectrographs, analyse and play back the sounds of animals with high fidelity. Computer-assisted manipulations of signals and interactive techniques were yet to come.

Table I. Practical uses of playback.

Attract animals	- capture / kill	**Census animals**	- crowded
	- marking		- dispersed
	- observation		- cryptic
	- recording		- rare
	- photograph		
		Repel animals	- roosts
Map territories			- crops
			- airfields, etc.

We should not forget that hunters have long used the actual sounds of animals as well as vocal and mechanically produced imitations to attract their quarry. This is another historical root of our method and tape recorders were rapidly seized upon for various practical purposes - to attract birds for photography, recording, capture or simply to see them, as well as to repel pests (Table I). Among the first applications of playback was the use of distress calls to disperse roosts of starlings (*Sturnus vulgaris*) and corvids by Frings, Busnel and their colleagues.

In 1956, I attended a conference arranged by Hubert Frings at Pennsylvania State University to establish the International Committee on Bioacoustics. Most of the participants at that meeting were preoccupied with recording or analysis but some of us had already begun playback experiments. The proceedings of the conference (somewhat augmented) were published under the editorship of R.-G. Busnel (1963). Two years after this first meeting, a symposium was organised by J.T. Emlen at the meeting of the AIBS at Indiana University. This was published under the editorship of W.E. Lanyon and W.N. Tavolga (1960). These two books, both of which cover the field broadly, provide an excellent overview of the state of bioacoustics by the end of the 1950's.

My own introduction to this field began through my friend Bill Gunn, who began tape recording in the late 1940's and produced some excellent records. He was also one of the first people to use playback in a study of American woodcock (*Scolopax minor*) in 1951. Bill demonstrated to me how, by playing even poor recordings, he could entice birds to approach the speaker and sing, where he could obtain a good recording. I had been interested in birds from my youth but my training was in experimental biology and population ecology. I was becoming interested in territoriality and it struck me that playback provided an excellent way to map a bird's territory and investigate its behaviour. At that time, territories were mapped by plotting points where a bird sang, which was inefficient because few of the points represented boundaries. I wanted to pull a bird to the edge of its territory, using playback as a simulated intrusion. By 1956, when Bill Gunn and I attended the bioacoustics conference, I had mapped the territories of ovenbirds (*Seiurus aurocapillus*) and Judith Stenger (later Weeden) and I had discovered neighbour-stranger discrimination in this species. I was soon to embark on studies of the components of song important to species recognition in ovenbirds and white-throated sparrows (*Zonotrichia albicollis*) using altered and artificial songs. At about the same time, Laszlo Szijj and I conducted interspecific playbacks to meadowlarks (*Sturnella* spp.). Once I had started along this path, there was no turning back.

The equipment we had in the early years was quite good but it was heavy. I had a tape recorder that used mains power so I had to have a car battery and convertor to operate it. This equipment was kept in a truck and we used speakers on cables 500 feet or more long. Later, when we had battery operated equipment, the "speaker method" of mapping territories became much more efficient (Dhondt 1966; Falls 1981).

Thus, as equipment improved, playback experiments became easier logistically but I don't think that had much influence on the nature of the questions asked. If our early studies were simple conceptually, it was because we were just starting and were interested in basic questions appropriate to that period. In ethology it was a time of releasing stimuli and IRM's (innate releasing mechanisms) but I was largely untouched by that and my interest continued to be in territorial behaviour. I suppose I did regard songs and responses as rather stereotyped. I had some background as a naturalist but didn't undertake detailed studies of the uses of song before I began my experiments. While I now advocate a more intimate combination of observation and experimentation, I don't think I would do my early experiments much differently.

Methods

I should like to make some observations about methods based on my experience with birds. I hope they may apply more broadly. Most of our experiments test the tendency of animals to discriminate between signals. For this purpose we use several types of presentations on which I shall comment briefly.

We may present each signal to a different sample of birds. This minimises intrusion for any one bird and we should take that into consideration. It also avoids carry-over of effects of one signal to another and is preferred by some investigators for that reason. However, individual variation in response is apt to be large even when we try to keep external variables within limits. This is a serious problem, because in most cases responses are more variable than signals, something that we will no doubt consider in our discussions of sampling. It is more usual to expose the same subjects to different treatments, making it possible to pair the data for each subject, so that the effect of individual variation on the final result is reduced. Several methods share this approach.

One method, often used in studies of insects and frogs, is to have two speakers broadcasting nearly simultaneously or alternately. It seems like a good system for choice and minimises differences in uncontrolled variables but has problems when used for large or mobile animals. If a bird gets closer to one speaker by chance, it may respond mainly to that signal. Birds are quick and may fly back and forth. The method may work well where one signal is relevant and the other is not. I suppose this was the case in Lanyon's (1963) studies of *Myiarchis* flycatchers. Otherwise, I think the effects of the two signals coming from different locations may interact, as illustrated by the experiments of Searcy et al. (1981) on swamp sparrows (*Melospiza georgiana*). These birds showed neighbour-stranger discrimination when a sequential presentation was made with one speaker but not when two speakers were used.

More often in bird studies we use a single speaker and present different signals to the same individuals in sequence. This may lead to carry-over of response from one signal to another, which we deal with by controlling for order of presentation. A long gap (a day or two) between presentations may minimise carry-over but allows considerable variation in the state of the subject and the environment. A short interval between successive presentations is more likely to result in some carry-over but reduces variation for other reasons. I have tried all the methods described so far and found that sequence presentation of signals with short intervals and one speaker was the most sensitive for my purposes.

The next method I want to mention consists of habituating an animal to one signal and then testing it with another. This seems good in principle, making use of a learning mechanism that operates importantly in nature. Nelson (1987) has used this method in studies of syllable discrimination by song sparrows (*Melospiza melodia*). It could be used either in the field or the laboratory.

Daniel Weary will tell us about the use of operant conditioning with captives. This method transfers the choice to an unusual context with food rewards. Does this tell us how a bird would respond in a normal context, say territorial or courtship behaviour? Or does it test the limits of the bird's perception as in a psycho-physical test? Clearly, the control here is very good and many extraneous variables are eliminated. I think it is important to pair this type of experiment with field studies.

Ten years ago, when I gave talks about playback studies, I said that they were good for immediate responses because they simulate what happens when a bird first hears a sound. I cautioned against putting much faith in what happened after the initial response. Long experiments might be appropriate to study habituation but they expose the subject to a disembodied, unresponsive intrusion. Since then, several investigators (e.g. Dabelsteen and Pedersen 1990) have experimented with interactive playback. This opens up new possibilities, but how far can one go in the absence of a visual stimulus? Torben Dabelsteen's paper will discuss his experience with the interactive method.

Now, I want to say a little about measuring response. In the field one cannot often observe subtle behaviours consistently across a sample of subjects. For that reason, I have tried to record simple attributes of response - locations and vocalisations at a series of times (real times or intervals). This allows one to measure latencies, closeness of approach, time near the speaker, and numbers of movements, songs or calls. One has to estimate distances but these measures call for little in the way of judgment and are not likely to be subject to bias. I find it helps to take or dictate notes with a tape recorder running. This records times and numbers of vocalisations and, if one wants to identify the song type, the recording can be analysed later, perhaps by someone else. Using tape recordings in this way, it is important to comment fully in order to get a two or three dimensional record of what happened. Indeed, without comments one can even confuse the playback with the subject.

Once we have the raw data, what should we do? I think we should at least summarise the individual measures to describe the nature of the response. If some of them show clear discrimination, that may be enough. However, individual measures are apt to be inter-correlated, so use of multivariate statistics allows us to use all the data in a more efficient way. I'll leave it to Peter McGregor to discuss that.

There is another way of looking at response, which is to construct a graded series of behavioural categories based on other studies (e.g. Falls and McNicholl 1979). This is a case where prior observation and experiment complement one another. There is more room for judgment here than when using the foregoing methods but the results may be easier to interpret. A problem with this method is that, if numerical values are assigned to behaviours, they will be arbitrary. Ranking may be more appropriate.

Before I leave methods I want to touch briefly on a more general point, namely that one must somehow find a level of response between zero and going flat out. If a subject is responding strongly, it may not make subtle discriminations - at any rate it may be hard to measure them. Let me give an example. When John Krebs, Peter McGregor and I experimented with great tits (*Parus major*) (Falls et al. 1982), we presented them with songs of neighbours and strangers as well as their own songs. Because we were interested in song matching, we used the data only from trials in which the subject sang. This eliminated some of the weakest responses. Although strangers received the strongest responses by most measures, differences were not significant. During the experiments, we noted whether or not the nearest neighbour sang. It turned out that when we played a subject its own song, the neighbour usually remained silent but when we played a stranger it almost invariably sang. In contrast to our inconclusive results with subjects, the neighbours showed very clear discrimination.

On a more personal note, I have tended to tackle small questions rather than large ones. I call this bricklaying, which in real life has proven to be a good way to build a solid edifice. Others might call it component analysis. If I understand Clive Catchpole's abstract, he may be a kindred spirit. This approach may seem limited but it has advantages. One can often get a clearer result if one studies a particular case or a particular component of a larger problem. I see a parallel here with theoretical modelling, C.S. Holling, who elaborates models from smaller components, refers to the trade-off between precision or realism on the one hand and generality on the other (Holling 1964). Perhaps that trade-off, which occurs throughout science, will be reflected in the discussions at this conference.

Uses of Playback

Now I want to review the varied uses of playback. This is going to be a very sketchy and biased survey - sketchy for lack of time and biased because I am not very familiar with work on taxa other than birds. Fortunately, we have experts on some of these other groups here and I hope they will bear with my ignorance and correct my mistakes. The literature on birds alone is enormous and I hope my ornithological colleagues will excuse me if I cite a number of studies from my own experience.

Please note that this is a survey of playback use, not a survey of our knowledge of bioacoustics - thus it is bound to be somewhat eclectic. Also, I do not intend to criticise the methods used or vouch for the results of the studies I cite; rather, I shall concentrate on the topics that have been investigated. That alone covers a lot of ground! I include references to books, reviews and individual papers that I have found useful but the list is not intended to be complete. For historical interest, I shall emphasise early studies, particularly those before 1960. I begin with insects and proceed through fish, amphibians, mammals, and birds in that order.

Insects

A recent book by Ewing (1989) provides an excellent overview of arthropod bioacoustics. Studies of insects were among the first to employ playback. In 1913, Regen transmitted the calling of a male cricket (*Gryllus campestris*) over the telephone to another room where a female approached the earphone. Thus, for the first time, the approach response was demonstrated using sound alone. Strictly speaking, this is not an example of playback but it does foreshadow what was to come. Since then, many investigators have observed the acoustic behaviour of insects, both in the field and in the laboratory. It was a short step to record and play back insect sounds when suitable equipment became available.

The first insect sounds to be used in playback experiments were the flight tones of female mosquitoes (*Aedes aegypti*). When the females mature, their flight sounds attract nearby males, setting off mating activities (Roth 1948). Kahn and Offenhauser (1949) were the first to use recordings of these sounds for mosquito control. Wishart and Riordan (1959) tested males with a variety of sounds and found that the fundamental frequency must lie within a narrow band to elicit the flight response.

Much of the research on insect sounds has dealt with Orthoptera. As early as 1951, Busnel and his colleagues in France investigated the properties of artificial sounds that induced calling and approach in male and female grasshoppers. Perhaps the first use of natural tape recorded songs was by Busnel and Busnel (1955), who demonstrated approach (phonotaxis) by female tree crickets (*Oecanthus pellucens*) to calls of males. Walker in 1957 was probably the first to report species discrimination, using the response of females to male calling songs of closely related, conspecific tree crickets (*Oecanthus* spp.). He also used artificial sounds to show that temporal pattern rather than spectral composition of the calls determined the female response. In the same year, Haskell (1957) demonstrated male response to playback of grasshopper (Acridinae) calls.

After 1960, the number of papers increased rapidly as did the subject matter addressed. The phonotactic response of females still dominated the field and several ingenious procedures were employed. Many studies involved the playback of signals from two speakers. In order to equalise the distance of the female from each source and still show a result, subjects were restrained in various ways. For example, a tethered insect could manipulate a bifurcated hoop, referred to as a "Y-maze globe" (see Ewing 1989). The most sophisticated apparatus in use today is the "Kramer Treadmill", a sphere on which the insect walks. An infrared beam reflected off the insect is used to control two servomotors which, by rotating the sphere, maintain the subject in a nearly constant position. At the same time, the signal fed to the motors provides a record of the insect's path. Huber and Thorson (1985) used this apparatus to investigate the properties of cricket (*Gryllus campestris*, *G. bimaculatus*) calls to which females respond, the neurological pathways involved in signal recognition, and the mechanisms by which these insects determine the direction of the sound source. In this and other recent work computer simulation of natural and altered songs has been employed.

Species discrimination as a means of reproductive isolation continued to be a subject of research on insects. For example, Morris and Fullard (1983) found that, while females of the katydid *Conocephalus nigropleurum* discriminated against a sympatric species (*C. brevipennis*), they responded to calls of allopatric species or even random noise, if the main energy peak was in the appropriate frequency range. Recently, Heady and Denno (1991) reported a study of planthoppers (Homoptera) in which males and females communicate by vibrations transmitted through the leaves of the host plant. They used a phonograph pickup to record these signals from leaves and an insect pin glued to the voice coil of a loudspeaker and touching a leaf to play them back. The signals are effective in mate location, attraction and mate choice. The insects discriminated between vibrations of two closely related species (*Prokelisia marginata* and *P. dolus*). Females displayed a variety of rejection behaviours to conspecific as well as heterospecific males, suggesting the possibility of female choice. Other studies have shown differential responses of female insects to intraspecific variation in loudness of songs, which may be related to the energy expenditure by males as well as their size. For example, higher amplitude synthetic calls attracted more females in mole crickets (*Scapteriscus acletus*, *S. vicinus*; Forrest 1980). In another case (Crankshaw 1979), female crickets (*Acheta domesticus*) preferred to approach playback of songs of dominant males over those of subordinates. These and other studies are discussed by Searcy and Andersson (1986) in their review of sexual selection and song evolution.

Some insects use song in male-male interactions involved in dominance and territorial behaviour. Alexander (1961) was able to "defeat" dominant male crickets (*Acheta* spp.) using models simulating "antennal lashing". Playback of aggressive song increased the effect. Again, playback of song of increasing amplitude will eventually silence a singing male bush cricket (*Teleogryllus oceanicus*; Cade 1981). (For other evidence, not necessarily involving playback, see Searcy and Andersson, 1986).

The danger of being attacked is a trade-off in the use of loud acoustic signals. Playback of insect calls revealed both risks and defensive measures. A number of cases have been reported of parasites using sound to locate insect hosts. For example, Cade (1975) found that the larviparous tachinid fly *Euphasiopteryx ochracia* could be induced to approach and larviposit on a loudspeaker broadcasting the calling song of a cricket (*Gryllus* spp.). Cricket species (*Teleogryllus* and *Gryllus* spp.) that are parasitised have different patterns of calling than congeners whose calls do not attract the parasite, suggesting that parasites exert selective pressures on these behaviours (Cade and Wyatt 1984).

The work of Roeder and his colleagues on the interaction of moths and bats is well known. As an example, certain arctiid moths (*Utetheisa* and *Rhodogastria* spp.) excrete a noxious foam when disturbed and are capable of producing a train of high frequency clicks. These sounds may have an aposematic function and bats avoid moths producing them. The moths produce the sounds in response to the cries of echolocating bats. Dunning and Roeder (1965) showed that bats trained to catch mealworms in the air avoided them if a recording of moth clicks was presented at the same time.

A common phenomenon in insects is the tendency of some species to sing in choruses. Playback studies dealing with synchrony and effects on females are reviewed by Ewing (1989).

To summarise, the use of playback in insect studies began early and has proliferated greatly. It will no doubt be extended to more taxa including other invertebrates. One cannot help but envy the ease with which responses of female insects can be measured. Although field studies are important, the scale of these animals and their habitats allows investigators to conduct laboratory experiments on insects that would be hard to arrange in larger animals.

Fish

When I was young there was a radio comedian whose act consisted of claiming that he was an expert on the sounds of fish. His fish imitations were silent. However, by the time of the bioacoustics conference in 1956, a number of fish sounds had been recorded and some had been played back. Indeed, a paper involving playback was published in that year. Moulton (1956) reported that when recordings of the staccato call of sea robins (*Prionotus* spp.) were played back in the murky waters of Woods Hole Harbour, the fish answered. At that time the behaviour of these fish when calling had not been observed and it was believed that the sounds were made by both sexes. Tavolga (1958) studied the reproductive behaviour of the goby *Bathygobius soporator*, which made its characteristic sounds in an aquarium. Playback of recordings of a male to an isolated female resulted in a temporary increase in activity. If a male was presented in a flask, the female would follow it but only while the sound was played. Apparently neither sight nor sound alone were sufficient to orient the female's movements.

More recent work on fish includes a variety of approaches. In two studies (Winn 1972; Myrberg et al. 1978), playback of altered signals showed that temporal characteristics were important in eliciting appropriate responses from conspecifics. Myrberg (1981) reviewed the literature and proposed a framework based on different sorts of receivers and benefits to senders. Under mating calls he tabulated a number of studies including six in which playback was used. These demonstrated attraction of females, interactions between the sexes and, in one case, arousal of gonadal activity in the female.

Myrberg's recent work on damselfish makes use of designs previously used in bird studies. Responses to underwater playback were monitored by observers with snorkels or SCUBA. In one study (Myrberg and Riggio 1985), sounds of neighbouring males were played to territorial bicolor damselfish (*Pomacentrus partitus*), both on the side occupied by the neighbour and on the opposite site. Responses were weak on the usual side compared to those on the opposite side, demonstrating individual recognition of neighbours by their sounds.

Another study (Myrberg et al. 1986) illustrated female responses. Speakers were placed behind a row of conch shells. In one trial, calls of two male bicolor damselfish of different size were broadcast. In another trial, calls of a bicolor male, calls of a sympatric damselfish of another species, and white noise were broadcast from different speakers. All the surrounding males were removed from the vicinity of the speakers. Females approached only calls of conspecific males. Significantly more of them approached the speaker playing calls of the larger male, which differed in frequency from those of the smaller male. Thus, this experiment demonstrated directional response and suggested species discrimination and female choice between conspecifics based on sound signals.

One of Myrberg's (1981) categories, interceptors, referred to the behaviour of "overhearers". For example, three cases were cited of males responding in various ways to playback of courtship calls of other males. Responses of companions to calls given during interactions with predators suggested "mobbing" behaviour (Winn 1967). Predators and prey respond to playback of each other's sounds but not all of these sounds can be considered signals.

In these few examples, one sees parallels to the questions investigated in insects and birds. My impression is that work with fish has progressed slowly and sample sizes tend to be relatively small, but I am also impressed with the difficulty of underwater research.

Amphibians

Nearly all of the acoustic research on amphibians has been done with frogs and toads. There are many parallels with insect studies, including the concentration on female responses to mating calls of males. I shall not attempt a detailed survey because I have not taken part in this work myself and we have two of the leading researchers in the field with us - Georg Klump and Carl Gerhardt, who will present papers about their work.

Playback to amphibians began in the late 1940's. In 1947, Bogert reported that playing a chorus of southern toads (*Bufo terrestris*) at the Archbold Station in Florida attracted both males and females. A decade later, Martof and Thompson (1958) showed that females were attracted to both natural and synthetic calls of male chorus frogs (*Pseudacris nigrita*). The following year, Littlejohn and Michaud (1959), using a two-speaker design, demonstrated that female chorus frogs (*Pseudacris streckeri*, *P. clarki*) could discriminate between calls of their own species and those of another sympatric species. A similar result was obtained with two closely-related allopatric species (*Pseudacris streckeri*, *P. ornata*) by Blair and Littlejohn (1960).

In subsequent studies, the response of females was used to analyse the effective components of male song (or synthetic copies), for example in bullfrogs (*Rana catesbiana*) by Capranica (1966) and in green treefrogs (*Hyla cinerea*) by Gerhardt (1974). An interesting case with parallels in other taxa, is the treefrog *Eleutherodactylus coqui* of Puerto Rico whose "coqui" call has different frequency components addressed to males (the "co" part) and females (the "qui"). Responses of each sex were obtained by Narins and Capranica (1976) in a series of playback experiments using partial and complete calls. The pulse rates of frog calls vary with temperature. Using two sympatric treefrogs (*Hyla versicolor*, *H. chrysoscelis*), whose calls differed in this parameter, Gerhardt (1978) showed that the responses of females were temperature dependent so that they tracked the changing pulse rate of the male calls of their own species.

Intraspecific variation in anuran calls has given rise to a number of studies using playback to investigate the possibilities of sexual selection through female choice and male-male encounters. Within a species, larger individuals tend to have deeper calls and several investigations have been based on this phenomenon. For example, Ryan (1981) showed that females of the tungara frog (*Physalaemus pustulosus*) prefer to approach calls of low, rather than high, fundamental frequency. The larger males that give deeper calls enjoy greater mating success than do smaller males. Searcy and Andersson (1986) review other studies in which female responses varied with calling rate or calling order of males. Gerhardt (1982) investigated sound pattern recognition in treefrogs (Hylidae) and has recently (Gerhardt 1991) reviewed the literature on female mate choice.

Bullfrogs have territorial calls that elicit aggression in other males (e.g. Wiewandt 1969). A case involving differential male responses was reported by Davies and Halliday (1978). Presented with playbacks of calls from large and small conspecifics, common toads (*Bufo bufo*) showed fewer and shorter attacks in response to the deeper "croaks" of the large individuals.

Two other topics have been investigated in treefrogs (*Hyla* spp.) using playback: the energetics of calling (e.g. Wells and Taigen 1986) and graded signalling (e.g. Wells and Schwartz 1984).

Another respect in which amphibia show similarity with insects is the tendency to call in large groups or choruses. Klump and Gerhardt will discuss research on this phenomenon.

Thus a wide range of acoustic behaviour has been investigated in amphibia using playback techniques. I could find few examples in reptiles and will not discuss this group.

Mammals

One of the earliest examples of playback was the demonstration by Garner (1892), nearly a hundred years ago, that rhesus monkeys (*Macaca mulatta*) would answer a disc recording of their food call. In view of the complex vocal repertoires and acoustic behaviour of some species, one might expect a large body of playback work with mammals. However, such does not appear to be the case. At the bioacoustics conference in 1956 there was much excitement about the work of Griffin and others on echolocation and new discoveries on cetacean vocalisations, but there were no reports involving playback. When the proceedings appeared, there was a long descriptive chapter on mammalian sounds by Tembrock (1963), who only briefly referred to two playbacks he had tried on foxes (*Vulpes vulpes*). One suggested recognition of his mate's call by a male; in the other case puppy sounds caused a male to collect and bring food to an empty box. The book edited by Lanyon and Tavolga (1960) had no chapter devoted to mammalian sounds.

From 1959 through 1963, my colleague Douglas Pimlott used recorded wolf howls in a study of the ecology of timber wolves (*Canis lupus*) in Algonquin Park, Ontario (Pimlott et al. 1969). Although negative results were hard to interpret, playback helped to determine the number, composition and movements of wolf packs. Also some insight was gained into seasonal and other factors affecting wolf howling. I can recall helping to obtain recordings of wild packs. On one occasion I preceded the truck carrying the playback equipment to a site where a pack normally howled. As the truck approached in the distance, the wolves began to howl and, by the time the driver got out, the howling was all over. It seemed that the wolves had become conditioned to the sound of the truck. This may have accounted for some negative results. As this study proceeded it was found that a human imitation of a wolf howl could be as effective as a recording in eliciting a response. This saved carrying heavy equipment in the bush. Since then, wolf "howling" has been a popular feature of the interpretive programme in Algonquin Park, with as many as 1,000 carloads of people in attendance. John Theberge and I carried out a descriptive study of howling by captives and their responses to imitations (Theberge and Falls 1967) but we were unable to use playback in the field for logistic reasons.

Playback studies have demonstrated the abilities of mammals to discriminate between sounds in a number of contexts. The ability of grey-cheeked mangabeys (*Cercocebus albigena*) to tell the "whoop gobbles" of strangers from those of their own band members was discovered by Waser (1975). Harrington (1986) showed that wild timber wolves discriminated between howls of adults and pups. Recognition by mothers of their own young in colonial or group situations has been demonstrated in a variety of species (e.g. reindeer (*Rangifer tarandus*) Espmark 1971; northern elephant seals (*Mirounga angustirostris*) Petrinovich 1974; Mexican free-tailed bat (*Tadarida brasiliensis mexicana*) Balcombe 1990). These studies parallel similar work on seabirds. Bottlenosed dolphins (*Tursiops truncatus*) have been shown to discriminate between "signature-whistles" of different individuals (Caldwell et al. 1969).

Discrimination is also apparent in predator-prey relationships. The widely quoted playback experiments of Cheney and Seyfarth (e.g. 1982) showed that a series of grunts of vervet monkeys (*Cercopithecus aethiops*) transmitted information about specific predators. Another well known paper is that of Tuttle and Ryan (1981) in which they demonstrated the ability of the tropical bat *Trachops cirrhosus* to discriminate between calls of large, small, poisonous and edible frogs, suggesting that predation may have influenced the evolution of calling behaviour in the frogs.

A few further examples will indicate the possibilities of using playback methods with mammals. Bats of several species (Microchiroptera) have been shown to respond to echolocation calls, mainly of conspecifics. This "eavesdropping" may help in finding resources (e.g. Barclay 1982; Balcombe and Fenton 1988). Playback of killer whale (*Orcinus orca*) screams have apparently intimidated other whales (e.g. white whale (*Delphinapterus leucas*) Fish and Vania 1971; grey whale (*Eschrichtius robustus*) Cummings and Thompson 1971). Tyack (1983) has obtained dramatic results playing back social sounds of humpback whales (*Megaptera novaeangliae*). Recent work on infrasounds of African elephants (*Loxodonta africana*) (Payne 1989) promises to help explain apparent long distance coordination in the movements of different groups. Bulls responded to playback of the call of an oestrous female from as far as two and a half miles away. I should also mention the work of Clutton-Brock and Albon (1979) on the roaring of red deer stags (*Cervus elephus*). Using playback of recorded roars, they showed that stags engage in vocal battles and avoid fights with opponents that they are unable to out-roar. More recently it has been found that roaring in this species influences ovulation and mate attraction in females (McComb 1987, 1991).

Despite the work I have cited, it appears that playback has been used relatively little in research on mammalian sounds. In their review in 1983, Hawkins and Myrberg describe most acoustic work on marine mammals as anecdotal and deplore the shortage of playback experiments. Perhaps more to the point, in the monumental compendium edited by Sebeok in 1977, half the book is devoted to mammals, including several authoritative, descriptive accounts of acoustical behaviour. Yet only two serious playback studies are mentioned and these dealt with topics investigated in birds two decades earlier. In contrast, many such studies are cited in the shorter chapters on other groups. No doubt there has been a considerable increase in mammalian work in the last decade, much of which I have missed but it has been late in coming.

Why has there been so little work involving playback with mammals? Many mammals are tuned more to visual and olfactory signals than to sounds and are not very vocal. Many are nocturnal and/or widely dispersed. Marine mammals are not difficult to record but it is hard to study their responses (Susan Cosens of Fisheries and Oceans, Canada informs me that the great expense and logistic difficulties of mounting playback experiments in the sea has discouraged their use). Moreover, the signals of mammals are often, at least partly, supersonic. All these things may have discouraged experimentation. In any case it appears that mammalogists have been interested in other subjects. Clearly there are great opportunities for anyone interested in applying playback techniques in mammalian studies. Karen McComb will describe her recent work with lions later in the conference.

Birds

Now I shall attempt a brief survey of the uses of playback in studies of birds. A rereading of Kroodsma et al.'s (1982) books suggests that playback either has been or could be used in most of the areas reviewed there. Indeed, there has been so much work and it is so varied that I have had difficulty organising this presentation. In any case, I can only touch on a few lines of investigation.

Although I have already mentioned some of the early work with birds, I should like to return to this subject briefly for comparison with other groups. Research before 1960 was concerned with both call notes and songs. Collias and Joos (1953), as part of a study of the calls of domestic fowl (*Gallus gallus domesticus*) played recordings to chicks to observe their effect. In 1954, Frings and Jumber experimented with distress calls of starlings (*Sturnus vulgaris*) in an effort to break up roosts; a year later, Busnel et al. (1955) reported similar work with European corvids. They soon collaborated in comparing responses of crows and gulls to recordings from both sides of the Atlantic (e.g. Frings et al. 1958).

In 1937, Arthur Allen described the response of a mockingbird (*Mimus polyglottos*) to playback of song in its territory. In 1951, Gunn used playback to capture woodcock. However, the first major experimental study, comparing responses of North American thrushes (*Hylocichla*, *Catharus*) to songs and models of their own and other species, was published by Dilger in 1956. This preceded investigations of song components important to species recognition that were soon to follow (e.g. Busnel and Brémond 1961; Abs 1963; Falls 1963). My students and I also investigated interspecific responses (Falls and Szijj 1959) and neighbour-stranger discrimination (Weeden and Falls 1959).

While these developments were taking place in North America, Thorpe (1958) in England, published the first account of song learning in the chaffinch (*Fringilla coelebs*) using playback as models. In the same laboratory, Hinde (1958) investigated factors influencing repertoire use in the chaffinch. Thus, by 1960, a wide range of studies involving playback were under way.

Responses of Females

I now turn to a review of the topics investigated in birds. Compared with insects, fish and frogs, wild birds have not shown obvious female responses to playback of male song. Only recently has female approach to song been demonstrated in pied flycatchers (*Ficedula hypoleuca*) (Eriksson and Wallin 1986). More females were captured in empty nest boxes with dummy males if song was broadcast. Much earlier, playback of male song to captive females had been shown to affect gonadal development in budgerigars (*Melopsittacus undulatus*) (Brockway 1965) and nest building behaviour in canaries (*Serinus canaria*) (Kroodsma 1976). Captive females of the brown-headed cowbird (*Molothrus ater*) perform precopulatory displays in response to playback of male song, providing a bioassay of features that affect the potency of song in this species (e.g. King and West 1983). Females of several other species, when implanted with oestradiol, have shown a heightened tendency to perform similar displays and this technique has been used to assess the effects of song parameters such as repertoire size on responses of females (e.g. song sparrows (*Melospiza melodia*) Searcy and Marler 1981). Catchpole used this same method in his studies of sexual selection on songs of *Acrocephalus* warblers (see Catchpole 1987 and references therein). William Searcy will be discussing this approach in his paper. One of our own studies (Dickinson and Falls 1989) investigated the effects of broadcasting female vocalisations to male and female western meadowlarks (*Sturnella neglecta*). Responses indicated conflict between members of a pair.

Responses of Males

Turning to male responses, much of the research involving playback of bird song has been done in the context of territorial defence in response to simulated intrusions. This has been used to demonstrate species discrimination and the parameters of song important to species recognition (see review of Becker 1982). These were among the earliest topics to be investigated with playback, and are still being pursued, for example by Nelson (1988) and Ratcliffe and her associates (e.g. Hurly et al. 1990). Laurene Ratcliffe's paper will illustrate her current approaches to this area.

Individual recognition of songs is closely allied to species recognition (see review of Falls 1982). Following our early demonstration of neighbour-stranger discrimination, experiments in which neighbours' songs were presented from unexpected directions showed that birds actually recognised the individual songs of neighbours. Further studies examined the structural basis of this recognition. It was also found that birds with repertoires performed less well than species with single songs in typical neighbour-stranger experiments. More work is needed to explain these results. Work on neighbour-stranger discrimination has continued, with at least 35 studies (including 22 species in 10 families) showing reduced response to neighbours (the "dear enemy" effect), as opposed to one in the opposite direction (Grove 1981). Recent work includes Weary et al. (1986) on the veery (*Catharus fuscescens*), Stoddard et al. (1990) on the song sparrow and the first case of neighbour recognition in a colonial seabird, the Adélie penguin (*Pygoscelis adeliae*) (Speirs and Davis 1991). Recently, McGregor and Avery (1986) have shown that great tits (*Parus major*) continue to learn neighbours' songs as they grow older but their results also suggest that a bird's memory for songs of other individuals may be limited.

Playback has proven useful in showing behavioural differences in singing and response to song between colour morphs of the white-throated sparrow (e.g. Lowther 1962). Discrimination among dialects by both males and females has also been studied using playback (see Baker and Cunningham 1985 for discussion and references) but the questions raised have not been resolved. Identification of kin based on songs has not been demonstrated but playback methods combined with molecular techniques to establish relationships, may make it possible to separate neighbour, dialect and kin recognition.

Several attributes of countersinging behaviour have been studied using playback. For example, one can investigate factors influencing the rate and timing of switching between song types in species with repertoires (Falls and d'Agincourt 1982; Kramer et al. 1985). Related studies have examined the effects on listeners of switching and contrast between successive songs (Horn and Falls 1988; Falls et al. 1990). Morton (1982) proposed that song repertoires may be used to disturb neighbours and thereby reduce their fitness. Predictions from this hypothesis could be tested using song playback.

Another attribute of countersinging in species with song repertoires is the tendency for a bird to match songs it hears if it has them in its repertoire. This is referred to as matched countersinging. The phenomenon was first observed by Hinde (1958) using song playback to chaffinches and has since been the subject of a number of studies (see references in Falls and Krebs 1975). It was suggested that matching represented a strong response, identifying the intended receiver. We attempted to test this hypothesis by measuring the frequency of matching in response to playback of strangers and neighbours and at the centre and edge of the territory. Both comparisons were expected to elicit responses of different strength. Our experiments were carried out using song types shared by the subjects, which were great tits (Falls et al. 1982) and western meadowlarks (Falls 1985). The frequency of matching did not vary with other measures of response strength but did increase in proportion to the similarity of the playback to the subject's own song.

Thus, in great tits, where neighbours' songs are similar, they were matched more than strangers. However, in meadowlarks, where neighbours' and strangers' songs were equally similar to a subject's song, more matching occurred to strangers' songs. It seemed that, other things being equal, familiarity reduced matching. We also found in meadowlarks that, when a bird did match, the times when it sang suggested that it had been entrained by the playback. Thus, the playback may have started a bout of singing, affecting the bird in the same way as its own previous songs. Todt (1981) has also examined timing by manipulating the onset of matching playback in relation to the subject's song in European blackbirds (*Turdus merula*). Since a bird will match songs of the same type as its own, we can play back songs that are slightly different to investigate how it classifies them. We have conducted experiments of this type with western meadowlarks (Falls et al. 1988) and great tits (Weary et al. 1990). Andrew Horn will discuss various approaches to the question of how a bird classifies conspecific songs. This brief excursion into matching shows how a simple observation has led to increasingly complex investigations. Clearly there is much more to be done before we understand this behaviour and I think playback will continue to be a useful tool in further research.

Some uses of playback have tackled the role of song in territoriality. The most direct approach has been to replace territory holders with speakers broadcasting song recordings and monitor the amount of intrusion (Göransson et al. 1974; Krebs 1977; Yasukawa 1981; Falls 1988). In the species where this has been tried, intrusions have been reduced or delayed, demonstrating a "keep out" (or at least a "keep quiet") meaning of territorial song. Krebs et al. (1978) showed that this effect increases in great tits if a repertoire of songs is played instead of a single type. Other studies have shown that the strength of a territory holder's response varies from the centre to the edge of the territory (e.g. Dhondt 1966). Playback can be used to define the shape of this "defence function" (Melemis and Falls 1982). Earlier, I described a method of mapping territories using playback. Recently, Jones (1987) used this method as well as telemetry to show that white-throated sparrows have more or less contiguous and non-overlapping song territories but move quietly over larger overlapping home ranges. Leary(1991) has used playback to enumerate the various types of intruders in *Junco* territories.

This may be an appropriate place to mention the possibility of using song and singing behaviour as a way of measuring male quality. To date there has only been a little work in this area involving playback (e.g. Weary et al. 1991). One of the active investigators, Marcel Lambrechts, will discuss this question in his paper.

A number of studies have examined interspecific responses, comparing results of playback in situations where closely related species are sympatric or allopatric (e.g. Falls and Szijj 1959; Emlen et al. 1975; Catchpole 1978). Interspecific responses have tended to occur in sympatry but not in allopatry, suggesting that these responses are learned. In a different situation, Reed (1982) found interspecific responses between chaffinches and great tits on Scottish islands, where resources were in short supply, but not on the mainland. Thus, playback has proven to be useful in showing the current relationships between species where they are in contact. Whether it is useful in elucidating phylogenetic relationships is another matter, and the possibilities and problems of this application are discussed by Payne (1986).

Ontogeny of Vocalisations

One area in which studies of bird song far exceed work with any other group concerns the ontogeny of vocalisations. Results of this research are taken to be a classic example of the interplay of inherited tendencies and environmental influences in the development of behaviour. While many experiments have used live tutors, playback of song to hand raised young began with the pioneer work of Thorpe (1958) and continued through the increasingly sophisticated studies of Marler and his co-workers. The defined signals provided by playback helped in measuring sensitive periods for song learning (see Kroodsma's 1982 review) and contributed to the recognition of stages in song development (e.g. Marler and Peters 1982). Douglas Nelson will discuss more recent developments in this area.

Sound Perception

Playback has also contributed to studies of sound perception in birds. Marler's (1955) proposals regarding locatability of different alarm calls led to the laboratory studies of Konishi (1977) and others. Richards' (1981a) demonstration that Carolina wrens (*Thryothorus ludovicianus*) discriminated between playback of degraded and undegraded songs (recorded at different distances from the source) led to further work on distance assessment. Gish and Morton (1981) showed that Carolina wren songs degraded less with distance in the habitats where they were sung than elsewhere. Other studies showed that birds had different vocalisations for long and short distance communication, with the former showing less degradation when played and recorded at different distances (e.g. Cosens and Falls 1984). McGregor et al. (1983) showed that great tits were only able to discriminate between degraded and undegraded songs if they or their neighbours had the songs in their repertoires. These and other studies (e.g. on song matching) suggest that birds use their own songs as standards in the way they perceive others. Research already mentioned suggests that song contrast enables listeners to notice switches between song types. Further studies of this type combining playback with environmental noise and measuring degradation with distance would help us to understand the structure and sequencing of song repertoires. Other playback experiments in this area are concerned with alerting components of song (e.g. Shiovitz 1975; Richards 1981b). I cannot leave the topic of perception without mentioning the neurological work of Margoliash and Konishi (1985) on cell populations in the brain that are sensitive to particular song elements. The combination of such laboratory research with field studies offers exciting prospects.

Call Notes

Finally, I should mention studies of bird calls other than song. The playback of distress calls by Frings, Busnel and their associates was among the first uses of the technique. These early workers were interested in dealing with pest problems but they also examined the structure of the calls and their interspecific as well as intraspecific effects. Later Stefanski and Falls (1972a, 1972b) followed both these lines of investigation with distress calls of three North American sparrows, (Emberizinae). Research on locating and non-locating alarm calls has already been mentioned. The effects of distress and alarm calls on prey and predator behaviour offer further possibilities for playback experimentation. Individual recognition between mates and between parents and young has been largely demonstrated using playback of calls (see Falls 1982). Much of this work has taken place in colonies of seabirds and swallows, where recognition poses serious problems. Studies with less crowded birds are few (e.g. Nowicki 1983).

In general, there is a dearth of research using playback to study the effects of stereotyped and graded calls on listeners. Possibilities in this area include testing Morton's (1982) structural-motivational rules.

Table II. Suggested topics for further playback studies.

Topics
Features of song related to male quality
Repertoires of songs with different meanings
Sequencing and timing of songs in interactions
Sequencing and contrast in relation to environment
Interspecific responses in relation to mimicry
Kin recognition through vocalisations
Habituation in the field
Disruptive effects of song (part of Morton's *ranging* hypothesis)
Variation in "stereotyped" signals
Meanings of call notes
Meanings of graded signals
Motivational / structural relationships (e.g. Morton's rules)
Physiological monitoring of responses in lab and field (telemetry)
Predator / prey interactions

Table II brings together my suggestions for topics that could profitably receive more attention using playback. The list is by no means complete. It was drawn up with birds in mind but may have some application to other groups. I feel sure that each of you could add to the list.

Concluding Remarks

This outline has been sketchy, incomplete and is undoubtedly biased towards my interests. However, I think it does show the variety of subjects that have been tackled and the considerable advances that have been made through the use of sound playback. We in this room are among the active practitioners of this technique. Now that it is well established, it seems appropriate that we should pause to consider problems and new developments in its use. There is still much to do and prospects for the future are very promising. I look forward to your papers and our discussions.

To end on a personal note, some of my most interesting and convincing results were obtained by accident. Scientific investigation can only partly be organised. As you and others continue to use playback experiments and to try and improve their design, I trust you will not lose the spontaneity, imagination and enthusiasm that characterised the early years of this study.

Acknowledgements

Without the work of my students and associates over the years, this account would have been considerably impoverished. I thank Glenn Morris and Brock Fenton for briefing me on work with insects and bats, respectively. My wife Ann and Rossana Soo assisted greatly with preparation of the manuscript. Peter McGregor and Xanthe Whittaker helped in the editing process.

References

Abs, M. 1963. Field tests on the essential components of the European nightjar's song. *Proc. Int. Ornithol. Congr.*, **13**, 202-205.

Alexander, R.D. 1961. Aggressiveness, territoriality, and sexual behaviour in field crickets (Orthoptera: Gryllidae). *Behaviour*, **17**, 130-223.

Allen, A.A. 1937. Hunting with a microphone the voices of vanishing birds. *Nat. Geographic*, **71**, 696-723.

Baker, M.C. and Cunningham, M.A. 1985. The biology of bird-song dialects. *Behav. Brain Sci.*, **8**, 85-133.

Balcombe, J.P. 1990. Vocal recognition of pups by mother Mexican free-tailed bats, *Tadarida brasiliensis mexicana*. *Anim. Behav.*, **39**, 960-966.

Balcombe, J.P. and Fenton, M.B. 1988. Eavesdropping by bats: the influence of echolocation call design and foraging strategy. *Ethology*, **79**, 158-166.

Barclay, R.M.R. 1982. Interindividual use of echolocation calls: eavesdropping by bats. *Behav. Ecol. Sociobiol.*, **10**, 271-275.

Becker, P.H. 1982. The coding of species-specific characteristics in birds sounds. In: *Evolution and Ecology of Acoustic Communication in Birds. Vol.I.* (Ed. by D.E. Kroodsma, E.H. Miller & H. Ouellet), pp. 213-252. Academic Press; New York.

Beer, C.G. 1982. Conceptual issues in the study of communication. In: *Evolution and Ecology of Acoustic Communication in Birds. Vol.II.* (Ed. by D.E. Kroodsma, E.H. Miller & H. Ouellet), pp. 279-310. Academic Press; New York.

Blair, W.F. and Littlejohn, M.J. 1960. Stage of speciation of two allopatric populations of chorus frogs (*Pseudacris*). *Evolution*, **14**, 82-87.

Bogert, C.M. 1947. A field study of homing in the Carolina toad. *Am. Mus. Novit.*, **1355**, 1-24.

Brockway, B.F. 1965. Stimulation of ovarian development and egg laying by male courtship vocalization in budgerigars (*Melopsittacus undulatus*). *Anim. Behav.*, **13**, 575-578.

Busnel, R.-G. (Ed.) 1963. *Acoustic Behaviour of Animals.* Elsevier; Amsterdam.

Busnel, R.-G. and Brémond, J.-C. 1961. Étude préliminaire du décodage des informations contenues dans le signal acoustique territorial du rouge-gorge (*Erithacus rubecula* L.). *Compt. Rend. Acad. Sci.*, **252**, 608-610.

Busnel, M.-C. and Busnel, R.-G. 1955. La directivité acoustique des déplacements de la femelle d'*Oecanthus pellucens* (Scop.). In: *L'Acoustique des Orthopteres.* (Ed. by R.-G. Busnel), pp. 356-364. Inst. Nat. de la Recherche Agronomique; Paris.

Busnel, R.G., Giban, J., Gramet, Ph. and Pasquinelly, F. 1955. Observations préliminaires de la phonotaxie négative des corbeaux à des signaux acoustiques naturels. *Compt. Rend. Acad. Sci.*, **241**, 1846-1849.

Cade, W. 1975. Acoustically orienting parasitoids: fly phonotaxis to cricket song. *Science*, **190**, 1312-1313.

Cade, W.H. 1981. Field cricket spacing, and the phonotaxis of crickets and parasitoid flies to clumped and isolated cricket songs. *Z. Tierpsychol.*, **55**, 365-375.

Cade, W.H. and Wyatt, D.R. 1984. Factors affecting calling behaviour in field crickets, *Teleogryllus* and *Gryllus* (age, weight, density and parasites). *Behaviour*, **88**, 61-75.

Caldwell, M.C., Caldwell, D.K. and Hall, N.L. 1969. An experimental demonstration of the ability of an Atlantic bottlenosed dolphin to discriminate between whistles of other individuals of the same species. *Los Angeles Co. Mus. Tech. Report*, **6**.

Capranica, R.R. 1966. Vocal response of the bullfrog to natural and synthetic mating calls. *J. Acoustical Soc. Am.*, **40**, 1131-1139.

Catchpole, C.K. 1978. Interspecific territorialism and competition in *Acrocephalus* warblers as revealed by playback experiments in areas of sympatry and allopatry. *Anim. Behav.*, **26**, 1072-1080.

Catchpole, C.K. 1987. Bird song, sexual selection and female choice. *Trends Ecol. Evol.*, **2**, 94-97.

Cheney, D.L. and Seyfarth R.M. 1982. How vervet monkeys perceive their grunts: field playback experiments. *Anim. Behav.*, **30**, 739-751.

Clutton-Brock, T.H. and Albon, S.D. 1979. The roaring of red deer and the evolution of honest advertisement. *Behaviour*, **69**, 145-169.

Collias, N. and Joos, M. 1953. The spectrographic analysis of sound signals of the domestic fowl. *Behaviour*, **5**, 175-188.

Cosens, S.E. and Falls, J.B. 1984. Structure and use of song in the yellow-headed blackbird (*Xanthocephalus xanthocephalus*). *Z. Tierpsychol.*, **66**, 227-241.
Crankshaw, O.S. 1979. Female choice in relation to calling and courtship song in *Acheta domesticus*. *Anim. Behav.*, **27**, 1274-1275.
Cummings, W.C. and Thompson, P.O. 1971. Gray whales, *Eschrichtius robustus*, avoid the underwater sounds of killer whales, *Orcinus orca*. *Fish. Bull.*, **691**, 525-530.
Davies, N.B. and Halliday, T.R. 1978. Deep croaks and fighting assessment in toads, *Bufo bufo*. *Nature*, **274**, 683-685.
Dhondt, A.A. 1966. A method to establish boundaries of bird territories. *Le Gerfaut*, **56**, 404-408.
Dickinson, T.E. and Falls, J.B. 1989. How western meadowlarks respond to simulated intrusions by unmated females. *Behav. Ecol. Sociobiol.*, **25**, 217-225.
Diehl, P., & Helb, H.-W. 1986. Radiotelemetric monitoring of heart-rate responses to song playback in blackbirds (*Turdus merula*). *Behav. Ecol. Sociobiol.*, **18**, 213-219.
Dilger, W.C. 1956. Hostile behavior and reproductive isolating mechanisms in the avian genera *Catharus* and *Hylocichla*. *Auk*, **73**, 313-353.
Dunning, D.C. and Roeder, K.D. 1965. Moth sounds and the insect catching behaviour of bats. *Science*, **147**, 173-174.
Emlen, S.T., Rising, J.D. and Thompson, W.L. 1975. A behavioral and morphological study of sympatry in the indigo and lazuli buntings of the great plains. *Wilson Bull.*, **87**, 145-179.
Eriksson, D. and Wallin, L. 1986. Male bird song attracts females - a field experiment. *Behav. Ecol. Sociobiol.*, **19**, 297-299.
Espmark, Y. 1971. Individual recognition by voice in reindeer mother-young relationship. Field observation and playback experiments. *Behaviour*, **40**, 295-301.
Ewing, A.W. 1989. *Arthropod Bioacoustics: Neurobiology and Behaviour.* Cornell University Press; Ithaca.
Falls, J.B. 1963. Properties of bird song eliciting responses from territorial males. *Proc. XIII Int. Ornithol. Congr.*, 259-271.
Falls, J.B. 1981. Mapping territories with playback: an accurate census method for songbirds. In: *Estimating the Numbers of Terrestrial Birds.* (Ed. by C.J. Ralph & J.M. Scott). *Stud. Avian Biol.*, **6**, 86-91.
Falls, J.B. 1982. Individual recognition by sounds in birds. In: *Evolution and Ecology of Acoustic Communication in Birds. Vol.II.* (Ed. by D.E. Kroodsma, E.H. Miller & H. Ouellet), pp. 237-278. Academic Press; New York.
Falls, J.B. 1985. Song matching in western meadowlarks. *Can. J. Zool.*, **63**, 2520-2524.
Falls, J.B. 1988. Does song deter intrusion in white-throated sparrows? *Can. J. Zool.*, **66**, 206-211.
Falls, J.B. and d'Agincourt, L.G. 1982. Why do meadowlarks switch song-types? *Can. J. Zool.*, **60**, 3400-3408.
Falls, J.B., Dickinson, T.E. and Krebs, J.R. 1990. Contrast between successive songs affects the response of eastern meadowlarks to playback. *Anim. Behav.*, **39**, 717-728.
Falls, J.B., Horn, A.G. and Dickinson, T.E. 1988. How western meadowlarks classify their songs: evidence from song matching. *Anim. Behav.*, **36**, 579-585.
Falls, J.B. and Krebs, J.R. 1975. Sequence of songs in repertoires of western meadowlarks (*Sturnella neglecta*). *Can. J. Zool.*, **53**, 1165-1178.
Falls, J.B., Krebs, J.R. and McGregor, P.K. 1982. Song matching in the great tits (*Parus major*): the effects of similarity and familiarity. *Anim. Behav.*, **30**, 997-1009.
Falls, J.B. and McNicholl, M.K. 1979. Neighbor-stranger discrimination by song in male blue grouse. *Can. J. Zool.*, **57**, 457-462.
Falls, J.B. and Szijj, L.J. 1959. Reactions of eastern and western meadowlarks in Ontario to each others' vocalizations. *Anat. Rec.*, **134**, 560.
Fish, J.F. and Vania, J.S. 1971. Killer whale, *Orcinus orca*, sounds repel white whales, *Delphinapterus leuca*. *Fish. Bull.*, **69**, 531-535.
Forrest, T.G. 1980. Phonotaxis in mole crickets: its reproductive significance. *Fla. Entomol.*, **63**, 45-53.
Frings, H., Frings, M., Jumber, J., Busnel, R.-G., Giban, J. and Gramet, Ph. 1958. Reactions of American and French species of *Corvus* and *Larus* to recorded communication signals tested reciprocally. *Ecology*, **39**, 126-131.
Frings, H. and Jumber, J. 1954. Preliminary studies in the use of a specific sound to repel starlings (*Sturnus vulgaris*) from objectionable roosts. *Science*, **119**, 318-319.
Garner, R.L. 1892. *The Speech of Monkeys*. New York; C.L. Webster.

Gerhardt, H.C. 1974. The significance of some spectral features in mating call recognition in the green treefrog (*Hyla cinerea*). *J. Exp. Biol.*, **61**, 229-241.
Gerhardt, H.C. 1978. Temperature coupling in the vocal communication system of the gray treefrog, *Hyla versicolor*. *Science*, **199**, 992-994.
Gerhardt, H.C. 1991. Female mate choice in treefrogs: static and dynamic acoustic criteria. *Anim. Behav.*, **42**, 615-635.
Gish, S.L. and Morton, E.S. 1981. Structural adaptations to local habitat acoustics in Carolina wren songs. *Z. Tierpsychol.*, **56**, 74-84.
Göransson, G., Högstedt, G., Karlsson, J., Källander, H. and Ulfstrand, S. 1974. Sangensroll für revirkallandet hos näktergal, *Luscinia luscinia* - nagra experiment med playback-teknik. *Var Fagelvärld*, **33**, 201-209.
Grove, P.A. 1981. The effect of location and stage of nesting on neighbor/stranger discrimination in the house wren. *unpublished* Ph.D. Thesis, City University of New York.
Gunn, W.W.H. 1951. *The Woodcock Program.* Ontario Dept. of Lands and Forests, Div. of Res., Wildlife Sec. 30 pp. (mimeographed).
Harrington, F.H. 1986. Timber wolf howling playback studies: discrimination of pup from adult howls. *Anim. Behav.*, **34**, 1575-1577.
Haskell, P.T. 1957. Stridulation and associated behaviour in certain Orthoptera, I. Analysis of the stridulation of, and behaviour between, males. *Brit. J. Anim. Behav.*, **5**, 139-148.
Hawkins, A.D. and Myrberg, A.A.,Jr. 1983. Hearing and sound communication under water. In: *Bioacoustics: a Comparative Approach.* (Ed. by B. Lewis), pp. 347-405. Academic Press; London.
Heady, S.E. and Denno, R.F. 1991. Reproductive isolation in *Prokelisia* planthoppers (Homoptera: Delphacidae): acoustic differentiation and hybridization failure. *J. Insect Behav.*, **4**, 367-390.
Hinde, R.A. 1958. Alternative motor patterns in chaffinch song. *Anim. Behav.*, **6**, 211-218.
Holling, C.S. 1964. The analysis of complex population processes. *Can. Entomol.*, **96**, 335-347.
Horn, A.G. and Falls, J.B. 1988. Response of western meadowlarks (*Sturnella neglecta*) to song repetition and contrast. *Anim. Behav.*, **36**, 291-293.
Huber, F. and Thorson, J. 1985. Cricket auditory communication. *Sci. Am.*, **253**, 60-68.
Hurly, T.A., Ratcliffe, L. and Weisman, R. 1990. Relative pitch recognition in white-throated sparrows, *Zonotrichia albicollis*. *Anim. Behav.*, **40**, 176-181.
Jones, J. 1987. Use of space by male white-throated sparrows (*Zonotrichia albicollis*). *unpublished* Ph.D. Thesis, University of Toronto.
Kahn, M.C. and Offenhauser, W.,Jr. 1949. First field tests of recorded mosquito sound used for mosquito destruction. *Am. J. Trop. Med.*, **29**, 811-825.
King, A.P. and West, M.J. 1983. Dissecting cowbird song potency: assessing a song's geographic identity and relative appeal. *Z. Tierpsychol.*, **63**, 37-50.
Konishi, M. 1977. Spatial localization of sound. In: *Dahlem Workshop on Recognition of Complex Acoustic Signals.* (Ed. by T. Bullock), pp. 127-143. Dahlem Konf.; Berlin.
Kramer, B. 1990. *Electrocommunication in Teleost Fishes. Behaviour and Experiments*. Springer-Verlag; Berlin.
Kramer, H.G., Lemon, R.E. and Morris, M.J. 1985. Song switching and agonistic stimulation in the song sparrow (*Melospiza melodia*): five tests. *Anim. Behav.*, **33**, 135-149.
Krebs, J.R. 1977. Song and territory in the great tit. In: *Evolutionary Ecology.* (Ed. by B. Stonehouse & C.M. Perrins), pp. 47-62. Macmillan; London.
Krebs, J.R., Ashcroft, R. and Webber, M. 1978. Song repertoires and territory defence in the great tit. *Nature*, **271**, 539-542.
Kroodsma, D.E. 1976. Reproductive development in a female songbird: differential stimulation by quality of male song. *Science*, **192**, 574-575.
Kroodsma, D.E. 1982. Learning and the ontogeny of sound signals in birds. In: *Evolution and Ecology of Acoustic Communication in Birds. Vol.II.* (Ed. by D.E. Kroodsma, E.H. Miller & H. Ouellet), pp. 125-146. Academic Press; New York.
Kroodsma, D.E., Miller, E.H. and Ouellet, H. 1982. *Acoustic Communication in Birds. Vol. I. Production, Perception, and Design Features of Sounds*. Academic Press; New York.
Kroodsma, D.E., Miller, E.H. and Ouellet, H. 1982. *Acoustic Communication in Birds. Vol. II. Song Learning and Its Consequences*. New York; Academic Press.
Lanyon, W.E. 1963. Experiments on species discrimination in *Myiarchus* flycatchers. *Am. Mus. Novit.*, **2126**, 1-16.

Lanyon, W.E. and Tavolga, W.N. (Eds.). 1960. *Animal Sounds and Communication.* AIBS; Washington.
Leary, J. 1991. Intruders on yellow-eyed *Junco* territories. *Wilson Bull.*, **103**, 292-295.
Littlejohn, M.J. and Michaud, T.C. 1959. Mating call discrimination by females of Strecker's chorus frog (*Pseudacris streckeri*). *Tex. J. Sci.*, **11**, 86-92.
Lowther, J.K. 1962. Colour and behavioural polymorphism in the white-throated sparrow, *Zonotrichia albicollis* (Gmelin). *unpublished* Ph.D. Thesis, University of Toronto.
Margoliash, D. and Konishi, M. 1985. Auditory representation of autogenous song in the song system of white-crowned sparrows. *Neurobiology*, **82**, 5997-6000.
Marler, P. 1955. Characteristics of some animal calls. *Nature*, **176**, 6-8.
Marler, P.R. and Peters, S. 1982. Subsong and plastic song: their role in the vocal learning process. In: *Evolution and Ecology of Acoustic Communication in Birds. Vol.II.* (Ed. by D.E. Kroodsma, E.H. Miller & H. Ouellet), pp. 25-50. Academic Press; New York.
Martof, B.S. and Thompson, E.F.,Jr. 1958. Reproductive behavior of the chorus frog, *Pseudacris nigrita. Behaviour*, **13**, 243-258.
Melemis, S.M. and Falls, J.B. 1982. The defense function: a measure of territorial behavior. *Can. J. Zool.*, **60**, 495-501.
McComb, K. 1987. Roaring by red deer stags advances the date of oestrous in hinds. *Nature*, **330**, 648-649.
McComb, K.E. 1991. Female choice for high roaring rates in red deer, *Cervus elephus. Anim. Behav.*, **41**, 79-88.
McGregor, P.K. and Avery, M.I. 1986. The unsung songs of great tits (*Parus major*): learning neighbours' songs for discrimination. *Behav. Ecol. Sociobiol.*, **18**, 311-316.
McGregor, P.K., Krebs, J.R. and Ratcliffe, L.M. 1983. The reaction of great tits (*Parus major*) to the playback of degraded and undegraded songs: the effects of familiarity with the stimulus song type. *Auk*, **100**, 898-906.
Morris, G.K. and Fullard, J.H. 1983. Random noise and congeneric discrimination in *Conocephalus* (Orthoptera: Tettigoniidae). In: *Orthopteran Mating Systems.* (Ed. by D.T. Gwynne & G.K. Morris), pp. 73-96. Westview Press; Boulder.
Morton, E.S. 1982. Grading, discreteness, redundancy, and motivation-structural rules. In: *Evolution and Ecology of Acoustic Communication in Birds. Vol.I.* (Ed. by D.E. Kroodsma, E.H. Miller & H. Ouellet), pp. 183-212. Academic Press; New York.
Moulton, J.M. 1956. Influencing the calling of sea robins (*Prionotus* spp.) with sound. *Biol. Bull.*, **111**, 393-398.
Myrberg, A.A.,Jr. 1981. Sound communication and interception in fishes. In: *Hearing and Sound Communication in Fishes.* (Ed. by W.N. Tavolga, A.N. Popper & R.R. Fay), pp. 395-425. Springer-Verlag; New York.
Myrberg, A.A.,Jr, Mohler, M. and Catala, J.D. 1986. Sound production by males of a coral reef fish (*Pomacentrus partitus*): its significance to females. *Anim. Behav.*, **34**, 913-923.
Myrberg, A.A.,Jr and Riggio, R.J. 1985. Acoustically mediated individual recognition by a coral reef fish (*Pomacentrus partitus*). *Anim. Behav.*, **33**, 411-416.
Myrberg, A.A.,Jr, Spanier, E. and Ha, S.J. 1978. Temporal patterning in acoustical communication. In: *Contrasts in Behavior.* (Ed. by E.S. Reese and F.J. Lighter), pp. 137-179. John Wiley; New York.
Narins, P.M. and Capranica, R.R. 1976. Sexual differences in the auditory system of the treefrog, *Eleutherodactylus coqui*. *Science*, **192**, 378-380.
Nelson, D.A. 1987. Song syllable discrimination by song sparrows (*Melospiza melodia*). *J. Comp. Psychol.*, **101**, 25-32.
Nelson, D.A. 1988. Feature weighting in species song recognition by the field sparrow (*Spizella pusilla*). *Behaviour*, **106**, 158-182.
Nowicki, S. 1983. Flock-specific recognition of chickadee calls. *Behav. Ecol. Sociobiol.*, **12**, 317-320.
Payne, K. 1989. Elephant talk. *Nat. Geographic*, **176**, 265-277.
Payne, R.B. 1986. Bird songs and avian systematics. In: *Current Ornithology. Vol. III.* (Ed. by R.J. Johnston), pp. 87-126. Plenum: New York.
Petrinovich, L. 1974. Individual recognition of pup vocalization by northern elephant seal mothers. *Z. Tierpsychol.*, **34**, 308-312.
Pimlott, D.H., Shannon, J.A. and Kolenosky, G.B. 1969. The ecology of the timber wolf in Algonquin Provincial Park. Ont. Dept. Lands and Forests Res. Rep. (Wildlife) No. 87.

Reed, T.M. 1982. Interspecific territoriality in the chaffinch and the great tit on islands and the mainland of Scotland: playback and removal experiments. *Anim. Behav.*, **30**, 171-181.
Regen, J. 1913. Über die Anlockung des Weibchens von *Gryllus campestris* L. durch telephonisch übertragene Stridulationslaute des Männchens. *Arch. Physiol. Menschen u. Tiere*, **155**, 193-200.
Richards, D.G. 1981a. Estimation of distance of singing conspecifics by the Carolina wren. *Auk*, **98**, 127-133.
Richards, D.G. 1981b. Alerting and message components in songs of rufous-sided towhees. *Behaviour*, **76**, 223-249.
Roth, L.M. 1948. A study of mosquito behavior. An experimental laboratory study of the sexual behavior of *Aedes aegypti* L. *Am. Midl. Nat.*, **40**, 265-352.
Ryan, M.J. 1981. Female mate choice in a neotropical frog. *Science*, **209**, 523-525.
Searcy, W.A. and Andersson, M. 1986. Sexual selection and the evolution of song. *Ann. Rev. Ecol. Syst.*, **17**, 507-533.
Searcy, W.A. and Marler, P. 1981. A test for responsiveness to song structure and programming in female sparrows. *Science*, **213**, 926-928.
Searcy, W.A., McArthur, P.D., Peters, S.S. and Marler, P. 1981. Response of male song and swamp sparrows to neighbor, stranger and self songs. *Behaviour*, **77**, 152-163.
Sebeok, T.A. 1977. *How Animals Communicate.* Indiana University Press; Bloomington, Indiana.
Shiovitz, K.A. 1975. The process of species-specific song recognition by the indigo bunting, *Passerina cyanea*, and its relationship to the organization of avian acoustical behavior. *Behaviour*, **55**, 128-179.
Smith, W.J. 1965. Message, meaning and context in ethology. *Am. Nat.*, **99**, 405-409.
Speirs, E.A.H. and Davis, L.S. 1991. Discrimination by Adelie penguins, *Pygoscelis adeliae*, between the loud mutual calls of mates, neighbours and strangers. *Anim. Behav.*, **41**, 937-944.
Stefanski, R.A. and Falls, J.B. 1972a. A study of distress calls of song, swamp and white-throated sparrows (Aves: Fringillidae). I. Intraspecific responses and functions. *Can. J. Zool.*, **50**, 1501-1512.
Stefanski, R.A. and Falls, J.B. 1972b. A study of distress calls of song, swamp and white-throated sparrows (Aves: Fringillidae). II. Interspecific responses and properties used in recognition. *Can. J. Zool.*, **50**, 1513-1525.
Stoddart, P.K., Beecher, M.D. Horning, C.L. and Campbell, S.E. 1990. Strong neighbor-stranger discrimination in song sparrows. *Condor*, **92**, 1051-1056.
Tavolga, W.N. 1958. The significance of underwater sounds produced by males of the gobiid fish, *Bathygobius soporator. Physiol. Zool.*, **31**, 259-271.
Tembrock, G. 1963. Acoustic behaviour of mammals. In: *Acoustic Behaviour of Animals.* (Ed. by R.-G. Busnel), pp. 751-786. Elsevier; Amsterdam.
Theberge, J.B. and Falls, J.B. 1967. Howling as a means of communication in timber wolves. *Am. Zool.*, **7**, 331-338.
Thorpe, W.H. 1958. The learning of song patterns by birds, with especial reference to the song of the chaffinch, *Fringilla coelebs. Ibis*, **100**, 535-570.
Todt, D. 1981. On functions of vocal matching: effect of counter-replies on song post choice and singing. *Z. Tierpsychol.*, **57**, 73-93.
Tuttle, M.D. and Ryan, M.J. 1981. Bat predation and the evolution of frog vocalizations in the neotropics. *Science*, **214**, 677-678.
Tyack, P. 1983. Differential response of humpback whales, *Megaptera novaeangliae*, to playback of song or social sounds. *Behav. Ecol. Sociobiol.*, **13**, 49-55.
Walker, T.J. 1957. Specificity in the response of female tree crickets (Orthop. Gryllidae Oecanthinae) to calling songs of males. *Ann. Entomol. Soc. Am.*, **50**, 626-636.
Waser, P.M. 1975. Experimental playbacks show vocal mediation of intergroup avoidance in a forest monkey. *Nature*, **255**, 56-58.
Weary, D.M., Falls, J.B. and McGregor, P.K. 1990. Song matching and the perception of song types in great tits, *Parus major. Behav. Ecol.*, **1**, 43-47.
Weary, D.M., Lambrechts, M.M. and Krebs, J.R. 1991. Does singing exhaust male great tits? *Anim. Behav.*, **41**, 540-542.
Weary, D.M., Lemon, R.E. and Date, E.M. 1986. Acoustic features used in song discrimination by the veery. *Ethology*, **72**, 199-213.
Weeden, J.S. and Falls, J.B. 1959. Differential responses of male ovenbirds to recorded songs of neighboring and more distant individuals. *Auk*, **76**, 343-351.

Wells, K.D. and Taigan, T.L. 1986. The effect of social interaction on calling energetics in the gray treefrog *Hyla versicolour*. *Behav. Ecol. Sociobiol.*, **19**, 9-18.
Wells, K.D. and Schwartz, J.J. 1984. Vocal communication in a neo-tropical treefrog, *Hyla ebraccata*: aggressive calls. *Behaviour*, **91**, 128-145.
Wiewandt, T.A. 1969. Vocalization, aggregation behavior and territoriality in the bullfrog, *Rana catesbiana*. *Copeia*, **69**, 276-285.
Winn, H.E. 1967. Vocal facilitation and the biological significance of toadfish sounds. In: *Marine Bioacoustics. Vol. 2.* (Ed. by W.N. Tavolga), pp. 283-303. Pergamon Press; New York.
Winn, H.E. 1972. Acoustic discrimination by the toadfish with comments on signal systems. In: *Behavior of Marine Animals, Current Perspectives in Research. Vol. 2, Vertebrates.* (Ed. by H. Winn & B.L. Olla), pp. 361-385. Plenum Press; New York.
Wishart, G. and Riordan, D.F. 1959. Flight responses to various sounds by adult males of *Aedes aegypti* L. (Diptera, Culicidae). *Can. Entomol.*, **91**, 181-191.
Yasukawa, K. 1981. Song repertoires in the red-winged blackbird (*Agelaius phoeniceus*): a test of the beau geste hypothesis. *Anim. Behav.*, **29**, 114-125.

INTEGRATING PLAYBACK: A WIDER CONTEXT

Clive K. Catchpole

Department of Biology
Royal Holloway & Bedford New College
University of London
Egham, Surrey, TW20 0EX
U.K.

Introduction

The scientific study of bird sounds has always been characterised and enriched by a variety of approaches. Classic works by Armstrong (1963) and Hartshorne (1973) were almost purely descriptive and yet both were pre-dated by Thorpe (1961) whose pioneering experimental approach to the development of bird song had far reaching implications influencing a whole generation of ethologists and experimental psychologists. The subject also reflects trends in scientific method and in recent years it has become fashionable to construct hypotheses and then design experiments which test their predictions. The hypothetico-deductive method advocated by Popper (1959), states a clear hypothesis then seeks to falsify it by experiment, whereas the inductive method (Platt 1964) tests alternative hypotheses by experiments which distinguish between them. Although the Popperian approach does not require statistical hypothesis testing, most biologists use inferential statistics in the design, analysis and interpretation of their experimental results.

However, Hurlbert (1984) has claimed that many published studies are flawed, in that they suffer from what he terms pseudoreplication. This is where inferential statistics are wrongly applied to test for treatment effects, where either the treatments are not replicated, or the replicates are not statistically independent. Kroodsma (1989, 1990) has recently reviewed playback experiments in avian bioacoustics and come to similar conclusions. His critical views have been contested by Searcy (1989) and Catchpole (1989a) and the whole area of playback experimental design was an important part of this NATO ARW. The consensus we reached is elaborated in the first chapter of this book and it seems clear from the other chapters that a variety of new playback techniques are emerging which will help to strengthen the rapidly developing field of bioacoustics. It seems to me that there is a danger of placing too much reliance upon one technique and that we should also use other techniques to test out predictions from our hypotheses.

In my view the playback experiment should not be seen as *the* test of a hypothesis, but one step (albeit a final one) in a whole chain of observations and experiments. Only at the end is it then possible to offer a sound interpretation based upon not one, but many, different strands of evidence, finally to accept or refute the main hypothesis in question. This idea is not a new one, but one which can easily be left behind in the rush to design the one perfect experiment immediately to confirm or refute a crucial prediction. The two major reviews of experimental design in ornithology both emphasise this point at some stage. James and McCulloch (1985), in dealing with the Popperian approach, state: "but it does not follow that advances are made only by hypothesis testing" and "The general field of statistics has been moving away from what is now viewed as an overemphasis on statistical hypothesis testing". Kamil (1988) also takes a very balanced view when he states: "there is still room for research based on simpler, less formal ideas. Progress can never take a science to the point where there is no room for the "What if?" type of research".

James and McCulloch (1985) outline a research strategy which starts with observing relationships in nature to help formulate simple verbal models. They also emphasise exploratory data analysis based on observations which they feel researchers should spend far more time on. It is only later that a move towards testing predictions by experiment is advocated. They draw a clear distinction between observational and experimental approaches, the latter being where the researcher controls the levels of certain variables. Kamil (1988) sees less of a dichotomy and more of a continuum, with purely descriptive field work at one end, where the researcher has no control over variables. Interestingly, Hurlbert (1984) considers this type of research to be "mensurative experiments". At the other end of Kamil's continuum is experimental research, where at least one variable is manipulated in true experiments. Hurlbert refers to these as "manipulative experiments".

I now return to my main theme, that an integrated variety of approaches are needed in order to test a hypothesis adequately. I will illustrate this with an example now well established in the literature - the song of the sedge warbler *Acrocephalus schoenobaenus*. Although I have worked on other *Acrocephalus* species, the sedge warbler and its extraordinary song have kept me occupied at intervals over twenty years. In many ways it also illustrates the main points that I am trying to get across.

A Field Observation

When I first watched and listened to a singing male sedge warbler I was struck by two things. First, the song seemed unusually long and complex compared to most other species and second, when the male obtained a mate he completely stopped singing. This struck me as decidedly odd, particularly as I had been told by my research supervisor that birds sang mainly to defend their territory from rival males, a view expressed by all the leading authorities of the day (e.g. Armstrong 1963). I quantified my observations by a sustained programme of field work on a population of colour-ringed sedge warblers (Catchpole 1973). I was able to confirm that as soon as a male attracted a female he indeed stopped singing, unless she left him or deserted, when he rapidly resumed song. But perhaps he sang briefly at dawn, dusk or even nocturnally. To check this I monitored song output from individual males over a full 24 hours before and after pairing. There was no doubt - I never found a ringed male in song when his female was with him on territory.

The diurnal plots of song from unpaired males revealed a tremendous investment in singing, with some males singing throughout all of the 24 hours, only pausing sporadically to feed. As soon as they attracted a female, the task was over and they never sang again for the whole breeding season. It seemed obvious to me then that there must be a firm link between male song and female attraction. If this was so, then perhaps it also explained the unusually long and complicated song structure. What I was listening to was perhaps the acoustic equivalent of the peacocks tail. My observations were already leading to the formulation of a simple verbal hypothesis which at the time seemed impossible to test. Stated simply - complex sedge warbler song is the product of runaway intersexual selection. With my observational evidence so far I took a first tentative step and was content to conclude that the main function of the song seemed to be in sexual attraction (Catchpole 1973).

A Field Experiment

Even then I was reluctant to publish without some experimental testing and so I set about designing and carrying out my first ever playback experiment. In reading my 1973 paper, I see that at least I made a reasonable start to my experimental career by sharply defining my hypothesis - long before I had ever read Karl Popper! "The main object was to experimentally test the hypothesis that paired male sedge warblers do not produce advertising song, even in territorial defence." Having decided that I could not yet test my ideas on intersexual selection directly, I was attempting to show that male song was not involved in a male-male intrasexual function. The experiment was certainly simple and primitive by today's standards I used a recording of continuous, spontaneous song taken from one male and played it back to ten male sedge warblers on territory. I merely noted whether they responded by approach and whether they sang or not. The results were certainly very clear: unpaired males interrupted their singing to approach the speaker silently and paired males also approached silently (Fig. 1).

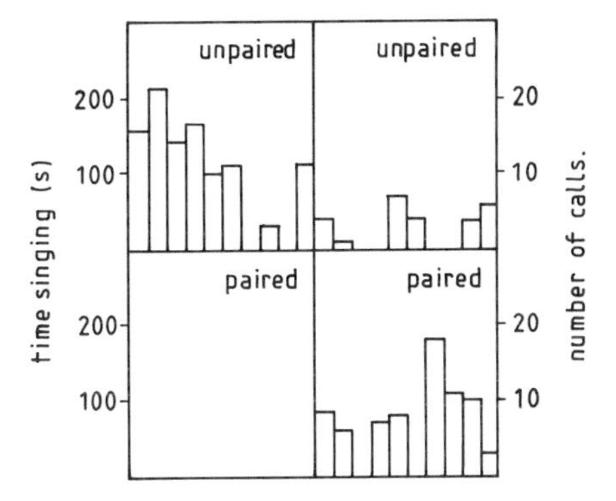

Figure 1. Time spent singing and number of calls given by ten male sedge warblers before and after pairing in response to playback of their species song (From Catchpole 1977). Reprinted by permission from Anim. Behav. **25**, 493-4. Copyright 1977 Academic Press Ltd.

Several years later I was able to confirm the results using better equipment, another recording, various controls and quantitative measurements on males both before and after pairing (Catchpole 1977). The results were just the same and so I now felt fairly confident that male sedge warblers did not use their song directly during territorial encounters with other males. Instead they use threat calls and visual threat displays as well as overt aggression to maintain territorial integrity. Although rival singing males are clearly conspicuous and may well avoid each other, I believe we are entitled to conclude at this stage that the song seems to function primarily in female attraction.

An Analytical Observation

So far I had avoided a detailed investigation into song structure. This was partly because, unlike most species, the long sedge warbler song did not fit conveniently onto the standard 2.5 seconds of Kay Sonograph paper. I also suspected that the song would be so variable and complex that there would be little to achieve by merely confirming this observation. What changed my view was another ethology fashion which swept through the seventies - sequence analysis. Bird songs were apparently marvellous material to demonstrate the mathematical rules thought to control sequences of behavioural events and a stimulating series of papers were produced by Lemon & Chatfield (e.g. 1971). When I eventually produced my own sequence analysis using the sedge warbler (Catchpole 1976) the complexity and variety it revealed exceeded my wildest predictions. Some of the songs were over a minute long and contained several hundred syllables, the sequencing of which was highly indeterminate. Apart from a few very general organisational rules, the songs are best thought of as unique compositions that are never repeated in exactly the same form. The analogy with the peacocks tail is certainly no exaggeration and may not do the sedge warbler justice, as the colours in the eye spots would need to be an ever changing light show to stand serious comparison.

As well as formally confirming the extraordinary elaboration of song structure, the sequence analysis was encouraging in other respects. The building blocks, or syllables, were repeated and easy to classify. Furthermore, the bird seemed to cycle through its syllable repertoire very quickly and plots of new syllable types rapidly became asymptotic after about 20 songs. Therefore a male with a larger repertoire of syllable types inevitably transmits a more complex pattern of song. This meant it was now possible to measure the main index of complexity - the number of different building blocks males can use to construct their songs. This is syllable repertoire size - which is not strictly comparable to song repertoire size used in studies of more conventional songbirds.

The Comparative Approach

Having a method to estimate repertoire size also opened up the possibility for a comparative study in the genus *Acrocephalus*. Kroodsma (1977) had published a comparative study of repertoire size in North American wrens and suggested that sexual selection was the driving force behind their evolution. In brief, he found that the most polygynous species had the most complex songs due to stronger sexual selection pressure in polygynous mating systems. When I looked at the six European *Acrocephalus* species (Catchpole 1980) I was able to test for the first time an important prediction from sexual selection theory, that more complex songs would be found in the polygynous species. What turned out was in fact the very reverse: monogamous species like the sedge warbler had long, continuous songs with high repertoire sizes. Conversely, the polygynous species had shorter, more conventional songs with lower repertoire sizes (Fig. 2).

This finding was not quite so surprising as I thought at first. Previous workers from Kroodsma to Darwin had commented upon examples of more exotic sexual displays and elaborate plumage dichromatism in monogamous birds. In *Acrocephalus* warblers the explanation was to be found in the relationship between female choice and investment in monogamous and polygynous systems, and how these influence the structure and function of song. In a polygynous system a female stands a high probability of being deserted and will have to raise the young alone. She is more concerned about territory than male quality and therefore selects males indirectly on the quality of their territories. Males in polygynous species defend large, resource-based territories to attract several females. Male song therefore evolves largely through male-male intrasexual selection pressure and songs develop a structure more suited to a territorial function. In monogamous systems the male is needed to feed the young and much food is collected outside the smaller territories. This switches the emphasis in female choice more directly onto male, rather than territory, quality. Songs therefore evolve largely through direct female choice and runaway intersexual selection pressure produces the long elaborate songs we now observe. Refuting the initial simple hypothesis has not led us to refute sexual selection, but allowed us to understand that different mechanisms of intra- and intersexual selection can result in the evolution of different song structures. It has also allowed us the opportunity to refine and sharpen our hypothesis before testing further predictions.

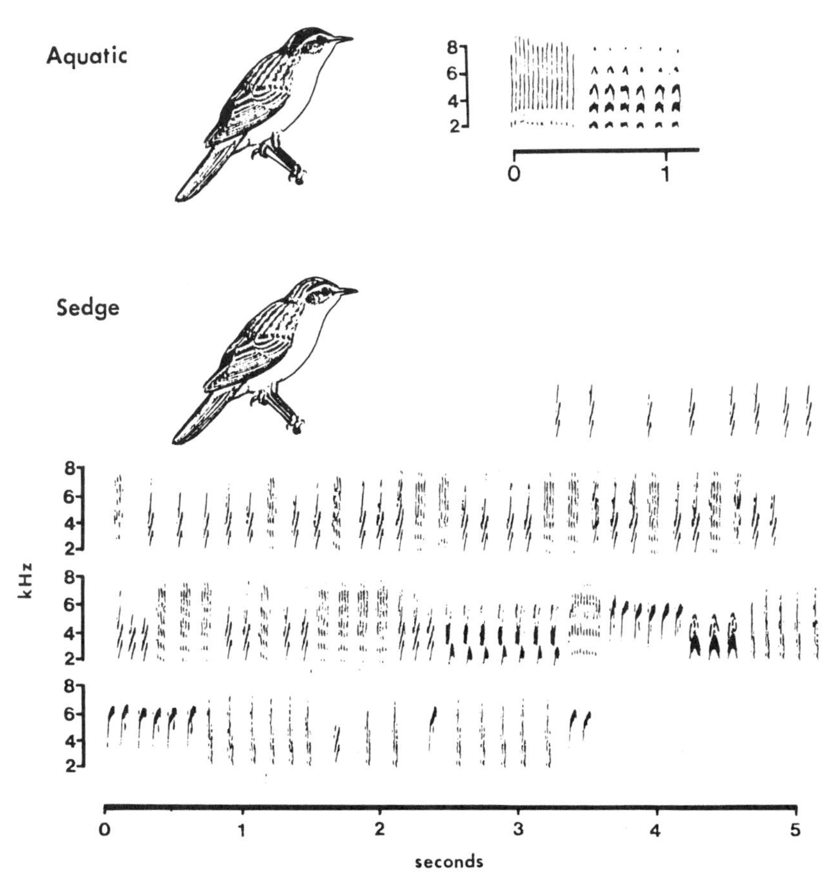

Figure 2. Sonagrams of typical songs from two *Acrocephalus* species. The polygynous aquatic warbler produces a short, simple song (here two syllable types) used mainly in territorial defence. The monogamous sedge warbler produces a longer, more complex song (here seven syllable types) used mainly for mate attraction (From Catchpole 1976 and 1980). Reprinted by permission from Behaviour **59** and **74**. Copyright 1976 and 1980 E.J. Brill Publishers Ltd.

Another Field Observation

Measuring repertoire size as an index of complexity now allowed me the chance to test a more crucial prediction from sexual selection theory, one which included female choice. Male sedge warblers with more complex songs should attract females before their rivals with simpler songs (Catchpole 1980). Although previously inhibited by the logistics of attempting a large scale integration of field and analytical work, I was inspired by the elegant study of Howard (1974) on the extremely complex songs of the mockingbird *Mimus polyglottos*. To do this, a whole population has to be monitored daily until they are paired and their songs recorded for later sonagraphic analysis. I duly completed this for a small population of sedge warblers and then estimated their repertoire sizes. When pairing date was plotted against repertoire size (Fig. 3), a highly significant inverse correlation was obtained ($P < 0.001$). Males with larger repertoire sizes were clearly obtaining females well before their rivals. There was no significant correlation between male settling pattern and pairing date, suggesting that females were not choosing the first males back (who might have selected the best territories). Such field data are encouraging and lend some support to our hypothesis that females do select males on the basis of song complexity.

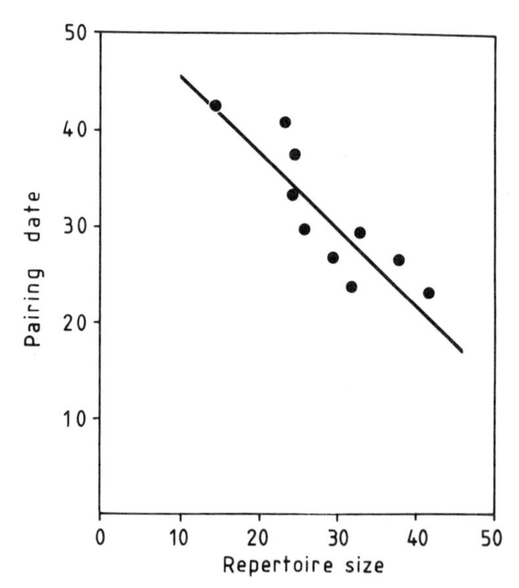

Figure 3. The relationship between syllable repertoire size and pairing date in a population of male sedge warblers. (From Catchpole 1980). Reprinted by permission from Behaviour **74**. Copyright 1980 E.J. Brill Publishers Ltd.

A First Laboratory Experiment

The problem with such field studies is that too many male and territory quality variables are beyond our control. One way round this is to control them in a laboratory experiment. For example, by playing back recorded songs to captive females it would be possible to eliminate totally all other male and territory quality variables from the system. At this stage there was no technique available to ask captive females which songs they preferred, until Searcy and Marler (1981) reported on a new technique. By implanting captive female songbirds with oestradiol, they became highly receptive to playback and signalled their interest with a quantifiable solicitation display. Since then it has become a standard technique applied to many different species (see chapter by Searcy in this volume) and it was clearly time to test female sedge warblers under more controlled conditions. My colleagues and I at the Max-Planck-Institut in Radolfzell decided to attempt an initial screening and control. We implanted a group of female sedge warblers and exposed each female only once to a four minute recording of spontaneous song from one individual of three different species (Catchpole et al. 1984). The presentations were made on different days and in random order to control for habituation and order effects and care was taken to standardise amplitude. The number of separate copulation solicitation displays was counted and summed for each individual subject over the four minute experiment. The three species used to make the experimental tapes were the sedge warbler; a related control, the reed warbler *A. scirpaceus*; and an unrelated control, the blackbird *Turdus merula*. Although the recorded sedge warbler male only had a repertoire size of 20, the reed warbler 80 and the blackbird 115, the females only gave full displays to the recording of their own species song. It seems that females are capable of discriminating their own species song and display to it in preference to the more complex songs of other species.

We were now ready to ask our implanted females a more interesting question in confirmation of our field study. Under controlled laboratory conditions, do females prefer males with more complex songs? To test this we used actual recordings from the field study. We took the male who paired first (repertoire size = 41), the male who paired last (repertoire size = 14) and another male from the middle of the range of repertoire sizes (repertoire size = 24). By doing this, our sample embraced the whole range of natural variation within our population. The results (Fig. 4) were very clear, the three males were ranked in the same order of response as they were in the wild: the females responded more than twice as frequently to the large repertoire male as to the small repertoire male and the difference in the number of displays was highly significant with the Wilcoxon matched pairs test ($P < 0.01$).

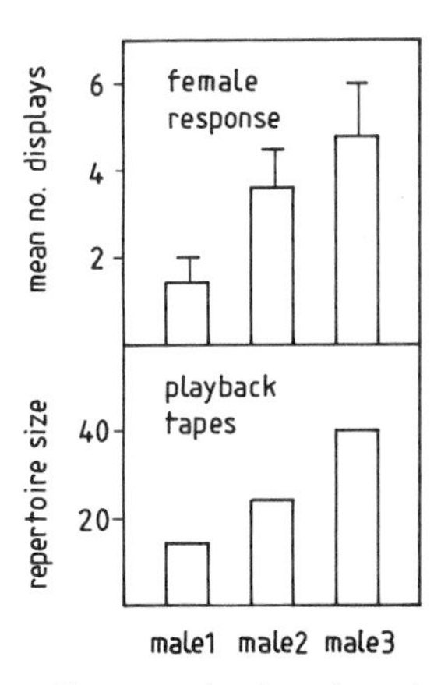

Figure 4. Responses of hormone implanted captive female sedge warblers to playback of songs recorded from three males with different syllable repertoire sizes (From Catchpole et al. 1984). Reprinted by permission from Nature **312**, 563. Copyright 1984 Macmillan Ltd.

Although we have effectively isolated confounding variables from other aspects of male quality, the songs themselves still encompass individual variation. For example, we are not controlling for variables other than repertoire size, such as singing rate or syllable sequencing. A more rigorous test of the hypothesis that repertoire size is the important variable requires us to attempt this.

Another Laboratory Experiment

The solution adopted was to eliminate individual variation entirely by using tapes prepared from only one individual male (Catchpole et al. 1984). Because sedge warbler songs are long, rambling permutations, with frequent repetition of syllable types, it was possible to construct many different tapes from a lengthy recording of a male with a large repertoire. We made seven such tapes with approximately equal numbers of syllables and the male cycling through his repertoire in a normal fashion. The only difference was that each male was allowed to do this with different sizes of the same syllable repertoire. The seven different tapes contained repertoires ranging from 2 to 55 syllables. As in the earlier experiments, each female was exposed only once a day to a four minute tape. Over seven days each female was therefore exposed to the seven different repertoire sizes in random order. A significant positive correlation ($P < 0.05$) was obtained between repertoire size and amount of display (Fig. 5). This particular experiment provides much stronger evidence for the hypothesis that repertoire size is indeed the most important variable in the system.

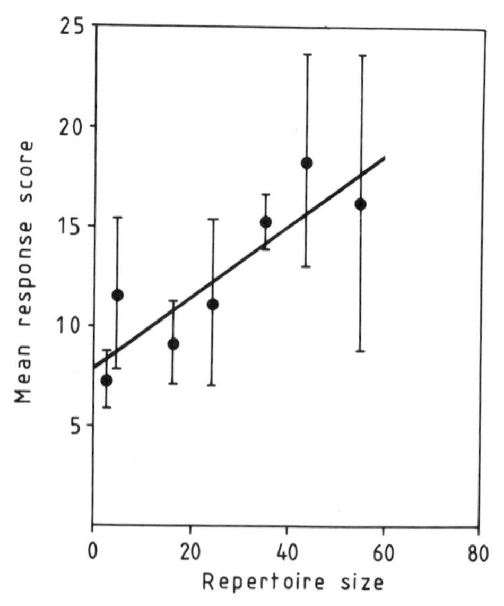

Figure 5. Responses of hormone implanted captive female sedge warblers to playback of one individual when syllable repertoire size is artificially manipulated (From Catchpole et al. 1984). Reprinted by permission from Nature **312**, 564. Copyright 1984 Macmillan Ltd.

Another Field Experiment

Even though there is considerable evidence that songs are not used directly in male-male interactions in sedge warblers, the possibility remains that repertoire size might have some effect upon rival territorial males. Having established a clear differential effect of repertoire size upon females, I decided to perform another playback experiment upon territorial males (Catchpole 1989b). This time I used the same tapes with seven different repertoire sizes which had been used on captive females. My hypothesis was that if repertoire size had any effect upon males then we might expect some correlations between repertoire size and various measures of approach. The experiments were carried out upon unpaired, singing males, but no such correlations were obtained with four different measures of approach. Although all the males responded strongly, it appears that repertoire size itself has no effect upon measures of approach.

At the start of each experiment the male was singing on territory and as in a natural situation stopped singing to search for the intruder. When the total amount of singing was considered there was a significant inverse correlation with time spent near the speaker ($P < 0.05$) and a significant positive correlation with nearest distance to the speaker ($P < 0.01$). Birds that sang most spent less time searching round the speaker and did not approach so closely. This confirms our earlier findings that singing in the unmated male is not directly related to territorial defence and that the amount of song during playback is an indirect consequence of time not spent in approaching and searching for intruders.

Discussion

The case of the sedge warbler illustrates the main point that I wish to make that an integrated series of observations, analyses, field and laboratory experiments are all needed before we can test a hypothesis adequately. The main hypothesis I have been testing is that the elaborate song of the sedge warbler is a product of intersexual selection. I do not claim to have tested all the many predictions that stem from this. For example, do male sedge warblers with larger repertoires not only attract females earlier, but follow this through by increasing their Darwinian fitness and leaving behind more offspring? To test this would certainly be difficult with a playback experiment, but it has been done in the field on another species, the great reed warbler *A. arundinaceus* (see review by Catchpole 1987). An integrated approach may not give all the answers, but I believe it has greater validity than a single playback experiment, however well designed and carried out. The problem of validity is one which Kroodsma (1989, 1990) frequently raises when suggesting improvements in the design of playback experiments. Campbell & Stanley (1963) suggested that there are two kinds of validity, internal and external (see also Kamil 1988). A well designed playback experiment has high internal validity, in that we can be confident that it has adequately tested our stated hypothesis under the conditions of that particular experiment. In his original criticism of the sedge warbler experiments Kroodsma (1989) suggested that we used only one repertoire and did not control for responsiveness to variation within song types. In fact our main experiment used three repertoires and many different variations of the same syllable type (sedge warblers don't have conventional song types). In his later paper, Kroodsma (1990) also feels that three repertoires is not an adequate sample to test for the effects of repertoire size. We used those three males because they encompassed the whole natural range in our field study, the two extremes and a median and we had already shown an effect in the field. The single male used in the next experiment was to control for individual variation and we had already used a fifth male in the initial screening. This brings me back to my main point, that the experiments and observations should be taken as a whole and criticised as such. They were never designed in isolation. An integrated approach does not mean that the individual experiments themselves should not be well designed, there is no excuse for sloppy design or bad practice. It is just that some questions may have to be tackled in a series of steps rather than all at once (see the "bricklaying approach" in chapter by Falls in this volume). Also, the experimental designs proposed by Kroodsma (1989, 1990) are really aimed at species with repertoires of song types, rather than the more variable singers like the sedge warbler and there are both technical and logistic difficulties in implementing them with such species. I also believe that an integrated approach confers higher validity - I feel much safer when experimental and field research support each other.

Kamil (1988) believes that the sort of integrated approach I have stressed could go some way to resolving the validity problem. "Laboratory work, with its potential for a high degree of internal validity, should be carried out in coordination with field research. In other words, one cannot maximise both internal and external validity in a single piece of research. Instead, highly experimental studies, which maximise internal validity and descriptive field studies, which maximise external validity, must be integrated". (The relationship between internal and external validity is discussed further in the consensus chapter in this volume.) At this point I should probably take my leave, as in some ways this is exactly what I have been getting at.

There is no doubt that Kroodsma's critical views have made us all think much more carefully about the design of our playback experiments. I feel quite sure that the design and execution of our playback experiments will continue to improve and there are indeed exciting developments ahead. With increasingly sophisticated computer technology now available, playback experiments will become even more fashionable and attractive to the experimental biologist. I hope that they will not entirely vacate what I have called "the wider context" and will attempt to integrate those exciting experiments into a firm base of relevant biology. After all, as Kamil (1988) reminds us: "Since the biological world is so complex, this means that a single experiment can only reveal a little bit of what we eventually need to know".

References

Armstrong, E.A. 1963. *A Study of Bird Song*. Oxford University Press, London & New York.

Catchpole, C.K. 1973. The functions of advertising song in the sedge warbler *Acrocephalus schoenobaenus* and the reed warbler *A. scirpaceus*. *Behaviour*, **46**, 300-320.

Catchpole, C.K. 1976. Temporal and sequential organisation of song in the sedge warbler *Acrocephalus schoenobaenus*. *Behaviour*, **59**, 226-246.

Catchpole, C.K. 1977. Aggressive response of male sedge warblers *Acrocephalus schoenobaenus* to playback of species song, before and after pairing. *Anim. Behav.*, **25**, 489-496.

Catchpole, C.K. 1980. Sexual selection and the evolution of complex songs among European warblers of the genus *Acrocephalus*. *Behaviour*, **74**, 149-166.

Catchpole, C.K. 1987. Bird songs, sexual selection and female choice. *Trends Ecol. Evol.*, **2**, 94-97.

Catchpole, C.K. 1989a. Pseudoreplication and external validity: playback experiments in avian bioacoustics. *Trends Ecol. Evol.*, **4**, 286-287.

Catchpole, C.K 1989b. Responses of male sedge warblers to playback of different repertoire sizes. *Anim. Behav.*, **37**, 1046-1047.

Catchpole, C.K., Dittami, J. & Leisler, B. 1984. Differential responses to male song repertoires in female songbirds implanted with oestradiol. *Nature*, **312**, 563-564.

Hartshorne, C. 1973. *Born to Sing*. Indiana Univ. Press, Bloomington Indiana.
Howard, R.D. 1974. The influence of sexual selection and interspecific competition on mockingbird song *Mimus polyglottos*. *Evolution*, **28**, 428-438.
Hurlbert, S.H. 1984. Pseudoreplication and the design of ecological field experiments. *Ecol. Monogr.*, **54**, 187-211.
James, F.C. & McCulloch, C.E. 1985. Data analysis and the design of experiments in ornithology. *Curr. Ornithol.*, **2**, 1-63.
Kamil, A.C. 1988. Experimental design in ornithology. *Curr. Ornithol.*, **5**, 313-346.
Kroodsma, D.E. 1977. Correlates of song organization among North American wrens. *Amer. Natur.*, **111**, 995-1008.
Kroodsma, D.E. 1989. Suggested experimental designs for song playbacks. *Anim. Behav.*, **37**, 600-609.
Kroodsma, D.E. 1990. Using appropriate experimental designs for intended hypotheses in song playbacks, with examples for testing effects of song repertoire sizes. *Anim. Behav.*, **40**, 1138-1150.
Lemon, R.E. & Chatfield, C. 1971. Organization of song in cardinals. *Anim. Behav.*, **19**, 1-17.
Platt, J.R. 1964. Strong inference. *Science*, **146**, 347-353.
Popper, K.R. 1959. *The Logic of Scientific Discovery.* Basic Books, New York.
Searcy, W.A. 1989. Pseudoreplication, external validity and the design of playback experiments. *Anim. Behav.*, **38**, 715-717.
Searcy, W.A. & Marler, P. 1981. A test for responsiveness to song structure and programming in female sparrows. *Science*, **213**, 926-928.

WHAT STUDIES ON LEARNING CAN TEACH US ABOUT PLAYBACK DESIGN

Irene M. Pepperberg

Department of Ecology & Evolutionary Biology
and Department of Psychology
University of Arizona
Tucson, Arizona
U.S.A.

Introduction

Communication must be studied as social interaction. If we are to understand how birds communicate, we need to examine not only the form of a given signal, *the song*, but also its meaning and the appropriate context for its use, *the singing behaviour* (see Smith 1991). Researchers who designed learning and playback experiments to study avian communication initially concentrated on form and sometimes meaning, but not the details of context. Recent studies on learning, however, have shown how birds are affected by the singing behaviour of other individuals: interactions that demonstrate how communication might occur, provide feedback on whether information has been transferred, or encourage the development of different forms of communication in different contexts (reviews in Pepperberg 1991; Pepperberg and Schinke-Llano *in press*). Playback experiments, in contrast, still generally concentrate on describing how birds respond to the form of a signal. But closer analysis reveals that playback studies, like those on song learning, actually do involve response to singing behaviour and not song alone (e.g. Smith 1988; Stoddard et al. 1988, 1990).

Specifically, the rationale behind playback experiments is that researchers can determine the meaning, or, more accurately, test a hypothesis about the meaning of a vocalisation by quantifying a bird's response to that set of sounds. Accurate testing of the hypothesis thus depends on observing a reaction by the subject that does indeed reflect upon the meaning of the playback and not upon extraneous variables (e.g. inappropriate sound levels) or variables that affect response but that are not directly related to meaning (e.g. the reproductive state of the playback subject). But more complex confounds also exist: The *meaning* of a signal is not conveyed by its physical form alone and the subject is being asked to interact with a stimulus that is not interactive. Such confounds involve response to singing behaviour, rather than to song.

Because these same confounds with respect to behaviour influenced the earliest studies on song learning, changes taken to improve the design of learning experiments

Playback and Studies of Animal Communication
Edited by P.K. McGregor, Plenum Press, New York, 1992

might similarly improve the quality of playback experiments. Researchers in learning, by determining the extent to which numerous variables influenced behaviour, designed new experiments with greater external validity (reviews in Pepperberg and Neapolitan 1988; Pepperberg 1991). Equivalent measures concerning singing behaviour can be adapted for playback studies. The purpose of this paper, therefore, is to review the variables known to affect avian vocal learning, suggest their potential effects on song playback experiments and discuss, briefly, the results of the few experiments that have taken such variables into account.

A Framework for Examining the Variables that Can Affect Experiments on Learning and Playback Response

Research on learning suggests that the most effective input is most *natural*; what psychologists term referential, contextually applicable and socially interactive. I have previously suggested (Pepperberg 1985, 1986a, 1986b, 1991; Pepperberg and Neapolitan 1988; Pepperberg and Schinke-Llano *in press*) that human social modelling theory (e.g. Bandura 1971, 1977) provides a framework for identifying the specific factors that make input effective. The theory consists of several principles, four of which apply to avian and human vocal learning (see Pepperberg 1991). Social modelling theory states that for learning to occur:

1) The level of competence of the student must be taken into account. The tutor/model therefore must constantly adjust the input to continue to challenge the student's increasing competence.
2) The tutor/model must show the student how the new material relates to present circumstances and the advantage that learning this material confers; i.e. the material must have *contextual applicability* and *functionality*.
3) The more intense the interaction between the student and model(s), the more effective is the training.
4) The more unlikely the situation for learning, the more important are the first three principles.

In sum, these principles suggest why one form of input, rather than another, affects how birds acquire song. Because these principles all involve responses to the singing behaviour of another individual, I propose that these principles are equally applicable to playback experiments. These principles should help identify and construct the forms of playback that will be effective in eliciting natural responses and thus allow us to deduce more accurately the meaning, to the receiver, of the stimuli used in playback.

Factors that Influence Learning and Are Likely to Affect Playback Effectiveness

Quality of Input

Social modelling theory emphasises three components of quality that input must have to be effective: optimality in *clarity*, *level* (nonacoustic) and *amount*. For playback studies, the importance of clear input is no longer an issue: researchers already appear sensitive to effects such as song degradation (e.g. Wiley and Richards 1982; McGregor and Falls 1984; Shy and Morton 1986) or of using a sound level appropriate to the environment (e.g. Dabelsteen 1981). Optimality in level and amount of input, however, are factors that have not always been taken into account.

Optimal Level of Input What constitutes an optimal level of input can differ for learning and playback, but the concept of optimality, as defined by social modelling theory, remains applicable: For any given situation, a particular level of input most effectively elicits the desired response. *Level* can refer to any interactive aspect of the input.

For human language learning, level usually refers to the level of competence, and the optimal level of input is just beyond the current level of competence (K. Nelson 1978; Krashen 1980; Kuczaj 1983; Scollon 1976). Thus speech that includes simple markers for *tenses* is optimal for a speaker who is just gaining understanding of how language can refer to the past and the future (see de Villiers and de Villiers 1978). Language-learners are also most likely to *respond* to input that is at or just slightly beyond their level of competence (Long 1981). The level to which and at which they respond demonstrates, in turn, their level of understanding of the meaning of the input.

Assessing optimal input for avian playback studies requires a somewhat different perspective, but, given other parallells between avian and human vocal behaviour (reviews in Pepperberg 1985, 1986a; Pepperberg and Neapolitan 1988; Pepperberg and Schinke-Llano *in press*), the existence of an optimal level of input is also likely. Because playback usually communicates information about an intrusion, optimality is likely related to levels of aggression. Some evidence does indeed exist for such a proposition: A recent study (Dabelsteen and Pedersen 1990; Dabelsteen chapter in this volume) suggests that playback that purportedly denotes a state at or just beyond the current level of aggression of the receiver is indeed likely to elicit the greatest (and, presumably, the most naturalistic) response (see also Kroodsma 1979). A signal that denotes too low a state might be ignored, whereas a signal that denotes a higher state might not (in the absence of a live sender) elicit a response that matches (and thus denotes) this purportedly higher state. In Dabelsteen and Pedersen's study (1990), a possible difference in the level of aggression represented by two vocalisations became apparent only when the vocalisations were scaled in an interactive playback.

Social modelling theory also predicts that the effectiveness of input in sustaining interaction and thus the possibility of determining the meaning of this input is increased if the stimulus level is constantly adjusted to reflect the current state of the receiver. For humans, input that is adjusted to the level of the receiver is indeed most likely to sustain naturalistic interaction at any level of competence (Krashen 1982; Scarcella and Higa 1982; Dore 1983; Furrow and Nelson 1986). Some birds might react similarly (see Petrinovich and Patterson 1980). Blackbirds (*Turdus merula*), for example, reacted for longer periods when playbacks simulated interactive escalation of threat than when playbacks randomly interspersed more and less aggressive vocalisations (Dabelsteen and Pedersen 1990). In contrast, input, at least for humans, that is too advanced may fail to elicit a response (see Shipley et al. 1969) or, if such input does maintain interaction, this interaction usually is not naturalistic: For humans, such a mismatch between encoder and receiver usually does not involve a transfer of information but rather involves an attempt to repair a *breakdown* in communication; e.g. encodes requests for clarification (see Long 1981). Application of such a possibility to the interpretation of playback experiments "... might explain why artificially emitted song in spite of great caution in the experimental design is sometimes "meaningless", or gives quite different meanings than those expected" (Dabelsteen 1985:202). Optimality of level thus appears to be an important variable in the appropriate design of playback experiments.

Optimal Amount of Input The amount of input is also likely to affect the strength of the response of the receiver and thus how the response is evaluated. Too little input, and the receiver may not perceive the stimulus; too much input of a particular type, and the receiver is likely to habituate and ignore the stimulus. Psychologists have devoted considerable effort to the study of habituation. Their data suggest "... that animals have attentional mechanisms that control the extent to which they process information about available stimuli" (Roitblat 1987:97). Animals will learn to ignore uninformative input, particularly input that repeats information that has already been processed (see Moran, Joch and Sorenson 1983; Davis 1984; Putney 1985; Davis and Bradford 1986; Pepperberg 1987). Many psychological studies specifically use a habituation/dishabituation paradigm to determine which aspects of input convey salient information by changing aspects of the input and measuring whether attention is reallocated (e.g. Swartz 1980; review in Roitblat 1987). The effect of habituation on all aspects of avian vocal behaviour would thus seem intuitively obvious.

Debate concerning habituation has, however, rather narrowly focussed on two areas. Some researchers have debated the relationship between habituation and repertoire organisation (e.g. Hartshorne 1956, 1973; Kroodsma 1982, 1990; Weary and Lemon 1988, 1990). Other researchers have adopted the psychological paradigm and actively used habituation as a tool to learn whether birds will react to distinctions that are perceived by humans, e.g. switches in song type, singing rate, or even individual syllables (see reviews in chapters by Horn and by Falls in this volume). Despite these studies, researchers have not often attended to the possible effects of habituation during other types of playback experiment.

For example, researchers in the field, with few notable exceptions (e.g. Petrinovich and Patterson 1979, 1980, 1981; Stoddard et al. 1988; chapter by Falls in this volume) do not take into account the potential for habituation when they replay the same song pattern on a loop for several minutes at a time. Although various reasons have been presented for discounting habituation effects in such circumstances (e.g. the assumption that minimising presentation time is sufficient to counter habituation), data from psychological studies do not support such assumptions. Interestingly, those studies cited above (see also Dabelsteen 1982; Dabelsteen and Pedersen 1990) that tested for habituation effects (and accounted for possible sensitisation effects of massed song presentation and individual differences among males) did indeed find either a heightened reaction or a slower rate of decline in response to playbacks with variability than to those without. Moreover, given the small likelihood of a bird in the wild reproducing even a single song in an *identical* manner in a given bout, artificial playback experiments need to determine the optimal amount of a given rendition of song to elicit the most natural responses in the receiver.

How Referentiality and Contextual Applicability of Input and Interaction Affect Playback

Why Input Should be Referential and Contextually Applicable

Referentiality and contextual applicability refer to the real-world use of a communication code. Referentiality describes the *purpose* or the *meaning* of an utterance: for humans, for example, whether the utterance is a request for a drink or a comment on the colour of a book (see de Villiers and de Villiers 1978); for birds, whether a song declares the occupancy of a territory or the lack of a suitable number of mates (e.g. Smith 1977).

Contextual applicability and functionality involves the particular situation in which an utterance is used. Talking about CS's, UCS's and VI schedules of reinforcement will draw blank looks at a NATO workshop such as this one, but is *de rigueur* at a Psychonomic Society meeting because of the specific information such jargon conveys. For birds, contextual applicability and functionality refer, for example, not only to whether a territorial song is used in a low *versus* a high intensity encounter, but also to whether the form of the song or the timing of the experiment is appropriate to the particular species' yearly cycle of territorial defence. Because most playback experiments are designed to provide information and test hypotheses about the referentiality or contextual applicability of a song, particular care must be taken to account for the subtleties of these effects.

Response to a territorial song that is played in an inappropriate context will likely be stronger than to a song that is played in the appropriate context; however, the researcher may not initially be able to determine the exact reason for the strong response. The researcher might make an unfounded conclusion as to the potency of the song itself. Maybe the location of the speaker playing a neighbour's song is inside the receiver's territory rather than on the border with the neighbour and the persistent singing of a male who should normally retreat triggers an unusually strong response (e.g. Stoddard et al. 1988; also Shy and Morton 1986). Maybe the neighbouring territory was taken over by a stranger shortly before the playback experiments began and the subject bird is thus responding appropriately to the real, but not the experimentally perceived, situation (Stoddard, *pers. comm.*). Maybe the neighbour is unmated and never quits intruding on the subject's territory; response to a playback of that neighbour's song might be unusually strong (Stoddard, *pers. comm.*). Although in some instances use of a well-known population and large sample size could counter the overall effect of these problems, for individual birds only additional experiments coupled with detailed observation could determine which aspect of the context is not being perceived appropriately by the experimenter, or whether the basic reference is inappropriate (see discussions in Petrinovich and Patterson 1979, 1980; Dabelsteen 1984; Stoddard et al. 1988).

Songs that are played back at a time of year or day that is not contextually appropriate may, in contrast, evoke a weaker response than is consistent with their intended function. If the existence of a diurnal or seasonal variation in response is not examined, data on playback responses can be misinterpreted (see Hoelzel 1986; see Brindley 1991). Simply monitoring fluctuations in song production may not, for example, indicate when *responses* will be greatest (see Logan 1988). Petrinovich and Patterson (1980) specifically note how the choice of pattern of stimulation in playback experiments can significantly affect the data and thus our understanding of how song is used.

In sum, social modelling theory (e.g. Bandura 1973, 1977) suggests that input must be referential and contextually applicable to elicit a socially appropriate response. Given that the underlying assumption of playback studies is that the referentiality or contextual applicability is being tested by whether a bird does respond in a way that is socially appropriate, contextual variables must be completely identified and fully understood so that they can be varied independently and systematically. Only through such systematic study can the proper choice of input be made to stimulate singing behaviour that provides information on the *single* variable under investigation.

Effects of Interaction on Referentiality and Contextual Applicability

Input must often be *inter*active as well as referential and contextually applicable if it is to be learned appropriately. Only through interaction can an emitting organism see how its vocalisations affect others and, by their responses, gauge how effective is its *mapping* of sound to meaning (Lieberman 1984; see also Snow 1979; Camaioni and Laicardi 1985). Such interactions are critical not only for an organism to learn how to use its vocalisation, but also for an investigator to learn about this pattern of behaviour.

As in the classic case of the linguist who tries to understand the meaning of a word in an unfamiliar language (Quine 1960), the scientist must "enter into communicative encounters with the natives, if only in order to go around pointing to things and asking "Gavagai?" in an attempt to figure out what "Gavagai" means" (Dennet 1990:275). The interaction, however, must be iterative: "The linguist tentatively adopts the hypothesis that rabbit is critical but...keeps watching for additional evidence that will confirm his hypothesis against still other alternatives" (Premack 1986:90). After several rounds of interaction, the linguist may still not be certain of the meaning of *gavagai*, but will have narrowed the field considerably. Because playback experiments are based on the identical testing paradigm, the need for interactive input seems apparent.

Until recently, interactive playbacks were difficult if not impossible to achieve (see Dabelsteen and Pedersen 1991; Capranica, *pers. comm.*). Yet only by fine-tuning the input of what can be considered a *model* bird to its live counterpart can we correlate the signals that are exchanged with actual behaviour, and thus determine what messages are indeed being transmitted (see Griffin 1981). As noted above, Dabelsteen and Pedersen's examination (1990) of the specific, interactive context in which their blackbirds used different song types was needed to clarify the extent of threat that the song types represented.

The Effect of Intensity of Interaction

Intensity of interaction (Petrinovich 1988; review in Pepperberg 1991) may make contextually appropriate, referential input more effective relative to other input. Intensity refers to degree of emotional arousal of the participants, not amount of input or opportunity for interaction, although those latter two factors (among others) will affect intensity. Bandura (1977) has shown that the extent of the influence exerted by a social tutor on the behaviour of human children is a function of the degree of arousal of the participants: Input that is identical in informational content will elicit stronger responses if the interactants are emotionally aroused. The interplay between increased interaction, arousal, attentional mechanisms and hormonal levels is far from clear (see discussion in Pepperberg 1991), and likely no single mechanism will be found to be responsible for effects of intense interaction. Even without knowing the mechanism, however, we must account for the effect.

In the wild, a bird's singing behaviour, as well as the song that it uses, may convey its degree of arousal; such meaning may be inferred from the reaction of the receiver. An organism's ability to respond to input that is thought to be graded with a graded range of behaviour is an indication of its ability to distinguish such graded series (see Bickerton 1990). Intensity (and arousal) may thus, for example, be reflected in an increased rate of song production in countersinging (e.g. Brindley 1991), an increase in amplitude (Kroodsma 1979; Morton 1982), the choice of song that is sung (Baptista 1985) and the addition of visual threats (Dabelsteen and Pedersen 1990). If a playback is to tell us anything about the use and meaning of a song, such factors must be included, in some way, into the design of the experiment.

In especially intense encounters, individuals may even produce a learned but rarely used code: that of a different dialect (Baptista and Morton 1982) or species (reviews in Pepperberg and Neapolitan 1988; Pepperberg and Schinke-Llano *in press*). A switch to another code can be a deliberate behaviour that "... highlights aggravation, mitigation and other discourse strategies..." (Zentalla 1990:85). Such a response to playback could considerably alter our interpretation of avian capacities, but such production may, however, occur only if the intensity has reached the appropriate stage in a natural manner; i.e., the organism may need time to re-access and reactivate the code from memory (see Berman 1979). Moreover, organisms may not, as noted above, respond to input that reflects another code if the context and the intensity of this input do not correlate appropriately (see Hatch 1983; McLaughlin 1984; also Dabelsteen 1985).

In sum, anyone who has observed avian interactions knows that the degrees of arousal of the participants vary continuously; a playback that does not permit such natural variation as input will not elicit natural behaviour from the recipient. Degree of arousal can be difficult to measure, but *changes* in states of arousal can be monitored (e.g. from physiological measurements as well as behavioural observations), and researchers can at least make first order correlations between singing situations and increased or decreased arousal. Such correlations could then be tested through interactive playback.

Contextual Variables that Make Natural Responses Unlikely

Effects of Differences in Relative Status between the Sender and Receiver

The social status (within the community) of the singer whose input is used can affect the results of a tutoring experiment (see Pepperberg 1985, 1988, 1991; Pepperberg and Schinke-Llano *in press*); I propose that status effects can similarly influence the results of a playback experiment. Research on human behaviour (e.g. Bandura 1973) has shown that the status of a potential model directly affects the amount of attention s/he receives from the learner and thus the amount of interaction that occurs. Thus a model of low status is ignored and the highest status model commands the greatest attention. Because dominance hierarchies exist in avian populations, similar responses should be expected from birds.

If the neighbour song chosen in a neighbour-stranger recognition study is, for example, especially effective at repelling intruders or is that of an particularly dominant individual, conceivably the results will not demonstrate the generally-observed weaker response to neighbour vs. stranger input. Recipients may have already learned that interactions with such an individual are costly, either in energy or in potential for injury. Responses to such songs thus would not reflect responses to natural interactions with other individuals in the area.

Conceivably, too, using the song of a subordinate bird in playback could be a source of confusion, both to the receiving bird and the researcher who attempts to analyse the data. Whatever the reaction of the dominant bird to the subordinate individual who now fails to back down after its initial challenge was met, the data is not likely to reflect the actual meaning of the song that was used or the actual context of the interaction (see Dabelsteen 1982). Given that the receiver is likely to judge the input on the basis of the singing behaviour that accompanies the song, experimenters should be careful not to judge the potency of a song solely on the response of the receiver to a static tape.

Effects of Individual Differences in Subjects

Early studies, in both humans and birds, rarely emphasised the effects of individual differences on communicative behaviour. Many studies on children, for example, tracked a single subject over the course of its development of one or two aspects of language (e.g. Bellugi 1967; Brown 1973). Although such studies provided important information that could be missed in short-term studies of many individuals, long-term studies on one or two subjects did not demonstrate how different children responded in an *overtly* different manner to the same input (Clifton and Nelson 1976; Bowerman 1978; Furrow and Nelson 1986). Petrinovich and Patterson (1980), however, noticed individual differences (termed *state effects*) in how white-crowned sparrows responded to playback: whereas some birds responded to the set of stimuli with a great deal of song, others responded hardly at all. Stoddard (*pers. comm.*) has similarly noted overt differences in the responses of song sparrows. In both birds and humans, determining whether such differences are specific to the individual (e.g. a chronic physiological condition) or to his environment is not always possible. However, observing these differences and taking them into account in the experimental design and subsequent data analysis is always possible (see chapter by Falls in this volume), and can be crucial for interpreting a response to a playback experiment (Petrinovich and Patterson 1980).

Summary and Conclusion

Song, like human language, is generally used in social contexts, be they affiliative or agonistic. Aspects of singing behaviour most likely to be affected by social factors thus are likely to involve choices that are made to achieve a specific communicative intent (Pepperberg and Schinke-Llano *in press*). If we are not interested in specific communicative intent, then much of what I have presented is not relevant. But, if we are to determine this communicative intent, which for me is the general rationale for playback experiments, then we must take into account the effect of social factors. I have tried to present a very brief review of the types of factors that might be important, based on the factors that affect another social phenomenon, that of learning.

Acknowledgements

Preparation of this chapter was supported by NSF grant BNS 91-9606. Travel to the NATO Workshop was supported in part by funds from the University of Arizona Foreign Travel Grant Program. I thank Philip Stoddard and Peter McGregor for comments on earlier versions of the manuscript.

References

Bandura, A. 1971. Analysis of modeling processes. In: *Psychological Modelling*. (Ed. by A. Bandura), pp. 1-62. Aldine-Atherton, Chicago.

Bandura, A. 1973. *Aggression: A Social Learning Analysis*. Prentice Hall, Englewood Cliffs, NJ.

Bandura, A. 1977. *Social Modeling Theory*. Aldine-Atherton, Chicago.

Baptista, L.F. 1985. The functional significance of song-sharing in the white-crowned sparrow. *Can. J. Zool.*, **63**,1741-1752.

Baptista, L.F. and Morton, M.L. 1982. Song dialects and mate selection in montane White-crowned sparrows. *Auk*, **99**, 537-547.

Bellugi, U. 1967. The Acquisition of Negation. *Unpublished* Ph.D. thesis. Harvard University, Cambridge, MA.

Berman, R. 1979. The re-emergence of a bilingual: A case study of a Hebrew-English speaking child. *Working Papers on Bilingualism*, **19**, 157-180.

Bickerton, D. 1990. *Language and Species.* University of Chicago Press, Chicago.

Bowerman, M. 1978. Words and sentences: Uniformity, individual variation, and shifts over time in patterns of acquisition. In: *Communicative and Cognitive Abilities - Early Behavioral Assessment.* (Ed. by F.D. Minifie & L.L. Lloyd), pp. 349-396. University Park Press, Baltimore.

Brindley, E.L. 1991. Response of European robins to playback of song: neighbour recognition and overlapping. *Anim. Behav.*, **41**, 503-512.

Brown, R.L. 1973. *A First Language.* Harvard University Press, Cambridge, MA.

Camaioni, L. and Laicardi, C. 1985. Early social games and the acquisition of language. *Brit. J. Devel. Psych.*, **3**, 31-39.

Clifton, R.K. and Nelson, M.N. 1976. Developmental study of habituation in infants: the importance of paradigm, response system, and state. In: *Habituation.* (Ed. by T.J. Tighe & R.N. Leaton), pp. 159-205. Lawrence Erlbaum Assoc., Hillsdale, NJ.

Dabelsteen, T. 1981. The sound pressure level in the dawn song of the blackbird (*Turdus merula*) and a method for adjusting the level in experimental song to the level in natural song. *Z. Tierpsychol.*, **56**, 137-149.

Dabelsteen, T. 1982. Variation in the response of free-living blackbirds (*Turdus merula*) to playback of song: I. Effect of continuous stimulation and predictability of the response. *Z. Tierpsychol.*, **58**, 311-328.

Dabelsteen, T. 1984. Variation in the response of freeliving blackbirds (*Turdus merula*) to playback of song: II. Effect of time of day, reproductive status and number of experiments. *Z. Tierpsychol.*, **65**, 215-227.

Dabelsteen, T. 1985. Messages and meanings of bird song with special reference to the blackbird (*Turdus merula*) and some methodology problems. *Biol. Skr. Dan. Vid. Slesk.*, **25**, 173-208.

Dabelsteen, T. and Pedersen, S.B. 1990. Song and information about aggressive responses of blackbirds, *Turdus merula*: Evidence from interactive playback experiments with territory owners. *Anim. Behav.*, **40**, 1158-1168.

Dabelsteen, T. and Pedersen, S.B. 1991. A portable digital sound emitter for interactive playback of animal vocalizations. *Bioacoustics*, **3**, 193-206.

Davis, H. 1984. Discrimination of the number three by a raccoon (*Procyon lotor*). *Animal Learning & Behavior*, **4**, 121-124.

Davis, H and Bradford, S.A. 1986. Counting behavior in rats in a simulated natural environment. *Ethology*, **73**, 265-280.

Dennett, D.C. 1990. *The Intentional Stance.* Bradford Books, The MIT Press, Cambridge, MA.

de Villiers, J.G. and de Villiers, P.A. 1978. *Language Acquisition.* Harvard University Press, Cambridge, MA.

Dore, J. 1983. Feeling, form, and intention in the baby's transition to language. In: *The Transition from Prelinguistic to Linguistic Communication.* (Ed. by R.M. Golinkoff), Lawrence Erlbaum Assoc., Hillsdale, NJ.

Furrow, D. and Nelson, K. 1986. A further look at the motherese hypothesis: A reply to Gleitman, Newport, & Gleitman. *J. Child. Lang.*, **13**, 163-176.

Griffin, D.R. 1981. *The Question of Animal Awareness.* Rockefeller University Press, New York.

Hartshorne, C. 1956. The monotony-threshold in singing birds. *Auk*, **73**, 176-192.

Hartshorne, C. 1973. *Born to Sing.* Indiana University Press, Bloomington, IA.

Hatch, E. 1983. *Psycholinguistics: A Second Language Perspective.* Newbury House, Rowley, MA.

Hoelzel, A.R. 1986. Song characteristics and response to playback of male and female robins, *Erithacus rubecula. Ibis*, **128**, 115-127.

Krashen, S.D. 1980. The input hypothesis. In: *Current Issues in Bilingual Education.* (Ed. by J.E. Alatis). pp. 168-180. Georgetown University Press, Washington, D.C.

Krashen, S.D. 1982. *Principles and Practices in Second Language Acquisition.* Pergamon Press, Oxford.

Kroodsma, D.E. 1979. Vocal dueling among males marsh wrens: evidence for ritualized expressions of dominance/subordinance. *Auk*, **96**, 506-515.

Kroodsma, D.E. 1982. Song repertoires: problems in their definition and use. In: *Evolution and Ecology of Acoustic Communication in Birds. Vol.II.* (Ed. by D.E. Kroodsma, E.H. Miller & H. Ouellet), pp. 125-146. Academic Press, New York.

Kroodsma, D.E. 1990. Patterns in songbird singing behavior: Hartshorne vindicated. *Anim. Behav.*, **39**, 994-996.

Kuczaj, S.A. 1983. *Crib Speech and Language Play.* Springer-Verlag, NY.

Lieberman, P. 1984. *The Biology and Evolution of Language.* Harvard University Press, Cambridge, MA.

Logan, C.A. 1988. Breeding context and response to song playback in mockingbirds (*Mimus polyglottos*). *J. Comp. Psych.*, **102**, 136-145.

Long, M.H. 1981. Input, interaction, and second-language acquisition. In: *Native Language and Foreign Language Acquisition.* (Ed. by H. Winitz). Vol. 379, pp. 259-278. Ann. NY Acad. Sci.

McGregor, P.K. and Falls, J.B. 1984. The response of Western Meadowlarks (*Sturnella neglecta*) to the playback of degraded and undegraded songs. *Can. J. Zool.*, **62**, 2125-2128.

McLaughlin, B. 1984. *Second-Language Acquisition in Childhood: Vol 1. Preschool Children.* Lawrence Erlbaum Assoc., Hillsdale, NJ.

Moran, G., Joch, E. and Sorenson, L. 1983, June: *The response of meerkats (*Suricata suricatta*) to changes in olfactory cues on established scent posts.* Paper presented at the annual meeting of the Animal Behavior Society, Lewisburg, PA.

Morton, E.S. 1982. Grading, discreteness, redundancy, and motivation-structural rules. In: *Evolution and Ecology of Acoustic Communication in Birds. Vol.I.* (Ed. by D.E. Kroodsma, E.H. Miller & H. Ouellet), pp. 183-212. Academic Press, New York.

Nelson, K.E. 1978, August. *Toward a rare-event cognitive comparison theory of syntax acquisition.* Paper presented at the 1st International Congress for the Study of Child Language, Tokyo. (Cited in Kuczaj, 1983.)

Pepperberg, I.M. 1985. Social modeling theory: A possible framework for understanding avian vocal learning. *Auk*, **102**, 854-864.

Pepperberg, I.M. 1986a. Acquisition of anomalous communicatory systems: Implications for studies on interspecies communication. In: *Dolphin Cognition and Behavior: A Comparative Approach.* (Ed. by R.J. Schusterman, J.A. Thomas, & F.G. Woods), pp. 289-302. Lawrence Erlbaum Assoc., Hillsdale, NJ.

Pepperberg, I.M. 1986b. Sensitive periods, social interaction, and song acquisition: The dialectics of dialects? *Behav. Brain Sci.*, **9**, 756-757.

Pepperberg, I.M. 1987. Acquisition of the same/different concept by an African Grey parrot (*Psittacus erithacus*): Learning with respect to color, shape, and material. *Animal Learning & Behavior*, **15**, 423-432.

Pepperberg, I.M. 1988. The importance of social interaction and observation in the acquisition of social competence: Possible parallels between avian and human learning. In: *Social Learning: Psychological and Biological Perspectives.* (Ed. by T.R. Zentall & B.G. Galef, Jr.), pp. 279-299. Lawrence Erlbaum Assoc., Hillsdale, NJ.

Pepperberg, I.M. 1991. Learning to communicate: the effects of social interaction. In: *Perspectives in Ethology, Vol. 9.* (Ed. by P.P.G. Bateson & P.H. Klopfer), pp. 119-164. Plenum Press, NY.

Pepperberg, I.M. and Neapolitan, D.M. 1988. Second language acquisition: A framework for studying the importance of input and interaction in exceptional song acquisition. *Ethology*, **77**, 150-168.

Pepperberg, I.M. and Schinke-Llano, L. *in press*. Language acquisition and form in a bilingual environment: A framework for studying birdsong in zones of sympatry. *Ethology*,

Petrinovitch, L. 1988. The role of social factors in white-crowned sparrow song development. In: *Social Learning: Psychological and Biological Perspectives.* (Ed. by T.R. Zentall & B.G. Galef, Jr.), pp. 255-278. Lawrence Erlbaum Assoc., Hillsdale, NJ.

Petrinovich, L. and Patterson, T.L. 1979. Field studies of habituation: I. The effect of reproductive condition, number of trials, and different delay intervals on the response of the white-crowned sparrow. *J. Comp. Physiol. Psych.*, **93**, 337-380.

Petrinovich, L. and Patterson, T.L. 1980. Field studies of habituation: III. Playback contingent on the response of the white-crowned sparrow. *Anim. Behav.*, **28**, 742-751.

Petrinovich, L. and Patterson, T.L. 1981. Field studies of habituation: IV. Sensitization as a function of the distribution and novelty of song playback to white-crowned sparrow. *J. Comp. Physiol. Psych.*, **95**, 805-812.

Premack, D. 1986. *Gavagai!* Bradford Books, The MIT Press, Cambridge, MA.

Putney, R.T. 1985. Do willful apes know what they are aiming at? *Psych. Record*, **35**, 49-62.

Quine, W.V.O. 1960. *Word and Object.* The MIT Press, Cambridge, MA.

Roitblat, H.L. 1987. *Introduction to Comparative Cognition.* W.H. Freeman & Co., New York.

Scarcella, R.C. and Higa, C.A. 1982. Input and age differences in second language acquisition. In: *Child-Adult Differences in Second Language Acquisition.* (Ed. by S.D. Krashen, R.C. Scarcella, & M.H. Long), pp. 175-201. Newbury House, Rowley, MA.

Scollon, R. 1976. *Conversations with a One-Year Old.* University Press of Hawaii, Honolulu.

Shipley, E.S., Smith, C.S. and Gleitman, L.R. 1969. A study in the acquisition of language: free response to commands. *Language*, **45**, 322-342.

Shy, E. and Morton, E.S. 1986. The role of distance, familiarity, and time of day in Carolina wrens responses to conspecific songs. *Behav. Ecol. Sociobiol.*, **19**, 393-400.

Smith, W.J. 1977. *The Behavior of Communicating*. Harvard University Press, Cambridge, MA.

Smith, W.J. 1988. Patterned daytime singing of the eastern wood-pewee (*Contopus virens*). *Anim. Behav.*, **36**, 1111-1123.

Smith, W.J. 1991. Singing is based on two markedly different kinds of signaling. *J. Theor. Biol.*, **152**, 241-253.

Snow, C.E. 1979. The role of social interaction in language acquisition. In: *Children's Language and Communication.* (Ed. by Collins, W.A.), pp. 157-182. Lawrence Erlbaum Assoc., Hillsdale, NJ.

Stoddard, P.K., Beecher, M.D. and Willis, M.S. 1988. Response of territorial male song sparrows to song types and variations. *Behav. Ecol. Sociobiol.*, **22**, 125-130.

Stoddard, P.K., Beecher, M.D., Horning, C.L. and Willis, M.S. 1990. Strong neighbor-stranger discrimination in song sparrows. *Condor*, **92**, 1051-1056.

Swartz, K.B. 1980. A comparative perspective on perceptual, cognitive, and social development. *J. Human Evol.*, **11**, 315-320.

Weary, D.M. and Lemon, R.E. 1988. Evidence against the continuity-versatility relationship in bird song. *Anim. Behav.*, **36**, 1379-1383.

Weary, D.M. and Lemon, R.E. 1990. Kroodsma refuted. *Anim. Behav.*, **39**, 996-998.

Wiley, R.H. and Richards, D.G. 1982. Adaptations for acoustic communication in birds: Sound transmission and signal detection. In: *Evolution and Ecology of Acoustic Communication in Birds. Vol.I.* (Ed. by D.E. Kroodsma, E.H. Miller & H. Ouellet), pp. 131-181. Academic Press, New York.

Zentalla, A.C. 1990: Integrating qualitative and quantitative methods in the study of bilingual code switching. In: *The Uses of Linguistics.* (Ed. by E.H. Bendix). Vol. 583, pp. 75-92. Ann. NY Acad. Sci.

CONDUCTING PLAYBACK EXPERIMENTS AND INTERPRETING THEIR RESULTS

H. Carl Gerhardt

Division of Biological Sciences
University of Missouri
Columbia, Missouri 65211
U.S.A.

Introduction

The aim of this chapter is to discuss factors to be considered in the design, execution and interpretation of playback experiments. Most of these factors will be of general concern, even though my examples will be drawn heavily from two-speaker designs with frogs and toads as subjects. I recognise at the outset that many researchers will consider some of these concerns trivial, taking for granted that every researcher will have taken this or that precaution in designing and executing the experiment. My response is that every factor under consideration has detracted from some recent study that I have done myself, that has recently appeared in the literature, or that I have reviewed in manuscript form. More significantly, the precautions that one laboratory routinely takes into consideration are often not even mentioned in the Methods and Materials section of a paper. However, this does not mean that every researcher is aware of the pitfalls, nor does it mean that every laboratory routinely adopts procedures to avoid such pitfalls. I strongly recommend that we adopt the policy of adding enough information to our papers (probably only a few sentences) so that readers are not required to make any assumptions about methodology.

The more controversial part of my contribution is a short introduction to some alternative (Bayesian) ways of thinking about statistical treatments of the results of playback experiments with small sample sizes. I will argue that many of us have blindly accepted significance tests using small alpha criteria as the only legitimate approach, even though the probability of Type II statistical errors is very high. Furthermore, alpha-levels do not provide a direct quantitative estimate of the evidence against the null hypothesis, and even small P-values must be interpreted carefully when the sample size is large. I will argue further that parameter estimation, either using classical or Bayesian procedures, are more appropriate and more readily interpreted in a straightforward fashion. I will also discuss Bayesian critiques of classical procedures with regard to significance testing, showing that many researchers conduct and analyse playback experiments in ways that invalidate their significance tests in the classical framework, but would be appropriate for Bayesian procedures.

Playback and Studies of Animal Communication
Edited by P.K. McGregor, Plenum Press, New York, 1992

I. Preparation of Experimental Stimuli

Digital techniques have eliminated a great deal of the tedium and imprecision of tape editing and other procedures used in the past. Natural sounds can be digitised and edited easily with inexpensive commercial packages for most kinds of personal computers. However, the use of synthetic sounds avoids many of the design problems arising from the multidimensional nature of natural signals and variations in recording quality. In principle, an investigator can explore the behavioural relevance of the entire perceptual or preference space that is delimited by variation in a set or sets of acoustic signals of particular interest: within-population, between-population (dialects), between different signals in the repertoire and between species.

A first step is to generate a synthetic signal that is comparable in its behavioural effectiveness to a typical natural signal. Normally, such a signal would have properties with values equal to the estimated mean values in the set of signals of interest. Ideally, in tests against a series of natural exemplars, the standard synthetic signal is neither more nor less attractive. In frogs, I have always found that a synthetic call of equal attractiveness is acoustically simpler than a typical natural exemplar, which is, of course, a general result from a wide range of ethological studies that have explored all sensory modalities (e.g. Gerhardt 1983).

The second step is to develop criteria for choosing the amount of change in the value of a given parameter. In my view any of the three following criteria are appropriate, depending on the question being asked. First, change the value of a parameter by units equal to the standard deviation or some other measure of variance of the property in the natural set of interest. This procedure would be appropriate for estimating the proportion of signallers in a population that might be favoured by mate choice based on the property in question. Second, vary the value of the parameter by some constant percentage, guided perhaps by any existing psycho-acoustical data. Third, if the distributions of a parameter in two classes of natural signals do not overlap, choose a difference between the property that corresponds to the minimum observed difference between the two sets (species, populations, dialects, signals within a repertoire). If the animals discriminate, then the difference in that property at least is adequate for recognition of the two natural classes of signals. If the animals do not respond selectively, then the difference can be increased systematically until there is a differential response. This information could then be related to the natural variation in the two sets of signals to provide an estimate of the proportion of individuals that would be likely to be distinguished in natural situations. In psychophysical experiments, failures to show selectivity are more likely to reflect failures to perceive differences than in experiments with untrained animals because of the lack of control of the animal's motivational state in the latter situation. But for many evolutionary biologists, the proximate explanation for the failure of animals to respond selectively is less important than the evolutionary consequences.

There will be considerable practical difficulties in studying systems in which there are many acoustic properties of potential relevance, but I am hopeful that, as in anurans, the set of relevant properties will be some relatively small subset of the possible properties. The major complication is that even in anurans, the values of two or more properties affect the overall attractiveness of a signal in ways that might not be predicted from the results of varying one property at a time. A few studies have begun to tackle the problem experimentally (Doherty 1985; Nelson 1988; Gerhardt and Doherty 1988; Date et al. 1991). As the number of simultaneously varying properties being examined increases, the design, execution and interpretation of such experiments will probably warrant another workshop like the one that has generated this book.

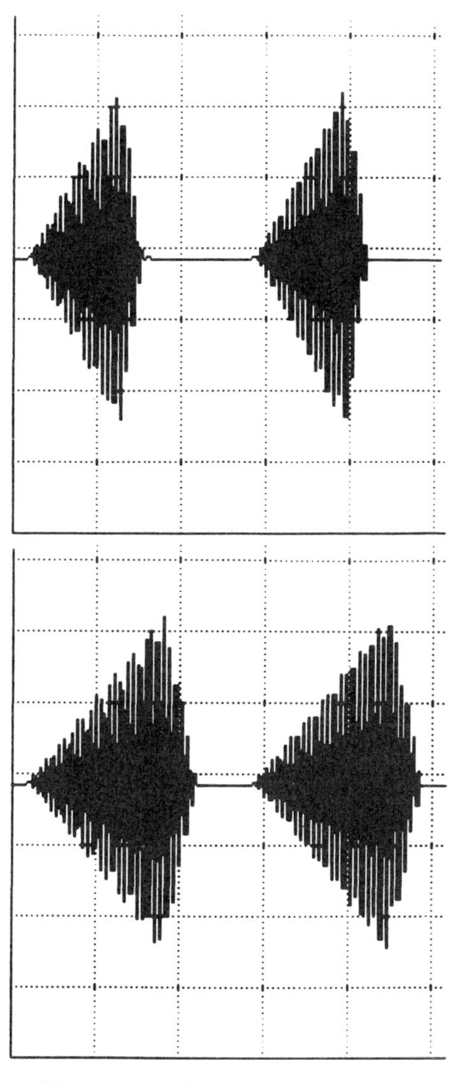

Figure 1. Oscillograms (amplitude *v.* time displays) of pulses of synthetic frog calls. The pulse period (time from the beginning of one pulse to the beginning of the second pulse) is the same, but the pulse duty cycle (ratio of sound to silence) differs because the pulses in the lower panel are longer. Adjusting the playback level of these two signals requires a consideration of the time-constant of the sound level meter and of the subject's auditory system. Each x-axis division equals 10ms.

Whereas varying two or more properties at the same time will be an important task for future studies, one major pitfall in generating signals in which the values of one specific property is varied is that sometimes additional differences, which may be salient to the animals, are also inadvertently generated. Even if the researcher is aware of these new variables, there needs to be an explicit rationale for dealing with them. I will mention two simple examples. Suppose that an animal produces signals that have two or more strong frequency components. In exploring the significance of varying the frequency of one component, while holding the frequency of the other one(s) constant, the waveform periodicity will necessarily vary. More specifically, the waveform periodicity will equal the largest common divisor of harmonically-related components and the waveform will be quasi-periodic or aperiodic if the components are not harmonically-related. More importantly, if two components differ relatively little (< 100Hz) in frequency, then obvious regular variations in the overall amplitude-time envelopes of the signals (e.g. beats) that are clearly audible will be introduced (see Greenewalt 1968 for a detailed discussion about the relationship between frequency and time-domain properties of bird song). Changes in the envelope of signals are used by females of the green treefrog to distinguish between the advertisement and aggressive calls of the male (Gerhardt 1978a, 1978b).

My second example concerns the gross temporal structure of signals. Many animals repeat basic acoustic units, notes or pulses, at relatively uniform intervals and organise them into trains. Suppose we want to explore the effect of varying the inter-pulse interval, holding all other properties constant. The fact is that we cannot hold all other properties constant. If we hold pulse duration constant, then we change the value of a derived but potentially important property: the ratio of sound-to-silence (the *duty cycle*; Fig. 1). If we hold duty cycle constant, then we obviously have to change pulse duration as well as the pulse interval. My recommendation is for researchers to examine how pulse duration varies with pulse interval in natural signals and then to model this variation in the synthetic signals. If pulse duration is held constant, so that the duty cycle varies appreciably, then the researcher must be careful in equalising the sound pressure levels of signals with different duty cycles. Equalising the maximum peak levels of the two sounds will result in a difference in their averaged (e.g. *fast* (about 100ms) root-mean-square) levels and vice-versa. Here it is important to have some knowledge about the time constants of the auditory system of the subject and in the absence of such information to use data from similar taxa. Another solution is to do the experiments both ways.

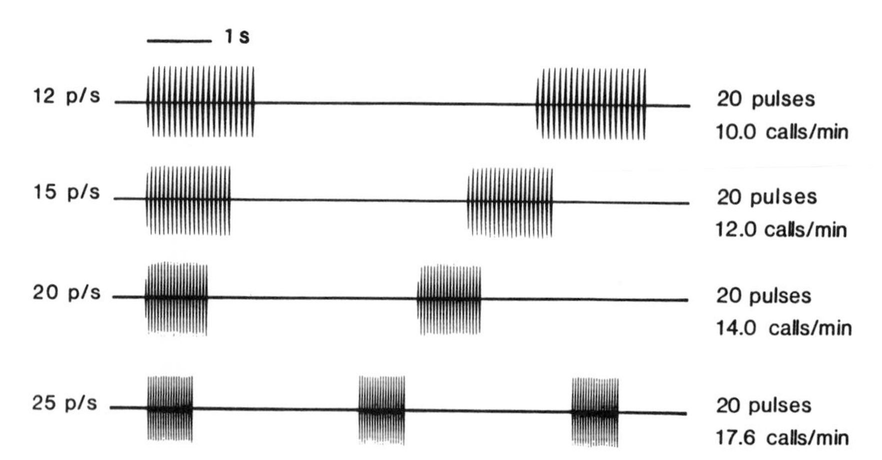

Figure 2. Oscillograms of synthetic frog calls. Varying one property of a sound sometimes necessarily causes changes in other variables. In this series of synthetic pulse trains, used to test females of the gray treefrog, variation in pulse rate (period) results in variation in pulse train duration because the number of pulses per train has been held constant. The time interval between pulse trains also varies, so that the pulse train duty cycle (ratio of sound to silence) is about the same. Decisions about which properties to vary and which to hold constant were based on how male gray treefrogs varied these properties as a function of their body temperature. These synthetic calls are similar to calls produced by males over a 12°C span of temperature. Modified from Gerhardt and Doherty (1988).

Similar complications arise at higher levels of organisation. Holding the number of pulses in a pulse train constant and varying pulse interval (rate) will obviously result in differences in the duration of pulse trains (Fig. 2; Gerhardt and Doherty 1988). If the interval between pulse trains is varied, then one has to decide whether or not to hold the duty cycle (in this case, of the pulse train) constant by appropriate adjustments of duration.

My final caveat about preparing playback stimuli is specific to designs in which animals are subjected to playbacks from two or more speakers at about the same time. In some amphibians and insects, the timing relationship and order of presentation may be very critical. For example, as Dyson and Passmore (1988) showed, the preference of female painted reed frogs for low- to high-frequency sounds was reversed when the high-frequency sound led the low-frequency sound rather than being separated in time by equal intervals of silence (see also the chapter by Klump and Gerhardt in this volume).

II. Playback Fidelity

Although there have been major advances in the ease of generating synthetic sounds and editing natural ones, the quality of playback experiments is still constrained by the analogue devices that reproduce the sounds, and by the acoustics of the testing arena, whether indoors or outdoors. *In my opinion, each study should include an assessment of the quality of the playbacks based on measurements of the test signals at the points at which the animal subjects are receiving them and making responses.* It is not sufficient to cite only the manufacturer's specifications for the playback equipment and loudspeakers. First, these specifications are usually derived from a statistical analysis of some sample of the products and may not apply to the equipment actually being used. Second, even an accurate estimate of, say, the frequency-response of the system is inadequate to characterise the effects of the acoustics of the playback environment.

I am not claiming that good experiments cannot be conducted without an excellent playback system and acoustics: the fidelity of reproduction required obviously depends on the questions being asked, i.e. the subtlety of acoustic differences that an animal may be required to discriminate. But without explicit statements about the fidelity of reproduction, the reader must assume that the researcher has made the appropriate measurements and taken steps to make sure that the acoustic differences to be tested can be adequately handled by the system.

In Figure 3 I show spectra and oscillograms of sound pulses that were synthesised digitally. Panel A shows displays of signals that were input directly from a digital-to-analogue interface board in a personal computer to the Kay DSP 5500 sonagraph. As designed, the signal has two spectral peaks, with the lower one having an amplitude of about -6dB relative to the upper component. The pulses have a duration of 25ms, with a linear rise and inverse exponential decay. Panel B shows the spectrum and oscillogram of the same signal after having been recorded on a high-quality cassette recorder, played back through a second cassette deck, amplifier and speaker, and then re-recorded at the point in a semi-anechoic chamber where females are initially released in a playback trial. Note that the relative amplitude difference between the two major peaks is nearly identical to that of the spectrum in the first panel, but that now there is a third component that has an amplitude that is approximately -35dB relative to that of the second peak. This third component is caused by harmonic distortion of the first peak. If the frequency range of the spectrum had been expanded, there would also be another component at about 6.6kHz, representing the third harmonic distortion component of the second peak.

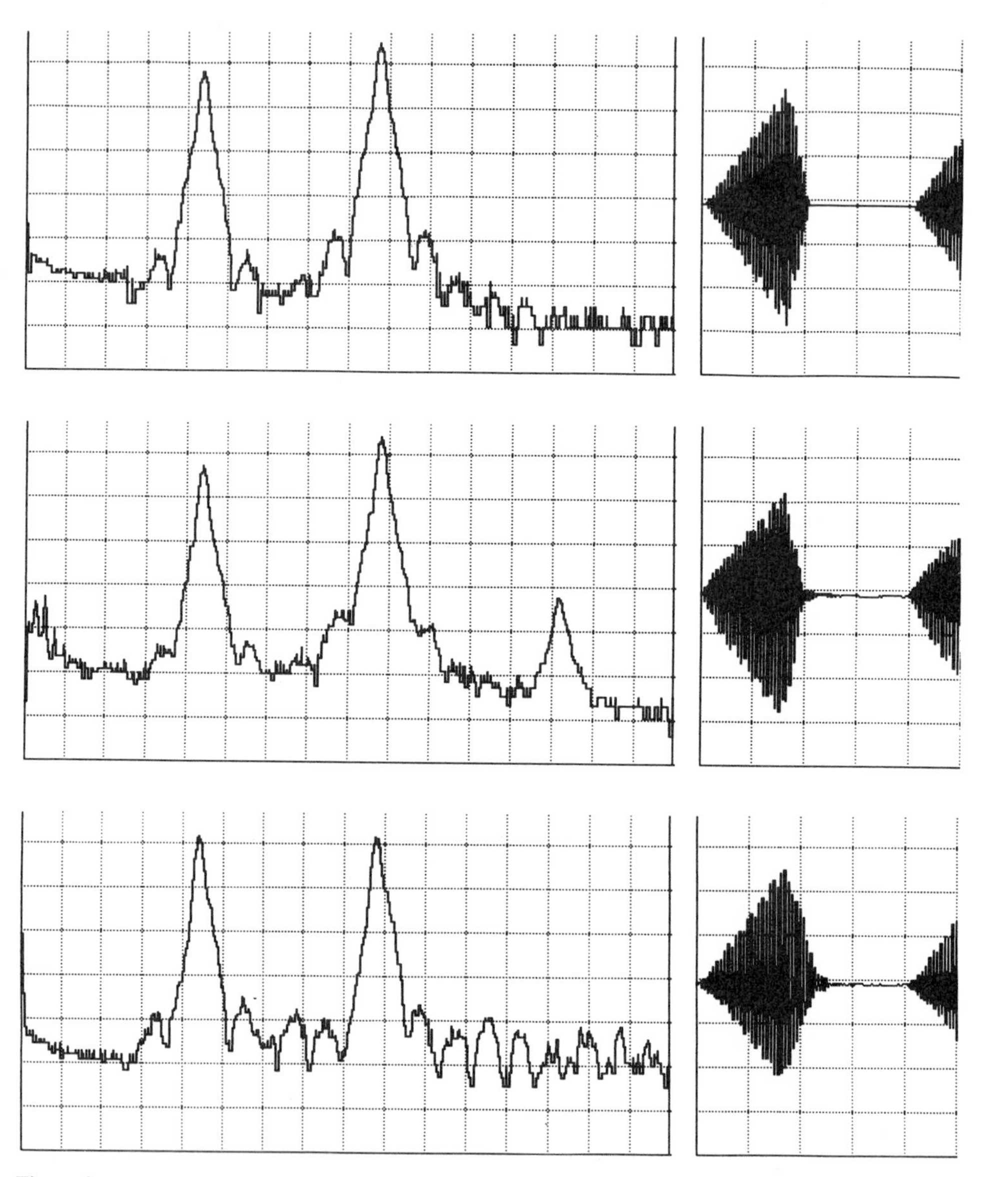

Figure 3. First column: amplitude *v.* frequency (power) spectra of synthetic frog calls (Frequency scale: each division = 250Hz; amplitude scale, each division = 5dB; second column: oscillograms of representative pulses, each division = 12.5ms. In **A** the sound has been input directly to a DSP 5500 Sonagraph from a digital-to-analogue board in a personal computer; in **B** the same sound has been recorded on a cassette recorder, played back in a semi-anechoic chamber and re-recorded at the point where frogs are released individually prior to phonotaxis; in **C** the recorded sound has been routed through an equaliser adjusted to make the relative amplitude of the two spectral peaks the same. Note the third harmonic distortion in **B** and the pulse duration and shape distortion in **C**. See the text for details.

Third harmonic distortion is usually generated by overdriving some analogue component, or by some poor quality analogue component. In fact, harmonic and non-harmonic (intermodulation) distortion can be introduced at any point in the chain: D/A board to cassette recorder, cassette recorder to amplifier, amplifier to speaker, microphone to recorder, recorder to DSP sonagraph. For this example, I deliberately introduced it by overdriving the speaker. Again, depending on the question being posed, harmonic distortion of this magnitude is unlikely to be a confounding variable. If, however, a two-speaker design is being used, I recommend verifying that the same magnitude of distortion occurs from the alternative loudspeaker (second channel of the playback system).

Now compare the oscillograms of the pulses in Panels A and B. The linear shape and duration of the direct signal are fairly well duplicated in the signal that has been broadcast and re-recorded. There are differences in the symmetry of the high-frequency representation of the waveform (the dark parts) that are caused by phase shifts. Such shifts can occur simply by recording a signal on one tape recorder and playing it back on another one, and they are unlikely to be perceived by humans or animals. The other difference between the two oscillograms is the presence of a small pulse of sound immediately after the end of the re-recorded signal. This represents a sound reflection, probably from some of the playback equipment in the semi-anechoic chamber. Its peak-to-peak amplitude is about 4% (about -29dB) of the maximum peak-to-peak amplitude of the pulse. I judge this to be an acceptable value, but again it is important to know that the magnitudes of reverberations generated by any pair of alternative stimuli are about the same. This is particularly true if the frequency content of alternative stimuli is very different because the magnitude of reflections is frequency- (wavelength-) dependent. For the same reason, it is also a good idea to check for position-dependent variations in reflections for signals that have two or more frequency components.

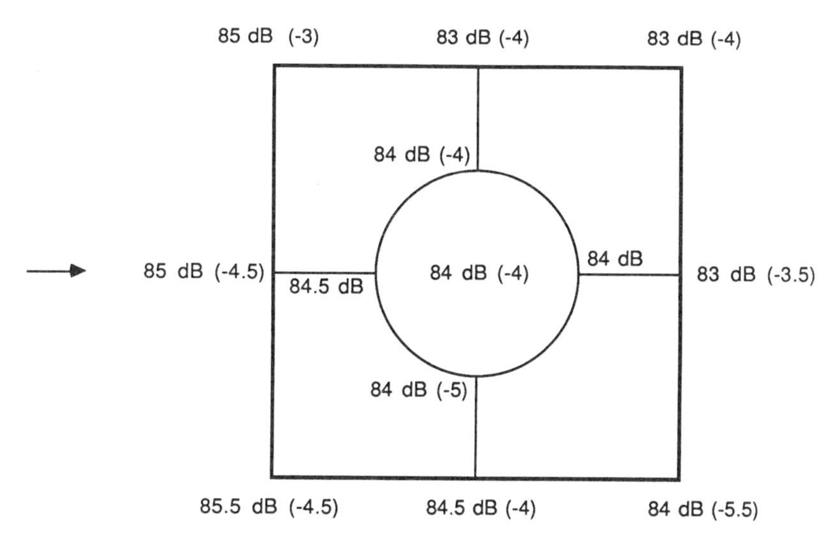

Figure 4. Diagram showing the results of measurements made in the immediate vicinity of the release point of female treefrogs in a semi-anechoic chamber. The sound pressure level was adjusted to 84dB SPL at ground level in the centre of the circle, which represents a hardware cloth releasing cage position in the centre of four reference squares (each side = 10 cm). The arrow indicates the direction from which the sound arrived from a speaker located 1m from the centre of the circle. The overall SPL is indicated for a sound of 1.1 + 2.2kHz. The amplitude of the 1.1kHz component relative to the 2.2kHz component (= 0dB) is shown in parentheses. See the text for discussion.

In Figure 4 I show the results of a set of measurements made near female release positions in experiments conducted in a semi-anechoic chamber. The maximum difference in the overall sound pressure level (*SPL*) was 2dB; the maximum variation in the relative amplitudes of frequency components of 1.1 and 2.2kHz was about 3dB.

Position- (wavelength-) dependent reflections and interference can be reduced by using an absorbent substratum, but then rather large differences in relative amplitude will occur because when both signals and receivers are situated on or just above such a substratum, there will be a severe attenuation of the high-frequency components relative to the low-frequency components. For example, I found a dramatic (about 6dB/octave) attenuation of frequencies above about 3kHz when a broad-band tonal signal was played back over a distance of just 1 metre over a low-cut pile carpet. If both source and receiver are elevated, then position (wavelength) effects will be non-monotonic, depending on the frequency, distance and phase-shifts introduced by the substratum on the indirect sound wave. These phenomena are described fully by Embleton et al. (1976), and Piercy and Daigle (1991) provide tables that show estimates of ground attenuation as a function of frequency (in octave bands up to 4000Hz), substratum, distance (from 10 to 100m) and source height. Piercy and Daigle (1991) also provide details for measurement techniques that take into account the effects of foliage, wind and temperature.

Panel C of Figure 3 illustrates a problem that may be introduced by compensating for non-linearities in the playback system by electronic filtering. Here I routed the signal depicted in Panel A through a Bruel and Kjaer graphic level equaliser and adjusted the relative amplitudes of the two major peaks to be equal. The most serious result is that the duration of the pulse is now noticeably longer than that of the original signal and its fall-time and shape are distorted. Such effects are caused by frequency-dependent phase shifts that will always occur when a signal is filtered (actively) strongly. One easy solution to the problem of non-linearity in the playback system is to measure it, and then compensate for deficiencies during the synthesis of signals by adjusting the relative amplitudes and phases before adding them. The dramatic effects of strong filtering on the temporal waveform of short pulses of sound are also evident in a series of synthetic frog calls that I used in an early study (Gerhardt 1978).

Many other problems arise when playback experiments are conducted outside. Wind effects, temperature gradients and background noise are all likely to be present and to vary (see Wiley and Richards 1978 for an extensive review). All of these factors may have important effects on the fidelity of sound reproduction and perception by the animal. It may also be difficult to eliminate visual stimuli that may influence an animal's response. For example, in two-speaker experiments with female frogs I found that when offering signals that differed relatively little in their attractiveness, females tended to move toward the speaker that had less moonlight behind it (Gerhardt 1981). For animals such as frogs and insects that move along the ground, slopes within the playback area may also be a factor. These directional biases can be eliminated in the laboratory, but there may still be some undetected directional factor. The usual solution is to switch the sounds between the two loudspeakers from one subject to another. The magnitude of a side bias can then be estimated and taken into account. The problem is that if the bias is strong, the females will all go to the same speaker, regardless of the signal being emitted. Thus, switching speakers will reveal the bias but not any subtle preference that may exist. If preferences are strong and robust, then side biases have very little, if any, effect on female responses.

Of course, animals do not communicate in semi-anechoic chambers, but in situations where acoustic fidelity will be compromised by many factors. While I appreciate attempts to test in the field for effects demonstrated in the laboratory, I think that a better approach is to add additional variables to the laboratory situation systematically, so that each effect can be tested separately. For example, additional speakers can be used to represent a small chorus, or background noise with specific spectral properties can be introduced. Studies with frogs have shown that females do not reliably move toward the source of advertisement calls of a single conspecific male if the SPL at her release point is not at least equal to that of chorus background noise (e.g. Gerhardt and Klump 1988). Changing the positions of the sources of signals and noise, so that the sounds arrive from 90 degrees apart, improved detectability only slightly and discrimination not at all (Schwartz and Gerhardt 1989).

III. Procedures and New Designs

In testing female anurans in the conventional two-speaker design with the female released midway between two speakers, I find that is important to restrain the animal in a small, acoustically-transparent cage until I have played back several repetitions of both stimuli. Otherwise, highly motivated frogs will hop immediately toward the speaker that first emits a sound. Once the female is released, her phonotactic behaviour is unequivocal. The frog usually turns her head or body immediately after the playback of a sound, and she usually moves to and touches the speaker. With the usual arrangement of speakers, then, the possibility of experimental bias is negligible.

One price for this objective criterion is that the design is too simplistic. Animals must seldom begin an assessment of two signalling males from midway between them. Rather, the female normally encounters an array of signallers at different distances and in a complex spatial array. The conventional two-speaker design also fails to provide an opportunity to learn if the female might actively avoid some signals, rather than merely being more attracted by one. I am currently experimenting with designs in which a female would have to pass within close proximity of a nearby speaker, which emits the signals of a heterospecific or very unattractive conspecific, to reach a speaker broadcasting very attractive conspecific signals (Fig. 5). Deviations from a more or less straight-line path to the attractive stimulus can be quantified and interpreted in terms of consequences in nature. For example, in many kinds of frogs and toads, males may stop calling and chase frogs that they detect nearby.

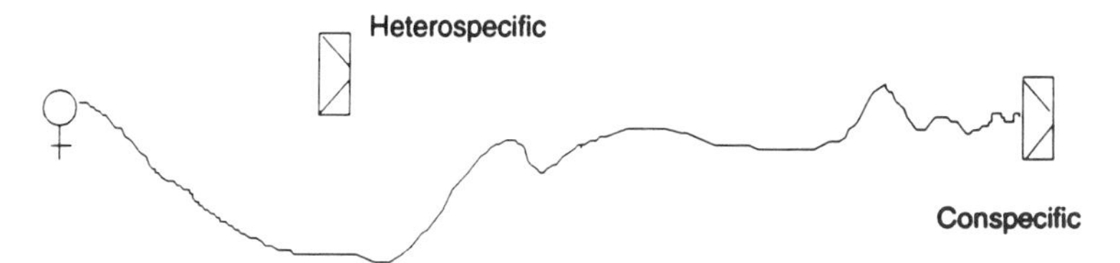

Figure 5. Design of a playback experiment to test for avoidance of heterospecific signals by female treefrogs. The diagram shows the positions of two speakers relative to the female release point. I would interpret the path shown to the conspecific speaker as avoidance of the nearby source of heterospecific calls. The experiments would be conducted with the speakers hidden, or with a second, inactive speaker to control for visual cues. A female treefrog would not hesitate to pass closely by or even hop on, a silent speaker that was adjacent to the shortest route to an active speaker that emitted attractive calls.

Another question addressed by these experiments concerns the costs that females might be willing to incur in mate choice. Extensive movements within choruses increase the chances that a female will be captured by a bullfrog, alligator, or other predator. Moving past a nearby male to reach a distant one will also involve some energetic costs. The effect of costs on the evolution of female mate choice has been termed direct selection and has been the subject of several new models (e.g. Maynard Smith 1991). The proposed new design also offers the female cues that exist in natural situations, but which are eliminated in the conventional two-speaker design, even if the sound pressure levels of the alternative stimuli are adjusted to be unequal at the female's release point midway between them. That is, when the speakers are placed at different distances from the female, she can obtain information about the absolute and relative difference in the distances of the sound sources and about their absolute levels by exploring the sound gradient (change in SPL per unit distance). There are also likely to be differences in the degree of degradation of the sounds emitted from the nearby and distant speakers. To my knowledge only one study of anurans has investigated the question of whether a difference in SPL created by moving speakers to different distances from the female is equivalent to that created by adjusting the relative SPL of signals emitted from speakers that are equidistant from the female (Arak 1984).

If, as in the design just described, phonotactic behaviour is being used to estimate female preferences in the context of sexual selection theory, then investigators must have some kind of estimate of the probability that an approach to a male will result in a subsequent mating with that male. In most treefrogs, the approach of a female to the close proximity of a calling male almost always leads to mating. Because unpaired males rarely contest paired ones, there is little doubt that the female's phonotactic approach in the laboratory represents a choice that would lead to fertilisation of her eggs by the male she chooses. In many other kinds of animals, including other kinds of frogs and toads, movements to the proximity of a male do not necessarily lead to a mating. Another possibility is that the female could later be taken over by a male to which she had not been attracted. In birds, copulation solicitation (e.g. Searcy and Brenowitz 1988; see chapter by Searcy in this volume) is an example of a much less ambiguous indicator of a female decision to mate than merely recording the time spent near a speaker.

IV. Limits of Generality

An important impetus for this book has, of course, been Don Kroodsma's critiques of the limited generality of many playback designs (references in consensus chapter). One of Kroodsma's main examples (see the consensus chapter for details) highlights a design in which a series of individuals are each tested with different sets of exemplars, where the number of exemplars is the main sampling unit. Here I am concerned about some other levels of analysis: within- and between- individuals and between populations.

In most of the research with selective phonotaxis in frogs and toads, researchers have based a judgment about a female's preference on one response in a given two-stimulus situation. For myself, the reason was to avoid non-independence of the data. That is, I was concerned that a female's experience in one trial would influence her preference in a second or third trial with the same pair of alternative stimuli. By recording only the initial responses of a series of females, one can test for a within-population preference. Now, however, I am concerned with questions that can only be answered by multiple testing of individual females in the same two-stimulus test. One of the major assumptions of nearly every model of the evolution of female mating preferences is that there exists within-population variation in preference. That is, in the simplest case, some proportion of the females prefer character state X and the rest prefer Y. The only way in which such variation can be estimated is to assess the consistency of the preferences of individual females. If all females do not make the same first choice, this could occur simply because individual females also have no particular preference. Indeed, this is the kind of result one would like to get in testing representative natural calls against one's initial standard synthetic calls (i.e. a call in which call properties have average values; see Gerhardt 1974a).

Before I became concerned about non-independence of responses, I conducted multiple trials with individual females (e.g. Gerhardt 1974b; Oldham and Gerhardt 1975). If the choice was an easy one (heterospecific *versus* conspecific), I found that nearly all females chose the same stimulus in their first trial and that each individual was consistent in her subsequent choices. However, I did obtain data from multiple tests of individual females in one experiment in which females as a group did not prefer one stimulus to the other: female green treefrogs offered a choice between calls of a conspecific and a hybrid involving the barking treefrog. As detailed in Gerhardt (1991), the majority of females were consistent in their preferences, a result that is consistent with the hypothesis that females within the population vary in their selectivity. (Breeding experiments are required to confirm the hypothesis). I think it is important to obtain multiple responses from individual females in any test in which first responses indicate that a large majority of females do not prefer one of the sounds. I have experimental evidence to suggest that carry-over effects from one trial to another are unlikely to be important, at least in one species of treefrog (Gerhardt 1981). Experiments to corroborate and generalise this result are badly needed.

Just as we know that there is geographic variation in the properties of male signals, there must certainly also be geographic variation in female selectivity. For example, I think that one reason for the apparent rarity of reproductive character displacement is that researchers have typically examined only geographic patterns in male signals. In fact, there are a number of studies that have shown differences in male or female discrimination that fit the expected pattern (e.g. Waage 1975; Wasserman and Koepfer 1977; Ratcliffe and Grant 1983). In addition to character displacement, there are many other evolutionary mechanisms that should give rise to geographic differences in the receiver's preferences or discriminatory abilities. I emphasise that the proper sampling unit for a study of geographic variation in female selectivity is the number of populations. For example, a comparison of two populations represents a sample of one, regardless of the number of females tested. I also think, however, that reasonable numbers of females should be tested from as many populations as possible to estimate simultaneously variability within populations.

V. Common Sense and the Statistical Analysis of Playback Experiments with Small Samples

Here I will argue against one of the aims of the workshop that generated the consensus chapter in this book: the design of experiments so that statistical tests of (point) null hypotheses will be valid. The problem of pseudoreplication that destroys the validity of significance tests does not disappear in the alternative, Bayesian view that I will advocate. We still need to identify the appropriate sampling unit for the hypothesis being examined. However, Bayesian and classical perspectives and procedures are very different with respect to how hypotheses are viewed in probabilistic terms and in how experiments should be planned and executed.

Playback experiments with anurans, and, indeed, with many other kinds of animals will often be limited by small sample sizes. One of my motives for questioning the usual way of statistically analysing the results of such experiments was the realisation that I and others, in routinely adopting a P = 0.05 alpha-level for the null hypothesis, have probably quite frequently committed Type II statistical errors. This is a simple consequence of the fact that significance tests with alpha-levels of this magnitude and lower are biased to protect against Type I errors. In the Bayesian approach, the investigator uses procedures that assign probability statements directly (as opposed to the indirect method of estimating error probabilities, α and ß) to competing hypotheses, including what we usually consider the null hypothesis (here, no preference). Finally, the Bayesian approach seems far more compatible with the way I think most researchers conduct behavioural research and interpret results than with classical procedures. In fact, some common and sensible ways of conducting experiments invalidate the use of classical inferential statistics.

To illustrate the contrast between classical and Bayesian approaches and interpretation, I will work through a simple example involving binary data from a two-speaker playback experiment with females of the green treefrog. Extending this approach to parametric data (e.g. response latencies, etc.) would be simple. Bayesian methods for multivariate problems (parametric and non-parametric) are also well established, and at the end of the chapter I discuss some of these references.

Females were given a choice between a synthetic call with a duration of 160ms (about average in the population) and an alternative of 120ms in trials conducted over a three-year period. I reported that 18 females chose the stimulus of 160ms and 7 females the alternative; a two-tailed binomial test of the null hypothesis of no preference allowed me to reject the null hypothesis at the 5% level (Gerhardt 1987). However, within the classical framework, this significance test was invalid because I did not plan (or specify) the sample size in advance. More significantly, the results I obtained during the first part of the experiment, and upon which I based my decisions to continue testing additional females at various times, illustrate one of the main reasons why the classical statistician insists on declaring sample size ahead of time.

In many experiments with green treefrogs, the first ten animals tested often choose the same alternative. But in this case, eight females chose the long call and two the short call. I then tested five more females, three of which chose the long call. Both of these results: 8:2 and 11:4 are not significant statistically, even using a one-tailed binomial test of the null hypothesis (P = 0.0547 and 0.0592, respectively; the two-tailed probability is just two times these values). Nevertheless, I tested ten more females and first achieved a *significant* P-value at N = 20 (15:5; P = 0.041, two-tailed). As stated above, the final results were 18:7 (P = 0.043). Obviously, I could be rightly accused of basing my decision to stop on the results that gave me a significant P-value, that is, I was acting as a *persistent investigator* (Pollard 1986). The more technical reason for stating and sticking with a sample size determined prior to the experiment is that the P-values of significance tests depend on knowing the outcome space of the experiment. That is, the possible ways (including all possible orders of choices) that a particular number of frogs could have responded depend on fixing the sample size or specifying a more complicated rule for when to stop collecting additional data. Indeed, the P-values I have calculated at various stages of the experiment are false ones in that they assume that the sample size was fixed in advance. The classical P-value is simply the probability that, given the null hypothesis, one would obtain the results observed and more extreme results (i.e. more against the null hypothesis) if the experiment were repeated very many times.

Here is how a simple Bayesian analysis might be applied to the same experiment, with the aim of estimating the proportion of females in the population that initially choose the long call to the short call and deciding if a preference exists. Because I had no idea about the existence of a preference beforehand, I would use a non-informative prior distribution for the probability density of a preference. Bayesian techniques require some kind of prior probability assessment. The necessity of some kind of subjective assessment (even saying that I had no idea which stimulus the frogs would prefer is a subjective judgment) is the main target of criticisms of Bayesian statistics, and the choice of an appropriate prior distribution is a main source of contention among Bayesian statisticians. In fact, final conclusions (the posterior probability density) are seldom affected significantly once a reasonable amount of data accumulates. This issue is fully addressed in several of the references discussed at the end of the chapter.

For the binomial case, one commonly used non-informative prior distribution is a uniform beta distribution, i.e. ß (1,1). The beta distribution is a continuous, natural conjugate for the binomial distribution. Indeed, the Natural Bureau of Standards binomial tables were derived from tables of the incomplete beta function (Novick and Jackson 1974). Using the formula and procedures from Pollard (1986), I obtain the probability density function for the results at each stage of the experiment (Fig. 6). These curves are the full Bayesian description of the probability of the unknown parameter. Following Novick and Jackson (1974), I might summarise the results by reporting the mean and mode. To express my degree of confidence that the females did show a preference, I might report the lower endpoint of the 90% (conservative) or 50% (less conservative) credible interval, which can be obtained using tables in Issacs et al. (1974). Credible intervals include 90% and 50% of the areas of the curves that contain the highest probability density (usually they are called highest density regions or HDRs).

Note that as the total sample size increases (compare Panel A to Panel D in Fig. 6), the curve more nearly approximates a normal distribution. Indeed, when there are at least 10 successes and 10 failures, a normal approximation can be used to assess credible intervals (Pollard 1986). In Bayesian terminology, I used the results obtained at stage one as my prior probability density for stage two and so forth. But the final conclusion (Stage 4; Panel D) would be exactly the same if I had waited until all trials had been completed before doing any analysis. More importantly, the Bayesian interpretation of the experimental result does not depend on any intentions I might have had. Nor does it consider any results I might have observed but did not, but rather solely on the choices actually observed. The final lower 90%-bound (0.55) on the proportion of females preferring the long call corresponds to the magnitude of a preference that I judge to be only marginally meaningful in terms of sexual selection. That is, if 55% were the true proportion of females showing the preference in the laboratory, I would not expect such a preference to be observable in field studies because there are many additional acoustic and other variables that would undoubtedly reduce a female's ability and propensity to choose a mate on the basis of a 40ms difference in call duration (e.g. Gerhardt 1982).

Figure 6. Posterior probability density functions for the results of a two-speaker choice experiment (long *versus* short call) at various stages of completion. The distributions are beta functions computed using a formula provided by Pollard (1986) and represent the full Bayesian description of the experimental data using a noninformative (uniform) prior distribution. The lower bound on the 90% high density region (or credible interval) is also indicated. Note that the inference that about 70% of the females prefer the long call and the lower bound on the 90% credibility interval do not change very much as the sample size is increased by 5 in each of the three stages after the initial sample of 10. By contrast, the significance level of a two-tailed binomial test of the null hypothesis of no preference changes from 0.1 to 0.04 when the sample size is doubled, i.e. from stage 1 (Panel A) to stage 3 (Panel B). In fact all of the P-values would be invalid from the standpoint of classical inferential statistics unless they corresponded to that associated with a sample size specified in advance. See the text for discussion. ⟶

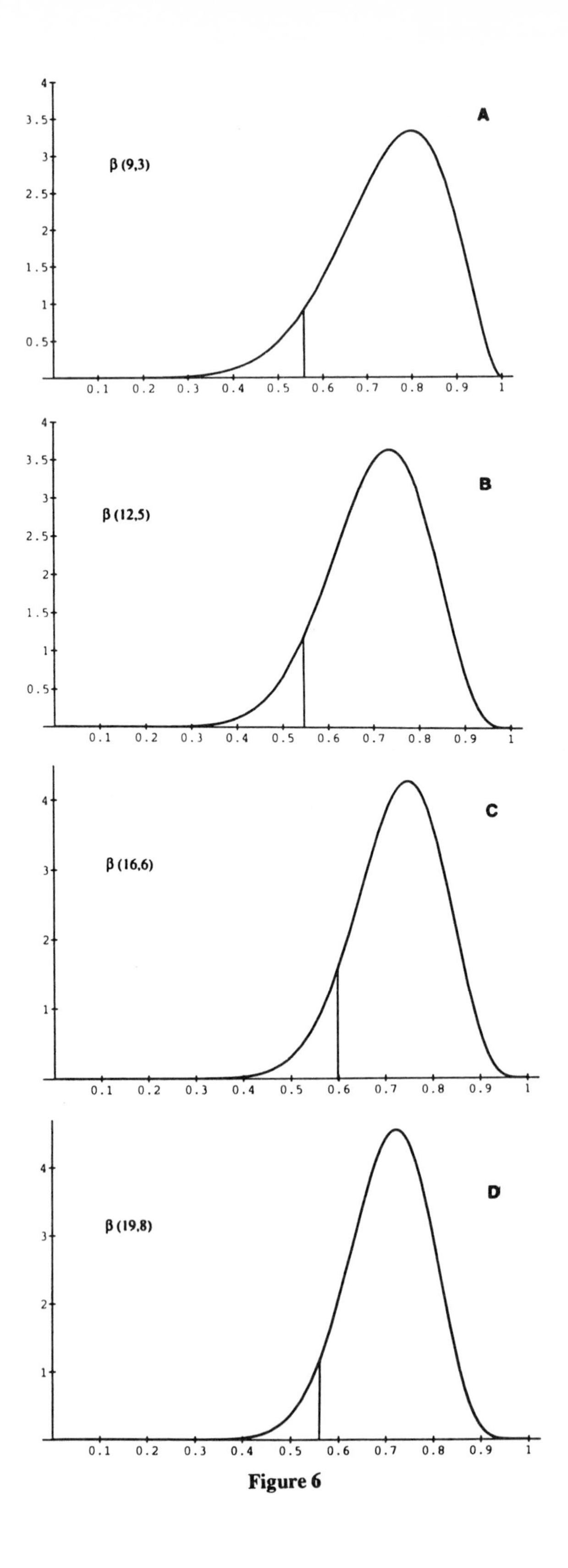

Figure 6

An alternative view is that, all things being equal, such a small preference may affect mating choices enough to have an evolutionary consequence over many generations.

Comparing the results from the first and third stages of the experiment provide a striking contrast with the classical approach. Even with a sample size of ten, the probability density function suggests that a preference exists, and the lower 90% credible interval is 0.62. This is obvious even within the classical approach if one computes the Type II error probability, which is 0.32 assuming that the true proportion of females preferring the long call were 80%, the observed proportion. After the initial stage, the maximum proportion of females preferring the long call was observed in the third stage (75%); here the lower 90% credible interval is 0.58. Contrast this with the change in the Type I error probability: from a non-significant 0.109 when N = 10 to a significant 0.041 when N = 20. Consider further the contrast if the next five females had chosen the long call: the lower 90% credible interval would have changed to 0.71, while the P-value would have dropped to 0.004 (two-tailed). Within the classical framework, the error probability is not a direct probability statement about a hypothesis (i.e. obtaining a P-value of 0.004 does not mean that the probability of the null hypothesis is 0.004, or that the probability of a preference in the population is 0.996). However, P-values (or significance levels) are often (understandably) interpreted as quantifying the evidence against the null hypothesis.

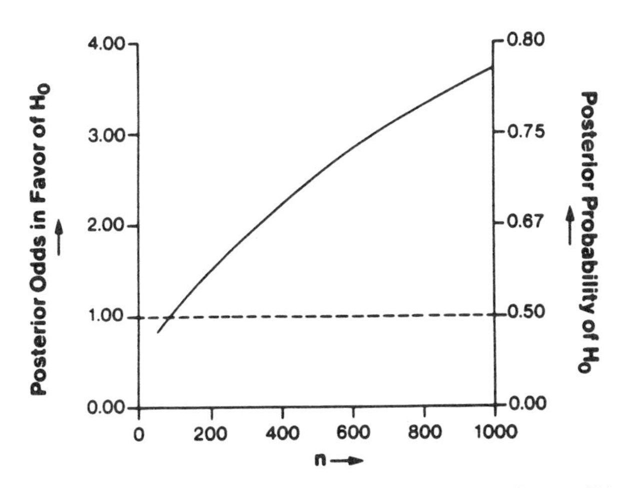

Figure 7. Posterior (Bayesian) probabilities (odds for) for a null hypothesis (solid line) as a function of sample size. A classical test of a point null hypothesis is just significant at the 0.05-level over the entire range of sample sizes, but a Bayesian analysis would conclude that the null hypothesis is about four times more likely to be true than the alternative hypothesis at a sample size of 1000. In this example, classical and Bayesian conclusions would be the same for small samples. Modified after Pollard (1986).

This contrast between Bayesian and classical procedures becomes even more pronounced as sample sizes become very large. Indeed, virtually any point null hypothesis can be rejected by a significance test if the sample size is increased sufficiently, whereas if an effect remains small, increasing the sample size will lead a Bayesian to favour the null hypothesis (Fig. 7; Pollard 1986; Berger 1985; Wilkinson 1990). In addition, the Bayesian approach explicitly demands that the investigator, who usually has a great deal of relevant information about the system within which the hypothesis has been generated, make judgments about the magnitude of significant effects not only relative to the precision with which measurements can be made (a classical concern as well) but also relative to the problem at hand. We all know that tiny effects, though statistically significant, are often biologically irrelevant.

It should be clear from the above example that if the sample size is small, then Bayesian procedures work no magic. But they do express more clearly the uncertainty that exists when N is small and the reasonableness of sampling in stages if the animal is hard to find or difficult to test. As stated by Edwards, Lindman and Savage (quoted by Berger 1985, p. 505), "Many experimenters would like to feel free to collect data until they have either conclusively proved their point, conclusively disproved it, or run out of time, money, or patience."

In Table I, I provide some credibility intervals and confidence intervals for a hypothesis of no preference (50% of the animals chose one of two alternative stimuli) for different sample sizes. If the proportion of animals choosing between alternative stimuli falls within these limits, then this would constitute evidence for the null hypothesis, and the choice of intervals (50%, 90% or some other interval) would depend on the investigator's assessment of the magnitude of a biologically significant effect and the accuracy of the behavioural data. Another, more conservative, procedure would be to estimate upper credible intervals above the proportion (close to 0.5) that would reflect the researcher's judgment about the magnitude of preferences that are meaningful for the problem at hand. Obviously, observed proportions that fall outside of these intervals would constitute evidence for a preference. Expressing results as an observed proportion (or the mean or mode of a probability density function) and estimating credible intervals (or interpreting confidence limits in the Bayesian fashion) gives the researcher and his colleagues the freedom to draw their own conclusions, rather than blindly submitting to alpha-level dictatorship.

Table I. Credible intervals and confidence intervals for a null hypothesis of no preference at four sample sizes (N). Confidence intervals were computed using Systat (Wilkinson 1990) and are generally more conservative than corresponding intervals for the binomial distribution, which can be computed easily using tables in Burstein (1971).

	50%		90%	
N	**credible interval**	**confidence interval**	**credible interval**	**confidence interval**
10	0.39 - 0.61	0.29 - 0.73	0.25 - 0.75	0.17 - 0.81
20	0.42 - 0.58	0.36 - 0.66	0.32 - 0.68	0.26 - 0.72
30	0.44 - 0.56	0.38 - 0.62	0.35 - 0.65	0.31 - 0.68
40	0.45 - 0.55	0.40 - 0.61	0.33 - 0.66	0.33 - 0.66

Suggestions for Further Reading

Historically, the Bayesian approach pre-dates the so-called classical school, originating with a posthumously published paper by Thomas Bayes (1702-1761), which is reprinted in Press (1989). Nowadays, Bayesian analyses are used to address many kinds of scientific and practical questions, ranging from foraging and game theory to genetic counselling (e.g. McNamara and Houston 1980; Young 1991). Maximum likelihood techniques, which are central to many Bayesian analyses (though these procedures may be arrived at and interpreted in non-Bayesian terms), have also been used extensively in analysing evolutionary problems and have recently been applied to analyses of phylogeny (e.g. Lynch 1991).

The conviction that the Bayesian approach to scientific reasoning is relevant to my research is mainly based on arguments in the book by Howson and Urbach (1989; see also recent short commentary by Howson and Urbach 1991 and a paper by Burger and Berry 1988). Howson and Urbach (1989) pursue an uncompromising, critical analysis of Popper's philosophy of science and classical statistical inference as founded by R.A. Fisher and extended by Neyman and Pearson. Moreover, their criticisms extend to the realm of experimental design. For example, Howson and Urbach's critique of Fisher's ideas about randomisation echoes that of Hurlbert (1984): randomisation is a poor substitute for designs that control explicitly for known factors that could confound treatment effects. One aim of Bayesian analysis is to incorporate prior (expert) knowledge about the system within which a hypothesis is being tested into the design of experiments and the interpretation of results.

In addition to the book by Howson and Urbach (1989), I recommend the books by Novick and Jackson (1974) and Pollard (1986) to those with limited mathematical and statistical background. These authors provide many worked examples of Bayesian analyses of practical research problems. The text by Lee (1989) is more demanding mathematically, but also provides worked examples that compare Bayesian and classical approaches to hypothesis testing. He also provides a method for computing credible intervals for beta distributions using tables of the F distribution. Press (1989) provides a list of available computer packages and programs for doing Bayesian statistics. Berger (1985) provides a rigorous mathematical treatment of Bayesian and related analyses, but he also offers clear verbal arguments against the so-called classical school, while at the same time taking the position that some classical methodology may be justified if given a Bayesian interpretation. Good (1983) provides a personal and often amusing outline of his long career as a Bayesian advocate and specifically calls for a Bayes/non-Bayes compromise.

Acknowledgements

I thank Georg Klump and Joshua Schwartz for comments on the manuscript. I am grateful to George P. Smith for introducing me to Bayesian statistics and for his comments on the last part of this chapter. My research has been funded generously by the National Science Foundation and the National Institutes of Mental Health.

References

Berger, J.O. 1985. *Statistical Decision Theory and Bayesian Analysis.* 2nd Edition. Springer-Verlag; New York.

Berger, J.O. and Berry, D.A. 1988. Statistical analysis and the illusion of objectivity. *Am. Sci.*, **76**, 159-165.

Burstein, H. 1971. *Attribute Sampling. Tables and Explanations.* McGraw-Hill; New York.

Date, E.M., Lemon, E.R., Weary, D.M. and Richter, A.K. 1991. Species identity by birdsong: discrete or additive information. *Anim. Behav.*, **41**, 111-120.

Dyson, M.L. and Passmore, N.I. 1988. Two-choice phonotaxis in *Hyperolius marmoratus* (Anura: Hyperoliidae): the effect of temporal variation in presented stimuli. *Anim. Behav.*, **36**, 648-652.

Embleton, T.F.W., Piercy, J.E. and Olson, N. 1976. Outdoor sound propagation over ground of finite impedance. *J. Acoust. Soc. Am.*, **59**, 267-277.

Gerhardt, H.C. 1974a. The significance of some spectral features in mating call recognition in the green treefrog (*Hyla cinerea*). *J. Exp. Biol.*, **61**, 229-241.

Gerhardt, H.C. 1974b. Vocalisations of some hybrid treefrogs: acoustic and behavioral analyses. *Behaviour*, **49**, 130-151.

Gerhardt, H.C. 1978a. Discrimination of intermediate sounds in a synthetic call continuum by female green tree frogs. *Science*, **199**, 1089-1091.

Gerhardt, H.C. 1978b. Mating call recognition in the green treefrog (*Hyla cinerea*): the significance of some fine-temporal properties. *J. Exp. Biol.*, **74**, 59-73.

Gerhardt, H.C. 1981. Mating call recognition in the green treefrog (*Hyla cinerea*): importance of two frequency bands as a function of sound pressure level. *J. comp. Physiol. A.*, **144**, 9-16.

Gerhardt, H.C 1982. Sound pattern recognition in North American treefrogs (Anura: Hylidae): implications for mate choice. *Am. Zool.*, **22**, 581-595.

Gerhardt, H.C. 1983. Acoustic communication in treefrogs. *Verh. Dtsch. Zool. Ges.*, **1983**, 25-35.

Gerhardt, H.C. 1987. Evolutionary and neurobiological implications of selective phonotaxis in the green treefrog, *Hyla cinerea. Anim. Behav.*, **35**, 1479-1489.

Gerhardt, H.C. 1991. Female mate choice in treefrogs: static and dynamic acoustic criteria. *Anim. Behav.*, **42**, 615-635.

Gerhardt, H.C. and Doherty, J.A. 1988. Acoustic communication in the gray treefrog, *Hyla versicolor*: evolutionary and neurobiological implications. *J. comp. Physiol. A.*, **162**, 261-278.

Gerhardt, H.C. and Klump, G.M. 1988. Masking of acoustic signals by the chorus background noise in the green treefrog: a limitation on mate choice. *Anim. Behav.*, **36**, 1247-1249.

Good, I.J. 1983. *Good Thinking: The Foundations of Probability and its Applications.* University of Minnesota Press; Minneapolis.

Greenewalt, C.H. 1968. *Bird Song: Acoustics and Physiology.* Smithsonian Inst. Press; Washington.

Howson, C. and Urbach P. 1989. *Scientific Reasoning: the Bayesian Approach.* Open Court Publ.; LaSalle, IL.

Howson, C. and Urbach, P. 1991. Bayesian reasoning in science. *Nature*, **350**, 371-374.

Hurlbert, S.H. 1984. Pseudoreplication and the design of ecological field experiments. *Ecol. Monogr.*, **54**, 187-211.

Issacs, G.L., Christ, D.E., Novick, M.R. and Jackson, P.H. 1984. *Tables for Bayesian Statistics.* University of Iowa (Iowa Testing Programs, Linquist Center for Measurement, Iowa City).

Lee, P.M. 1989. *Bayesian Statistics: An Introduction.* Oxford University Press; New York.

Lynch, M. 1991. Methods for the analysis of comparative data in evolutionary biology. *Evolution*, **45**, 1065-1080.

McNamara, J. and Houston, A.I. 1980. The application of statistical decision theory to animal behaviour. *J. Theor. Biol.*, **85**, 673-690

Maynard Smith, J. 1991. Theories of sexual selection. *Trends Ecol. Evol.*, **6**, 146-151.

Nelson, D.A. 1988. Feature weighting in species song recognition by the field sparrow (*Spizella pusilla*). *Behaviour*, **106**, 158-181.

Novick, M.R. and Jackson, P.H. 1974. *Statistical Methods for Educational and Psychological Research.* McGraw Hill; New York.

Oldham, R.S. and Gerhardt, H.C. 1975. Behavioral isolation of the treefrogs *Hyla cinerea* and *Hyla gratiosa*. *Copeia*, **1975**, 223-231.

Piercy, J. E. and Daigle, G.A. 1991. Sound propagation in the open air. In: *Handbook of Acoustical Measurements and Noise Control.* (Ed. by C.M. Harris), pp. 3.1-3.26. McGraw-Hill; New York.

Pollard, W. 1985. *Bayesian Statistics for Evaluation Research: An Introduction.* Sage Publ., Beverly Hills.

Press, S.J. 1989. *Bayesian Statistics: Principles, Models and Applications.* Wiley and Sons; New York.

Ratcliffe, L.M. and Grant, P.R. 1983. Species recognition in Darwin's finches (*Geospiza*: Gould). II. Geographic variation in mate preference. *Anim. Behav.*, **31**, 1154-1165.

Schwartz, J.J. and Gerhardt, H.C. 1989. Spatially mediated release from auditory masking in an anuran amphibian. *J. comp. Physiol. A.* **166**, 37-41.

Searcy, W.A. and Brenowitz, E.A. 1988. Sexual differences in species recognition of avian song. *Nature*, **332**, 152-154.

Waage, J.K. 1975. Reproductive isolation and the potential for character displacement in the damselflies, *Calopterix maculata* and *C. aequabilis* (Odonata: Calopterygidae). *Evolution*, **33**, 104-116.

Wasserman, M. and Koepfer, H.R. 1977. Character displacement for sexual isolation between *Drosophila mojavensis* and *D. arizonensis*. *Evolution*, **31**, 812-823.

Wiley, R.H. and Richards, D.G. 1978. Physical constraints on acoustic communication in the atmosphere: implications for the evolution of animal communication. *Behav. Ecol. Sociobiol.*, **3**, 69-94.

Wilkinson, L. 1990. *Systat: The System for Statistics.* Systat Inc.; Evanston, IL.

Young, I.D. 1991. *Introduction to Risk Calculation in Genetic Counseling.* Oxford University Press; Oxford.

QUANTIFYING RESPONSES TO PLAYBACK: ONE, MANY, OR COMPOSITE MULTIVARIATE MEASURES?

Peter K. McGregor

Behaviour and Ecology Research Group
Department of Life Science
University of Nottingham
University Park
Nottingham NG7 2RD
U.K.

Introduction

Acoustic communication is one of the better studied areas of animal behaviour (Halliday and Slater 1973). Part of the reason for the large number of studies in this area is the relative ease with which sounds can be recorded and broadcast. This has permitted experiments in the laboratory and in the field in which sound is played to animals and their responses noted. Such *playback* experiments have been conducted on a large number of species, notably songbirds and anuran amphibians (see chapters by Falls and by Gerhardt respectively in this volume), but including a wide diversity of taxonomic groups; from whales (e.g. Clark and Clark 1980) to wolves (e.g. Harrington 1989), from reef fish (e.g. Myrberg and Riggio 1985) to rodents (e.g. Cherry 1989), and from deer (e.g. Clutton-Brock and Albon 1979, McComb chapter in this volume) to *Drosophila* (e.g. Bennet-Clark and Ewing 1969). As argued by Catchpole (this volume), playback is now regarded as an important part of any integrated investigation of a communication system.

There is much to be said for gathering as many, and as varied, measures of an animal's response to playback as is possible. If the response is recorded by video or tape recorder a large number of such measures can be noted. These measures are unlikely to be statistically independent, and even if it were statistically valid to present all of the measures, the sheer number of them could make it difficult to interpret the result unambiguously. The purpose of this chapter is to assess the relative merits of solutions to the problem of the embarrassment of riches resulting from a well documented playback experiment. There are three potential solutions:

1) choosing to collect data on just one measure from the many possible, termed *the single-measure* approach;
2) presenting all of the many measures collected, termed *the many-measures* approach;
3) using a multivariate technique, principal components analysis, to generate a single composite measure, termed *the multivariate-measure* approach.

Attention will be concentrated on the multivariate-measure approach, because although it has considerable potential, it is currently the rarest in the playback literature, perhaps because most workers are reluctant to become involved with unfamiliar multivariate statistics. To try and overcome this reluctance, I will illustrate the method by using a common statistics package to re-analyse a set of playback data that was originally analysed by separately considering the eight variables measured.

A. The Single-Measure Approach

The Rationale

This approach assumes that the behaviour elicited by playback is adequately described by a single measure of response.

Examples

A.1 A common technique to determine the attractiveness of, or preference for, a sound to anuran amphibians is to place the test animal midway between two loudspeakers, each playing a different stimulus. The measure used in such studies is simply which loudspeaker is approached. An example is the study of female natterjack toads *Bufo calamita* by Arak (1988); pairs of synthetic calls differing in various features such as mean frequency, relative sound pressure level and absolute sound pressure level were played simultaneously from two speakers and female preference was inferred from which loudspeaker the female approached.

A.2 Corn buntings (*Miliaria calandra*) show clear local dialects in song (e.g. McGregor 1980; McGregor et al. 1988), but the details of a local dialect change in a concerted fashion from year to year, with all males adopting the changes (McGregor and Thompson 1988; McGregor and Shepherd *in prep*). We wanted to see if males present in 1987 and also in 1990 responded to playback of the 1987 local dialect by replying with the 1987 local dialect rather than their current (1990) song. Therefore, in 1990, we recorded the songs elicited by playback of 1987 local dialect and used a real-time spectrum analyser to classify the songs as 1987 or 1990 local dialects, thus generating a single measure (McGregor and Kitwood *unpublished*).

A.3 An example of a case where there was the potential to collect data on a number of measures was a study of intra- and inter-specific response to playback of a song of a great reed warbler (*Acrocephalus arundinaceus*). The experimenters chose to measure one variable, the time the test male spent within 1m of the loudspeaker (Catchpole and Leisler 1986). The rationale for choosing such a measure was that approaching, and staying close to, a singer is an important component of aggression in *Acrocephalus* warblers, and the question addressed by playback was whether song playback elicited an aggressive reaction from conspecifics and congenerics.

Advantages

The single-measure approach has three main advantages.

First, it forestalls the criticism of non-independence of variables, since only one variable is measured. However, as in *Example B.1* below, the single aspect of response noted may give rise to a series of interrelated variables, which negates the single variable advantage.

Second, it can be difficult to note responses accurately during the somewhat hectic business of managing the playback apparatus and keeping track of the test animal's response. Collecting only one piece of information during the response makes accurate data gathering more likely.

Third, it is easy to present the results clearly, although interpretation may not be as easy (see *Disadvantages* below).

Disadvantages

The first and arguably the only major disadvantage of the single-measure approach is that it relies on the single variable being an adequate description of the animal's response. Choosing this variable is crucial to the validity of the experiment and ease of interpretation. In some cases the variable to choose is obvious (e.g. *Example A.2* above), but in most instances the choice is based on an extensive knowledge of the species concerned and its pattern of interactions, although this information is rarely presented in the same paper as the playback result, if it is presented at all. This requirement for such detailed background knowledge will sometimes render the single-measure approach inappropriate.

A second disadvantage is that some types of single measure may not change in a monotonic fashion. A good example is the relationship between song and strength of aggressive response. In many species, no song is associated both with a close, silent approach and with no aggressive response. In other words, playback has either elicited the strongest aggressive response, with the test bird silently hopping around the speaker probably displaying visually, or the weakest, with the test bird ignoring playback. Deciding between these alternative extremes would require another measure, such as approach distance. If some song is elicited by playback then the aggressive response is intermediate between these extremes.

A more insidious disadvantage is that sometimes a number of measures have been noted during response to playback, but only the one significant measure is presented and no mention is made of the others. This is completely unacceptable, but very difficult to detect either when refereeing the manuscript or when reading the published paper.

Suggested Guidelines for Use

1) Explain why the single measure noted is considered to be an adequate description of the behaviour of interest
2) State that no other measures of response were noted

Summary

This approach has a role to play in measuring response to playback, particularly when the question posed is expected to have a rather straightforward answer. However, it can be constrained by the need for a detailed knowledge of the study species' response and by the possibility that other measures have been noted, but not presented because they did not reach significance.

B. The Many-Measures Approach

The Rationale

This approach tacitly assumes that a single measure is not adequate to describe the range of responses of interest that are likely to be elicited by playback.

Examples

B.1 A study by McGregor and Krebs (1984) of the response of 32 territorial male great tits (*Parus major*) to the playback of songs varying in degree of degradation and familiarity measured eight variables of aggressive response. This study is the data set for the multivariate-measure approach and is explained in more detail below. The justification for recording and presenting eight measures of aggressive response to playback was that there is no adequate single measure of such a response in this species.

B.2 In a study of the length of song (*strophe length*, see Lambrechts and Dhondt 1987, and chapter by Lambrechts in this volume) elicited by playback of long and short strophes to great tits, each test male's singing response was recorded (McGregor and Horn *in press*). The number of phrases per strophe, a measure of strophe length, was counted for each song. A number of measures (the mean, maximum, minimum and first strophe length) were then derived from this data. It was argued that maximum, minimum and first strophe length elicited were likely to illustrate differences in response between individuals that would not be readily apparent from a comparison of mean strophe length on its own.

B.3 Interactive playback (see Dabelsteen chapter in this volume) with great tits (McGregor et al. *in press*), allowed the strophe length of playback to be changed song by song, and for this reason various measures of the test bird's strophe were related to the playback strophe immediately preceding it. The use of such relative measures and presenting the mean, maximum, minimum and first value for each measure led to a total of 26 variables being presented. Our justification for presenting this apparently inordinate number of measures of response was that the responses elicited by interactive playback were complex and needed this number of measures to describe them adequately.

Advantages

The main advantage of presenting a number of measures of responses to playback is that a number of different aspects of response can be investigated. This is likely to be most important where there are no strong *a priori* reasons for expecting playback to elicit a particular pattern of response. A good example of this is the analysis of the interactive playback experiment with great tits (*Example B.3*). One of the variables gave us an unexpected insight into the way song is used in counter-singing in great tits, thus vindicating our argument for the use of so many measures. The measure was relative delay, defined as the time from beginning of playback strophe to the beginning of bird's response strophe, divided by playback strophe duration. One variable derived from this was minimum relative delay. If minimum relative delay was less than unity, then the bird had overlapped playback at least once. We found interesting differences between birds and treatments in the tendency to overlap playback, suggesting a new interpretation of counter-singing and leading to an experiment to investigate the idea (Dabelsteen et al. *ms*).

Disadvantages

The disadvantage most commonly noted for the many-measures approach is that the measures are likely to be correlated. However, even if the measures are significantly correlated, it is statistically valid to present them as measures of different aspects of the response elicited by playback. It is not statistically valid to combine the measures. This is commonly done in two ways. First, tests such as the sign test are applied to the direction of difference of the measures. A fictitious example is the statement 'none of the eight measures showed significant differences, but there was a significant difference when combined (six of the eight showed a greater response to own species song, $P < 0.05$ 2-tailed sign test)'. If there are significant correlations between the eight measures, then they are not independent observations and cannot be used as such in subsequent analyses. The second way in which measures are combined is not valid regardless of whether there are significant correlations between the measures. This is combining variables using a form of weighting that is arbitrary, or justified by the experimenter's preconceptions about the relationships between measures. Such composite scores are commonly found in studies involving playback to oestradiol-implanted females. The precopulation solicitation postures elicited vary in the extent of 'completeness' and a composite score is generated either by simply adding all degrees of posture (implicitly giving all postures equal weight), or by giving categories of degree different weights (e.g. number of instances of some evidence of posture x1, number of semi-complete postures x5, number of full postures x10). Since the final score depends on the weightings chosen, it may be possible to generate a significant difference between stimuli simply because of the magnitude of weights. For this reason the composite score should be generated in a statistically appropriate way (see section **C**), or the three measures presented and analysed separately.

A second disadvantage of the many-measures approach is the inelegance of presenting the results (e.g. 38 of the 61 column inches of the Results section of McGregor and Krebs (1984) were tables of data). This has more than aesthetic consequences. The experiment was designed to test a possible mechanism of degradation assessment (Morton 1982); however, the result of the test has been ignored in the relevant literature (e.g. Morton et al. 1986), probably because of the difficulty of clearly presenting (and therefore of the reader easily interpreting) the analyses based on eight separate variables.

Suggested Guidelines for Use

1) Never use second order statistics (e.g. sign test) on significantly correlated variables
2) Never generate a composite measure by weighting variables according to the experimenter's preconception of how variables are interrelated
3) It may be useful to present a table of correlation coefficients and probabilities in an Appendix, so that readers can see for themselves whether the measures are likely to be measuring different aspects of the response to playback

Summary

Despite the difficulties of clear presentation and subsequent interpretation that are raised by using the many-measures approach, it is often important to show the way in which the different measures vary with playback treatment, even when a multivariate approach is also used.

C. The Multivariate-Measure Approach

The Rationale

This approach attempts to combine the many variables originally measured into a smaller number of composite measures. The multivariate methods employed to achieve this are principal components analysis or factor analysis. The aim of principal components analysis is to produce new variables that are statistically independent of each other, that is, each component represents an axis that is orthogonal to the other component axes. Factor analysis has the same aim, to describe the original variables in terms of a smaller number of factors and therefore discover the relationship between these factors. It differs from principal components analysis because it assumes a particular statistical model (constant correlation ratios between the rows of the correlation matrix) (Manly 1986). Unless one has a particular model of the underlying causes of the correlations between variables, it is best not to use factor analysis (Reyment et al. 1984). Therefore, only principal components analysis will be considered for the rest of this chapter.

Examples

Although there are examples of this approach in the literature, for example Huntingford (1976) and McGregor and Avery (1986), I am going to use one study to look at the approach in detail.

The experiments (McGregor and Krebs 1984, see also *Example B.1*) were designed to investigate the ability of territorial male great tits to judge the distance of a source of song using information on the degree of sound degradation or distortion (song amplitude was held constant). Each male was played a song in a more, and in a less, degraded form - the *degraded* song (made by re-recording from 100m) and the *undegraded* song (re-recorded from 5m), respectively. The songs were played from the same position, 25m inside the territory boundary of the test male. Each male was played two different songs in degraded and in undegraded forms; one song was of a song type which was either sung by the test male, or by a neighbour, or by both (but note that all playback songs were recorded from males >500m but <6km from the test male). This song was called the *familiar* song. The other song was of a song type not sung by the test male, nor by any male within 500m of the test male: this was called the *unfamiliar* song. Therefore, each of the 32 test males was played 4 stimuli; *familiar undegraded*, *familiar degraded*, *unfamiliar undegraded* and *unfamiliar degraded*. The eight measures describing strength of response to playback were chosen because they could be noted relatively accurately in the field and because they had proved useful in previous playback experiments with great tits. The first measure was termed total time responding (*TTR*), defined as the time spent singing, calling, approaching, staying within 20m of the loudspeaker, or any combination of these features. There were two measures of song output: number of strophes (termed bursts (*BST*) after Krebs 1976), and seconds of song (*SSO*). The pattern of approach generated two measures: the time spent within 20m (*SCL*), and the closest approach to the loudspeaker (*MIN*). Three measures assessed the rapidity of response, or latency: latency to first vocalisation or approach (*LAT*), latency to song (*LATS*), latency to closest approach to the loudspeaker (*LMIN*).

In the next section, this data set is subjected to principal components analysis by the large, common statistics package SPSS, in this case the PC$^+$ version.

Principal Components Analysis of the Example

In many large statistics packages (including SPSS), principal components analysis is found under the heading Factor analysis. This is because principal components analysis is used as a first stage in factor analysis to determine the number of factors for subsequent factor analysis. Although details of command lines and output vary between statistics packages, the sections of analysis needed to describe response to playback as a single measure are fundamental to the analysis and are therefore common to all major packages.

Table I shows the form of the lines of control information needed to perform principal components analysis using SPSS.PC$^+$.

Table I. The session control commands used to analyse McGregor and Krebs (1984) data to generate a single principal component measure of response to playback. The commands are explained in the text.

```
FACTOR /VARIABLES TTR BST SSO SCL MIN LAT LATS LMIN
       /PRINT UNIVARIATE INITIAL CORRELATION EXTRACTION
       /CRITERIA FACTORS (1)
       /SAVE REGRESSION (1 PC).
```

The command FACTOR invokes factor analysis, the first stage of which is the principal components analysis we require. The variables to be used in the analysis (the eight original variables in this example) are defined with the /VARIABLES command. The information generated when carrying out the analysis is controlled by /PRINT: UNIVARIATE gives means, standard deviations and number of valid cases for the analysis (Fig. 1); INITIAL gives initial estimates of the variance in each original variable explained by all principal components extracted (this is referred to as communality) and % variation explained for each component (Fig. 3); CORRELATION gives the correlation matrix (Fig. 2); EXTRACTION gives revised communality for each original variable (Fig. 5). The analysis is restricted to generating only the first principal component by the command /CRITERIA FACTORS (1). The principal component scores, generated by the regression method (see section 4 below), for each case are written to the file containing the original variables by the /SAVE REGRESSION (1 PC) command. The number of principal component variables generated and their variable names are determined by the information in brackets. In this example, only the first principal component score will be saved as the variable PC1. If more than one principal score had been requested these would have been labelled PC1 to PCn, where n is the number of scores requested (see Appendix 2).

Principal components analysis has four stages:

1) checking that the data are suitable for principal components analysis
2) computing the principal components (component extraction)
3) interpreting the first principal component
4) computing principal components scores (a new variable used in analyses of response to playback).

```
Page    1                          SPSS/PC+                            7/23/9

                 - - - -   F A C T O R   A N A L Y S I S   - - - -

Analysis Number  1  Listwise deletion of cases with missing values

                     Mean       Std Dev    Label

TTR             210.45313     118.18555    TOTAL TIME RESPONDING
BST              20.62500      13.58380    BURSTS OF SONG
SSO             139.85938      93.98758    SECS SONG
SCL             123.03906     131.25526    SECS IN 20M
MIN              23.20313      26.46030    CLOSEST DISTANCE
LAT              70.62500      99.38837    LATENCY
LATS             95.40625     126.15597    LATENCY TO SONG
LMIN            145.47656     106.99187    LATENCY CLOSEST

Number of Cases  =      128

-----------------------------------------------------------------------------
```

Figure 1. The first page of printout generated by SPSS.PC^{+} when analysing McGregor and Krebs (1984) data to produce the first principal component (see Appendix 2 for the production of further components): Summary statistics for the original variables. This information is useful to check that all the cases you expect to be included in the analysis have been (use the mean values and the number of cases).

```
Page    2                          SPSS/PC+                            7/23/9

                 - - - -   F A C T O R   A N A L Y S I S   - - - -

Correlation Matrix:

              TTR        BST        SSO        SCL        MIN        LAT        LATS

TTR         1.00000
BST          .62323    1.00000
SSO          .63554     .95019    1.00000
SCL          .73579     .20204     .17383    1.00000
MIN         -.57181    -.31021    -.27649    -.62402    1.00000
LAT         -.58047    -.51090    -.47105    -.38012     .70656    1.00000
LATS        -.49441    -.62716    -.59207    -.27276     .53463     .81942    1.00000
LMIN        -.30298    -.34454    -.33019    -.12913     .42093     .71015     .54531

             LMIN

LMIN        1.00000
-----------------------------------------------------------------------------
```

Figure 2. The second page of printout: the correlation matrix. The original variables are arranged in the same order as that in which they were read (see /VARIABLES line of control information in Table I).

1) *Checking suitability of data* The very fact that principal components analysis is being considered is probably the best indication that the data are suitable for such analysis, that is, there are good reasons to think that the variables are correlated to some extent - they are all measuring aspects of strength of response to playback. Most packages give a range of more formal indications of suitability, ranging from a correlation matrix (e.g. Fig. 2) to tests designed to detect significant differences from lack of any interrelationships. Appendix 1 looks at these in more detail for the sample data set, which is suitable for principal components analysis on all criteria. At the publication stage it is probably adequate to state that the correlation matrix has been inspected.

2) *Component extraction* The next stage is to carry out a principal components analysis to construct a number of orthogonal axes through an n-dimensional scatter of data points (where n = number of original variables). The first axis is calculated (by weighting the original variables) to pass through the multi-dimensional cloud along the direction of maximum variation: it is exactly equivalent to the major axis in bivariate correlation. Further axes are then chosen successively, each constrained to be uncorrelated with the preceding ones. Each new axis therefore accounts for less variation than the previous one. The printout of this stage (Fig. 3) is confusing because the original variables are arranged alongside the list of components ('Factor') extracted. The line of stars between the two parts of this section of printout are meant to indicate that the two parts have no relation to each other. The most important columns in this section are 'Eigenvalue' and 'Pct of Var'; any factor with an eigenvalue greater than 1 is explaining more variation in the data than a single standardised original variable (Manly 1986, p.65), and the other column lists the percent of variation explained by the factor. In the sample data the first principal component (factor 1) is explaining nearly 56.7% of the variation in the data set and it has a considerably bigger eigenvalue (4.54) than the next two factors (1.35 and 1.13).

The next page of printout (Fig. 4) shows how each of the original variables contributes to the first principal component extracted. The communality of each original variable is shown on the next page of the printout (Fig. 5).

Many packages standardise the values for variables (express deviation from the mean in units of standard deviations) because principal components analysis is rather sensitive to the assumption that all variables have roughly equal absolute values, although it is relatively robust with regard to the assumption of parametric data (Hope 1968). (Standardisation is not considered to be appropriate with morphometric data because this will have the effect of transforming shape. Most packages allow standardisation to be bypassed, so if you are using a package designed for morphometric analyses, check that your data will be standardised.) As well as removing any problems caused by variables with differing absolute magnitudes, standardisation also means that the correlation matrix is the same as the covariance matrix. Therefore, the option offered by many packages to do principal components analysis on one or the other is irrelevant if the data are standardised.

The effect of standardisation on the original data is two-fold. First, all of the weightings (columns headed FACTOR 1) are in the range 1 to -1. Second, the sign of the weight simply groups variables changing in the same direction; in the sample data set the variables where small values indicate a strong response (LAT, LATS, LMIN, MIN) are positive, and where a large value indicates a strong response (TTR, SSO, BST, SCL) are negative.

3) *Interpreting the first principal component* An important part of component extraction is deciding whether principal components analysis has generated a *strength-of-response-to-playback* factor. The decision is based on two features. The first is the magnitudes of the component weights: in the sample data all eight of the original variables have similar weights and are therefore contributing about equally to the first principal component. The second is the sign of the weightings: as explained above, in the sample data original variables are grouped according to whether a large value indicates a strong response (e.g. TTR) or a weak response (e.g. LAT). The sample data clearly illustrate that the first principal components score is a strength of response factor, analogous to size in morphological data. By contrast, the second principal component (See Fig. 7 in Appendix 2) shows a wide range of weights (0.077 to 0.689) and the signs do not always correspond to groupings of original variables.

```
Page    3                         SPSS/PC+                              7/23/9

              - - - -  F A C T O R   A N A L Y S I S   - - - -

Extraction  1  for Analysis  1, Principal-Components Analysis (PC)

Initial Statistics:

Variable       Communality  *  Factor   Eigenvalue   Pct of Var   Cum Pct
                            *
TTR              1.00000    *     1       4.53929       56.7        56.7
BST              1.00000    *     2       1.34856       16.9        73.6
SSO              1.00000    *     3       1.12982       14.1        87.7

SCL              1.00000    *     4        .42206        5.3        93.0
MIN              1.00000    *     5        .28316        3.5        96.5
LAT              1.00000    *     6        .15376        1.9        98.5
LATS             1.00000    *     7        .08111        1.0        99.5
LMIN             1.00000    *     8        .04224         .5       100.0

   PC Extracted    1 factors.
-----------------------------------------------------------------------------
```

Figure 3. The third page of printout: details of the principal components analysis. **Factor** refers to the number of the principal component extracted (1 to n, where n is the number of original variables). Eigen values greater than 1 show principal components that explain more variance than a single original variable (PC1 to PC3 in this example). **Pct of Var** shows the percent of variance explained by each principal component. In this example, principal component 1 accounts for over half of the variance in the original data. **Communality** is the proportion of the variance in the **original variable** explained by all the principal components, in this table it should be very close to one. If it isn't, then it can be used to decide which of the original variables have little in common with the others and should therefore be dropped from further principal components analysis.

```
Page    4                         SPSS/PC+                              7/23/9

              - - - -  F A C T O R   A N A L Y S I S   - - - -

Factor Matrix:

              FACTOR  1

TTR            -.81997
BST            -.77712
SSO            -.75411
SCL            -.56151
MIN             .73252
LAT             .87163
LATS            .83068
LMIN            .62668

-----------------------------------------------------------------------------
```

Figure 4. The fourth page of printout: the factor matrix. This shows the weightings of the original variables (standardised) in the principal components extracted. Only one factor was extracted in this example.

```
Page    5                          SPSS/PC+                              7/23/9

              - - - -   F A C T O R   A N A L Y S I S   - - - -

Final Statistics:
```

Variable	Communality	*	Factor	Eigenvalue	Pct of Var	Cum Pct
		*				
TTR	.67236	*	1	4.53929	56.7	56.7
BST	.60391	*				
SSO	.56867	*				
SCL	.31529	*				
MIN	.53658	*				
LAT	.75973	*				
LATS	.69003	*				
LMIN	.39272	*				

Figure 5. The fifth page of printout: the final statistics. This page is similar to Fig. 3 in layout, but only shows the principal component requested. In this example, only the first component.

4) *Computing principal components scores* The aim of the multivariate-measure approach is to reduce the number of measures of playback to be considered and it does this by constructing orthogonal principal component axes through the original data. The eight values for each case in the original data set can be reduced by determining the position of the original data point in n dimensional (8 dimensional in this case) space relative to the major axis, principal component 1, or axes (other principal components with eigen values greater than one). These values can be saved for further analysis. In the example data set, only the first principal component scores were saved and used to investigate various aspects of the study (Fig. 6).

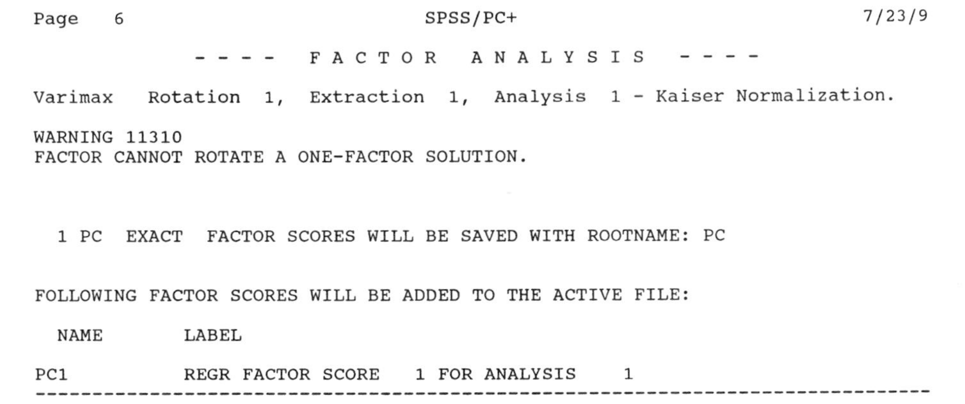

```
Page    6                          SPSS/PC+                              7/23/9

              - - - -   F A C T O R   A N A L Y S I S   - - - -

Varimax   Rotation  1,  Extraction  1,  Analysis  1 - Kaiser Normalization.

WARNING 11310
FACTOR CANNOT ROTATE A ONE-FACTOR SOLUTION.


  1 PC  EXACT  FACTOR SCORES WILL BE SAVED WITH ROOTNAME: PC


FOLLOWING FACTOR SCORES WILL BE ADDED TO THE ACTIVE FILE:

  NAME        LABEL

PC1           REGR FACTOR SCORE   1 FOR ANALYSIS    1
```

Figure 6. The sixth page of printout: factor analysis. This page first shows the package attempting to proceed into factor analysis (trying to perform a Varimax rotation). It is unable to do so because only one principal component was requested. If more than one principal component is to be extracted, factor analysis can be suppressed by replacing the /CRITERIA FACTORS (1) line in the control lines (see Table I) with /ROTATION=NOROTATE. The second part of this page confirms that the principal components scores (**exact factor scores**) will be saved. It also notes the new variable names and labels (in this example, **PC1** and **REGR FACTOR SCORE**).

Advantages

The difficulty of presenting and interpreting many variables is nicely illustrated by Table IIa, which shows the response to the four playback stimuli as measured by the eight original variables. The presentation problem is not solved by representing the data as a figure (see Fig. 2 in McGregor and Krebs 1984, which is also rather unclear). The main result of the study was that familiar song types elicited consistently stronger responses from the males when played in an undegraded form than when played in degraded form. All eight of the original measures showed significant differences. No such difference was apparent with unfamiliar song types. The presentation problem caused by the number of variables is exacerbated by the fact that a strong response results in a large value for four measures (TTR, BST, SSO and SCL) but a small value for the other measures (LAT, LATS, LMIN, and MIN). Presenting the single measure derived by principal components analysis solves both these problems; Table IIb clearly shows the difference in response to degraded and undegraded familiar song types and the lack of such a difference for unfamiliar song types.

Table IIa. The response to playback of four stimuli by 32 male great tits, as shown by the original eight variables (see text for details). Values are means ± 1se. Units are s, except BST (number) and MIN (m).

	Familiar		*Unfamiliar*	
	Undegraded	*Degraded*	*Undegraded*	*Degraded*
TTR	271.3±19.0	154.0±17.8	199.2±20.2	217.3±21.8
BST	024.1± 2.5	014.8± 2.0	022.8± 2.4	020.1±02.5
SSO	162.8±16.9	101.8±13.7	157.5±17.3	132.0±17.3
SCL	165.9±27.1	079.2±17.7	097.5±19.8	150.5±24.5
MIN	016.0± 3.2	025.4± 4.2	025.7± 4.8	020.2±04.5
LAT	029.9± 5.9	092.8±22.1	075.3±15.1	088.4±20.8
LATS	048.9±13.4	118.6±25.5	105.5±22.3	119.8±26.6
LMIN	089.4±11.3	173.4±20.4	151.1±16.6	165.1±20.9

The original data was also analysed by eight separate two-way analyses of variance (Table 1 in McGregor and Krebs 1984), seven of the eight variables showed a significant interaction term; that is, both familiarity (whether familiar or unfamiliar song type played) and degradation (whether undegraded or degraded version played) are important in determining response strength. The same analysis done with the single principal component score measure also gives a significant interaction term ($F_{1,31df} = 8.37$, $P < 0.005$).

Table IIb. The response to playback of four stimuli by 32 male great tits, as shown by the principal component variable derived from the above measures. Note that a large negative value indicates a strong response to playback. P values are for 2-tailed paired t-test.

	Familiar		*Unfamiliar*	
	Undegraded	*Degraded*	*Undegraded*	*Degraded*
	-0.479±0.104	0.417±0.182	0.032±0.175	0.030±0.201
P	0.0001		0.43	

The analyses clearly show that the method arrives at the same conclusions as the use of eight separate measures, but in a simpler, more easily interpreted form.

It was argued above that presenting many measures may have led to one of the main results of this set of experiments being ignored because of the complexity of information (see **B**. *Disadvantages*). McGregor and Krebs (1984) set out to test a suggested mechanism of degradation assessment (Morton 1982). Morton has proposed that the degree of degradation of a sound can only be assessed if the listener has an undegraded copy of the sound with which to compare what it hears. He further suggested that such undegraded internal standards are only present for songs which the bird sings itself.

The test by McGregor and Krebs consisted of playing three kinds of familiar song. A song of a song type sung by: 1) the test male and at least one neighbour, 2) the test male and no neighbours, and 3) neighbours of the test male, but not the test male. The rationale was straightforward, if Morton's idea was right, test males would respond differently to degraded and undegraded versions of songs of the same song type as they sang (1 and 2), but would show no difference if the song was of a song type only sung by neighbours (3). In the original paper the three kinds of familiar song were compared for each of the eight measures (Table 2 in McGregor and Krebs 1984) and led to the rather tentative statement "there is little evidence for category of familiar song having a marked effect on degradation discrimination".

The same analysis is presented for the principal components score in Table III. It is clear that there is a significant difference between response to degraded and undegraded song within each of the kinds of familiar song, but no such difference for unfamiliar songs. A one-way analysis of variance shows no significant difference between the three kinds of familiar song (analysis done on the undegraded value minus the degraded value; $F_{2,29df} = 1.66$, not significant). The analyses make it very clear that male great tits can assess the degree of degradation of songs which they do not sing, but which are of the same song type as a song sung by their neighbours. This result means that species capable of neighbour discrimination cannot acquire songs which their neighbours find unrangeable (cf. Morton 1982, Morton et al. 1986).

Disadvantages

One disadvantage of the multivariate-measure approach is that the sign of the principal components score is arbitrary (although its meaning is not), especially when the measures change monotonically in opposite directions (e.g. the meaning of large values is opposite for the TTR group of measures and for the LAT group). Although initially a bit disconcerting, this can be easily resolved by looking at the component weights shown in the factor matrix (Fig. 4). This disadvantage can be overcome when presenting the data by adding an arrow to the axis to indicate the direction of a strong response (e.g. Fig. 3 in Brindley 1991, Fig. 2 in McGregor and Avery 1986). Alternatively, the original measures can be transformed so that a large value always indicates a strong response. In the sample data set this would mean transforming the latency and closest approach measures by subtracting the actual value from the maximum value possible (420s and 100m respectively).

A second disadvantage is that although the first principal component may account for the most variation, it may not be the axis of most biological interest. This is a common problem in morphometric studies which include size and shape variables; size is almost always a major aspect of the variation, but often shape variation is the subject of study (Reyment et al. 1984). Once again, the component weights give a good indication of whether the principal component axes have sensible meanings in terms of response to playback.

Table III. The response to playback of unfamiliar stimuli and three kinds of familiar song (shown below in italics, see text for full details) as shown by the principal component variable derived from the original eight variables (see Table IIa). Note that a large negative value indicates a strong response to playback. There is a significant difference between undegraded and degraded versions of all three kinds of familiar song (2-tailed paired t-test) but no significant differences for unfamiliar songs.

	Familiar		*Unfamiliar*	
	Undegraded	*Degraded*	*Undegraded*	*Degraded*
Own+Neighbour	-0.418±0.16	0.219±0.38	-0.247±0.22	-0.138±0.26
Neighbour	-0.309±0.17	0.487±0.27	0.251±0.31	0.354±0.42
Own	-0.899±0.16	0.625±0.20	0.101±0.43	-0.283±0.17

Suggested Guidelines for Use

1) Standardise variables before carrying out principal components analysis
2) Use the correlation matrix to confirm that variables are interrelated to a reasonable extent
3) Check that the principal components extracted do represent a strength-of-response-to-playback measure (inspect the weights and signs of coefficients, see Appendix 1)
4) Use the minimum number of axes. If the first axis explains much more of variation than subsequent, only use this measure. In any case only use axes which explain more variation than single original variables (i.e. eigenvalue >1)
5) Give component weights of the principal components used, perhaps as footnote to Tables

6) Explain how the sign of the principal components score relates to strength of response and label figures accordingly

Summary

Principal components analysis seems to be the obvious solution to the problem of a playback experiment where a number of variables were measured because they were thought essential to adequately describe the responses, but where these original variables are likely to be interrelated and where it is difficult to maintain clarity in explaining the results if all the original variables are presented. Usually the analysis is used to derive one measure of response, although there may be instances were more are derived.

It deserves to be a more common tool for the analysis of playback results.

Acknowledgements

I thank: Mark Avery for being insistent about the merits of principal components analysis in relation to playback; John Krebs for permission to use data from joint research (funded by SERC); the Royal Society for funding equipment; Chris Barnard and Francis Gilbert for sharing the cost of the SPSS package (via University of Nottingham funding); the Behaviour and Ecology Research Group for general assistance and especially Francis Gilbert for improving the ms and statistical advice.

References

Arak, A. 1988. Female mate selection in the natterjack toad: active choice or passive attraction? *Behav. Ecol. Sociobiol.*, **22**, 317-327.

Bennet-Clark, H.B. & Ewing, A.W. 1969. Pulse interval as a critical parameter in the courtship of *Drosophila melanogaster*. *Anim. Behav.*, **17**, 755-759.

Brindley, E.L.C. 1991. Response of European robins to playback of song: neighbour recognition and overlapping. *Anim. Behav.*, **41**, 503-512.

Catchpole, C.K. & Leisler, B. 1986. Interspecific territorialism in reed warblers: a local effect revealed by playback experiment. *Anim. Behav.*, **34**, 299-280.

Cherry, J.A. 1989. Ultrasonic vocalizations by male hamsters: parameters of calling and effect of playbacks on female behaviour. *Anim. Behav.*, **38**, 138-153.

Clark, C.W. & Clark, J.M. 1980. Sound playback experiments with southern right whales (*Eubalaena australis*). *Science*, **207**, 663-665.

Clutton-Brock, T.H. & Albon, S.D. 1979. The roaring of red deer and the evolution of honest advertisement. *Behaviour* **69**, 145-170.

Falls, J.B. 1982. Individual recognition by sounds. In: *Evolution and Ecology of Acoustic Communication in Birds. Vol.II.* (Ed. by D.E. Kroodsma, E.H. Miller & H. Ouellet), pp. 237-278. Academic Press, New York.

Halliday, T.R. & Slater, P.J.B. 1983. *Animal Behaviour. Volume 2. Communication.* Blackwell Scientific Publications, Oxford.

Harrington, F.H. 1989. Chorus howling by wolves: acoustic structure, pack size and the Beau Geste effect. *Bioacoustics* **2**, 117-136.

Hope, K. 1968. *Methods of Multivariate Analysis*. University of London Press, London.

Huntingford, F.J. 1976. An investigation of the territorial behaviour of three-spined sticklebacks using principal components analysis. *Anim. Behav.*, **27**, 822-834.

Krebs, J.R. 1976. Habituation and song repertoires in the great tit. *Behav. Ecol. Sociobiol.*, **1**, 215-227.

Lambrechts, M.M. & Dhondt, A.A. 1987. Differences in singing performance between male great tits. *Ardea* **75**, 43-52.

McGregor, P.K. 1980. Song dialects in the corn bunting (*Emberiza calandra*). *Z. Tierpsychol.*, **54**, 285-297.
McGregor, P.K. 1991. The singer and the song: on the receiving end of bird song. *Biol. Rev.*, **66**, 57-81.
McGregor, P.K., & Avery, M.I. 1986. The unsung songs of great tits (*Parus major*): learning neighbours' songs for discrimination. *Behav. Ecol. Sociobiol.*, **18**, 311-316.
McGregor, P.K., Dabelsteen, T., Shepherd, M. & Pedersen, S.B. *in press* The signal value of matched singing in great tits: evidence from interactive playback experiments. *accepted* 6/91 *Anim. Behav.*,
McGregor, P.K. & Horn, A.G. *in press* Strophe length and response to playback in great tits. *accepted* 4/91 *Anim. Behav.*,
McGregor, P.K. & Krebs, J.R. 1984. Sound degradation as a distance cue in great tit (*Parus major*) song. *Behav. Ecol. Sociobiol.*, **16**, 49-56.
McGregor, P.K. & Thompson, D.B.A. 1988. Constancy and change in local dialects of the corn bunting. *Ornis Scand.*, **19**, 153-159.
McGregor, P.K., Walford, V.R. & Harper, D.G.C. 1988. Song inheritance and mating in a songbird with local dialects. *Bioacoustics* **1**, 107-129.
Manly, B.F.J. 1986. *Multivariate Statistical Methods. A Primer.* Chapman & Hall, London.
Morton, E.S. 1982. Grading, discreteness, redundancy and motivational-structural rules. In: *Evolution and Ecology of Acoustic Communication in Birds. Vol.I.* (Ed. by D.E. Kroodsma, E.H. Miller & H. Ouellet), pp. 183-212. Academic Press, New York.
Morton, E.S., Gish, S.L., & Voort, M. van der. 1986. On the learning of degraded and undegraded songs in the Carolina wren. *Anim. Behav.*, **34**, 815-820.
Myrberg, A.A., & Riggio, R.J. 1985. Acoustically-mediated individual recognition by a coral reef fish (*Pomacentrus partitus*). *Anim. Behav.*, **33**, 411-416.
Reyment, R.A., Blackith, R.E. & Campbell, N.A. 1984. *Multivariate Morphometrics.* 2nd Edition, Academic Press, London.
SPSS Inc. & M.J. Norusis. 1988. *SPSS.PC$^+$ Advanced Statistics V2.0 Manual.* p.43-45. SPSS, Chicago.

Using Information from the Correlation Matrix to Decide on the Suitability of the Data for Principal Components Analysis

The reason for producing a set of correlations between the original variables is that such a matrix allows you to see whether it is worth attempting principal components analysis. If the original variables are not, or only weakly, correlated, then they are already more or less orthogonal and principal components analysis will merely pick them in order of variability.

The correlations between each of the eight variables in the study are shown in Fig. 2). As expected, the highest correlations are generally between those variables which measure a similar aspect of the response, for example, the two measures of song output (BST and SSO, $r = 0.95$), and between TTR and its main constituents (SSO and SCL, $r = 0.64$ and $r = 0.74$). Although the two approach measures (MIN and SCL) are negatively correlated ($r = -.62$), this makes sense in terms of response because a close approach (small value of MIN) and a long time spent within 20m (large value of SCL) are both aspects of a strong response.

Two features of the correlation matrix allow a qualitative assessment of the suitability of the data for the multivariate approach. First, the proportion of the matrix with reasonably large correlation coefficients. In the example, over half the coefficients are greater than 0.5. Second, whether any single variable has low correlation coefficients with most other measures. The example data show that all variables have most correlations >0.3, with LMIN showing the weakest correlations with variables other than the remaining two latency measures (LAT and LSO) (Fig. 2). There do not appear to be any fixed rules governing when the proportion of low correlation coefficients is sufficiently high for principal components analysis to be inadvisable, nor for when a variable should be dropped from the data set because correlations with other variables are too low for this variable to contribute to a composite measure. However, the strength of relationships between variables is tested by two statistics in SPSS.PC$^+$. To generate these values in the printout use the line /PRINT ALL in place of the /PRINT line in the example control information (Table I). Bartlett's test of sphericity compares the observed correlation matrix with that expected when variables are completely unrelated (such a matrix would have leading diagonal terms of 1 and all other values would be 0). The sample data are significantly different from such an identity matrix (test value = 955.8, $P < 0.000001$). The second test is the Kaiser-Meyer-Olkin (KMO) measure of sampling adequacy, it is an index ranging from 1 (all correlations between pairs of variables can be explained by other variables) to 0 (no correlations can be explained in this way). Any value above 0.5 is acceptable (SPSS Inc. & Norusis, 1988). The sample data has a KMO value of 0.72.

All of this information is consistent with the idea that the multivariate approach is acceptable, that is, the original variables are interrelated to a sufficient extent in the sample data set for principal components analysis to be a used to generate a composite measure of response to playback.

Appendix 2

Part of the Printout Generated by SPSS.PC$^+$ when Analysing McGregor and Krebs (1984) Data and when the Analysis Is not Limited to the First Principal Component

```
Page                                  SPSS/PC+                               7/23/9

             - - - -   F A C T O R   A N A L Y S I S   - - - -

Factor Matrix:

             FACTOR  1      FACTOR  2      FACTOR  3

TTR          -.81997        -.20699         .41865
BST          -.77712         .50958         .30400
SSO          -.75411         .53240         .33424
SCL          -.56151        -.68869         .36363
MIN           .73252         .49734         .13445
LAT           .87163         .07720         .38673
LATS          .83068        -.16919         .25370
LMIN          .62668        -.07978         .62144

--------------------------------------------------------------------------------
Page                                  SPSS/PC+                               7/23/9

             - - - -   F A C T O R   A N A L Y S I S   - - - -

Final Statistics:

Variable      Communality  *  Factor   Eigenvalue    Pct of Var    Cum Pct
                           *
TTR               .89047   *     1       4.53929        56.7         56.7
BST               .95600   *     2       1.34856        16.9         73.6
SSO               .96384   *     3       1.12982        14.1         87.7
SCL               .92181   *
MIN               .80200   *
LAT               .91525   *
LATS              .78302   *
LMIN              .78528   *
--------------------------------------------------------------------------------
```

Figure 7. The fourth and fifth page of printout generated by the package when the number of factors is not limited to one.

The printout is identical to that shown in the main body of the text (Figs. 1-6), except for pages 4 and 5 (cf. Figs. 4 and 5), which are shown below (Fig. 7). The factor matrix now shows the components chosen by the package using the default criterion of an eigenvalue <1. This generates 3 principal components in this example which together explain over 87% of the variance in the original data. The factor matrix shows the weightings used to generate each of these components from the standardised original data.

The final statistics show the communality for each of the original variables when each of the three factors is included.

INTERACTIVE PLAYBACK: A FINELY TUNED RESPONSE

Torben Dabelsteen

Institute of Population Biology
University of Copenhagen
Universitetsparken 15
DK-2100 Copenhagen Ø
DENMARK

Types of Avian Vocal Interactions

Animal vocal communication is usually interactive when individuals are active at the same time; their acoustic signals influence each other's behaviour, sometimes resulting in complicated types of vocal co-ordination. The highly variable singing behaviours performed by male blackbirds, *Turdus merula*, at dawn are a good example of such effects (the general features of blackbird song necessary to understand this example are described in Box I).

Box 1. Blackbird song

The song of the blackbird consists of distinct sound sequences - songs. A full song starts with a *motif part* of two to five powerful and low frequency sound elements and ends with a *twitter part* of quieter, but more complicated, sounds covering a broad frequency range (Fig. 1). Both types of sound elements vary considerably in both frequency and amplitude modulation, and they occur together in fixed combinations, termed *motifs*. One to two motifs $\pm$ a few sounds per motif part; one to three motifs $\pm$ a few sounds per twitter part. Neighbours always share some motifs, especially motif part motifs, while other motifs are unique to the individual. The repertoire size of individuals varies from less than 10 to more than 30 motif part motifs. The repertoire size of twitter part motifs is unknown, but it is definitely greater than for motif part motifs. Males change motifs continuously, song by song. The pattern of singing can be divided into five different types, or intensities, of singing. Each type represents a section on a continuous scale starting with *motif songs* (**MS**) and *low intensity songs* (**LI**), through with *high intensity songs* (**HI**) and *transitional songs* (**TS**), and ending with *strangled songs* (**SS**). **MS** lacks twitter parts, but has relatively long (on average 1.3s) motif parts and inter-song pauses of about 4s. **LI** is similar to **MS**, but each song finishes with a short (0.6s) twitter part. **HI** has a relatively short (1.1s) motif part and pauses between motif sounds and between songs (2.7 s), but long twitter parts (0.9 s). Often, **HI** also has a high sound pressure level and the motif sounds have a high lowest fundamental frequency. **TS** is similar to **HI**, but the motif parts have more energy concentrated in the overtones, a higher number of well-modulated sounds and a higher degree of modulation. **SS** consist of combinations of twitter parts without motif parts at the start of the song. For further details see Fig. 1, Fig. 5 and Dabelsteen (1984), Dabelsteen and Pedersen (1990, *submitted*).

Blackbird song activity at dawn lasts for 30 to 90min. A single male often starts with a short period of motif songs (MS) followed by low intensity songs (LI), but relatively quickly it changes to high intensity songs (HI) (Fig.1), probably when it hears the song of other males. It continues with HI for 20-30min, sometimes longer. Now and then it changes song post, often approaching singing or calling rival males and/or females, and it often interrupts HI with a few transitional songs (TS) and/or strangled songs (SS) (Fig. 5). The changes between HI, TS and SS may be very frequent and rapid. For instance, a male often switches from HI to TS if a neighbour changes its behaviour such as by singing or changing song perch, and it then reverts to HI after a few songs. Should a rival male intrude into the territory, the owner usually changes to SS, flies towards the intruder and chases it while emitting SS. It reverts to HI if the intruder flees. Females are courted with SS in a similar manner. SS is usually only emitted at relatively short distances from conspecifics, while TS and, especially HI, are emitted at greater distances. For example, HI is often used in song interactions or *duels* with rivals over distances of 50 to >200m. Further examples of the use of the five different ways of singing are described in Dabelsteen (1984, 1985, 1988) and Dabelsteen and Pedersen (1988, 1990, *submitted*).

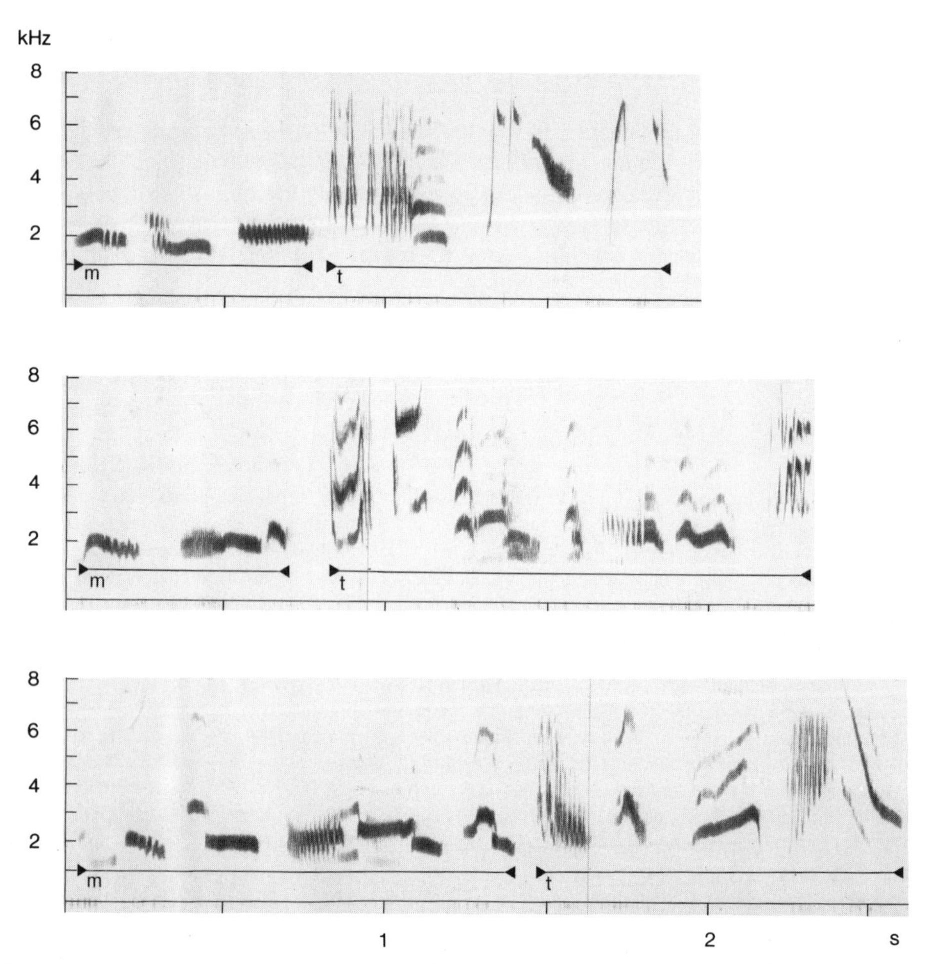

Figure 1. Sonagrams of three high intensity blackbird songs. Horizontal underlining indicates motif (**m**) and twitter (**t**) parts, respectively.

The singing behaviour of males does not seem to be finely coordinated during periods with frequent changes between HI, TS and SS, except that they often respond more or less immediately to each other's songs. The *changes* between HI, TS and SS seem to depend on the behaviour of conspecifics in a broad sense. On the other hand, during a song duel two males sometimes seem to coordinate singing song by song. The published data on such pair-wise singing interactions are still sparse, but three features are apparent: 1) the two males may *alternate* with each other or *overlap* each other's songs to different extents; 2) the two males may adopt *leader/follower roles* for various periods; and 3) males may try to *match* each other by replying to an opponent's song with a song containing a similar motif part (e.g. Todt 1968, 1970a). The description of such behaviour as a pair-wise interaction is, of course, a simplification. It ignores the fact that several individuals of the same species are within singing/hearing range of each other and are usually active at much the same time. The extent to which less actively singing or calling birds influence interactions between *pairs* of very actively singing males is unknown. In principal they could all act as senders and/or receivers and hence form a local communication network.

The song activity of blackbirds at dawn is complicated, but it may be typical of many species of songbird because one or more of the blackbird singing interactions have also been reported for other species, see e.g. Gompertz (1961), Lemon (1968a, 1968b), Kroodsma (1971, 1979), Smith and Norman (1979). It seems reasonable to assume that the various patterns of singing interaction are an aspect of communication between the individuals involved, but exactly how is still poorly understood. There are two main reasons for the relatively slow progress in our understanding of bird song phenomena such as switching between patterns and intensities of singing and switching between song types, alternating or overlapping song, adopting leader/follower roles, and song matching. First, the precise timing of these interactions have been difficult to determine, and second, they have been almost impossible to simulate in a natural way in interactive playback experiments.

The Timing of Vocal Interactions

In certain groups of animals, such as frogs, vocalising individuals usually interact over relatively short distances (a few metres in frogs, see chapter by Klump in this volume). In these species the relatively slow speed of sound transmission in air (330-345m/s) has negligible implications for the timing of the vocal interactions. However, in birds, interactions may occur over distances of hundreds of metres and up to several kilometres in mammals (see chapter by McComb in this volume). In such cases the sound delay imposes a constraint on a signaller's ability to alternate and overlap songs and also constrains the adoption of leader/follower roles during song duels. For example, a song of a male blackbird (male **A**) will take about 0.5s to reach another male blackbird (male **B**) if the distance between them is about 170m. An immediate response from **B** will take another 0.5s to reach **A**. Therefore **B** cannot achieve an overlap of the motif part of **A**'s song at the location of **A** (because motif parts last about 1s). The motif part is probably the only part of the song that can reach another male relatively undegraded at this distance (Dabelsteen et al. *submitted*). **A** will have finished most of its motif part when **B**'s response arrives, although **B** hears the song as though he is overlapping half of **A**'s motif part. If **B** matches **A**'s song, it will delay **B**'s response further, since **B** must wait to reply until it has received the part of **A**'s song which is necessary for recognition. A second consequence of the delay imposed by the speed of sound transmission is that each of the two males will hear itself as the leader and the other as the follower at their respective locations, if they begin songs simultaneously.

The fact that each male hears the timing of the interaction only at its own location does not necessarily prevent them knowing what happens at the location of a singing rival. Information from song degradation may be used by the males to assess distance to the opponent (e.g. McGregor and Krebs 1984) and therefore, in theory, the time the opponent produces and receives songs could be deduced. Theoretically, males could compensate for the delay imposed by the speed of sound transmission constraint during the process of co-ordinating singing with a rival. (This argument may seem far fetched, but a similar sort of prediction to overcome reaction time constraints has been proposed for the intricate coordination of wheeling in bird flocks (Potts 1984)).

At present most of our knowledge of the timing of interactions between singing birds is based on analyses which have not considered that the songs of the different individuals are delayed to different extents on their way to the human ear or to a microphone. If the delay is not considered in an analysis of a tape recording one may get very imprecise or misleading information on different aspects of timing. Some aspects may be correct, but others will always be wrong. Which will be right and wrong will depend on the location of the microphone in relation to the singers (Fig. 2 and Fig. 3).

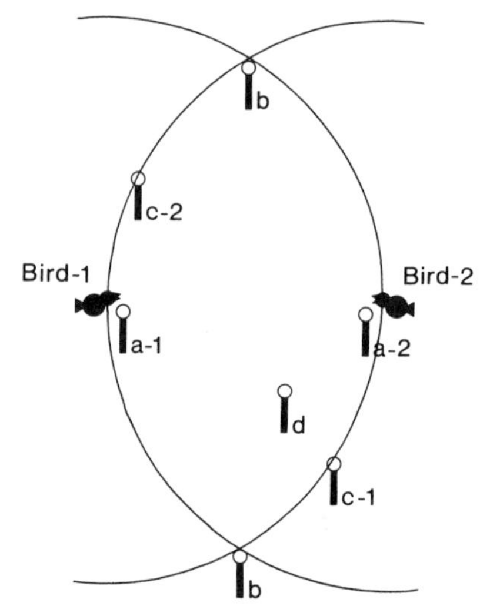

Figure 2. Examples of microphone positions (**a** - **d**) during recordings of singing interactions between two birds, **Bird-1** and **Bird-2**. The radii of the two arcs are identical and equal to the distance between the two birds. Microphone positions **a-1** and **a-2** give recordings corresponding to what **Bird-1** and **Bird-2** will hear, respectively. At the two **b**-positions the mutual timing of the two birds' song production will be recorded correctly, but the delay between each of the two birds' own songs and those of their opponent, as heard by the birds, will not. Positions **c-1** and **c-2** will give correct information on the time that **Bird-2** and **Bird-1**, respectively, receive the songs from their opponent, but not on the timing between the birds' own songs and those of their opponent as heard by the birds. Finally, position **d** gives incorrect information on arrival times and on the timing as heard by the birds.

A direct recording of what the birds involved actually hear requires that a microphone is placed at each bird's location, but this is usually impossible for practical reasons. Another possibility is to make the recordings from fixed positions, with known distances between microphone and birds and between birds and use these distances in conjunction with recordings to back-calculate the timing of the singing interaction as heard by each of the birds involved. Since birds often change song post during singing interactions, this will not be an easy task and it will be compounded by the problem that birds may be difficult to follow and locate during interactions involving many changes in song post over the whole territory. Furthermore, in some species, for example forest living blackbirds, males are usually disturbed by the observer's attempts to follow the birds during their interactions. So far I have outlined the problems involved with two individuals. The situation becomes even less tractable when more than two individuals are involved in the interaction.

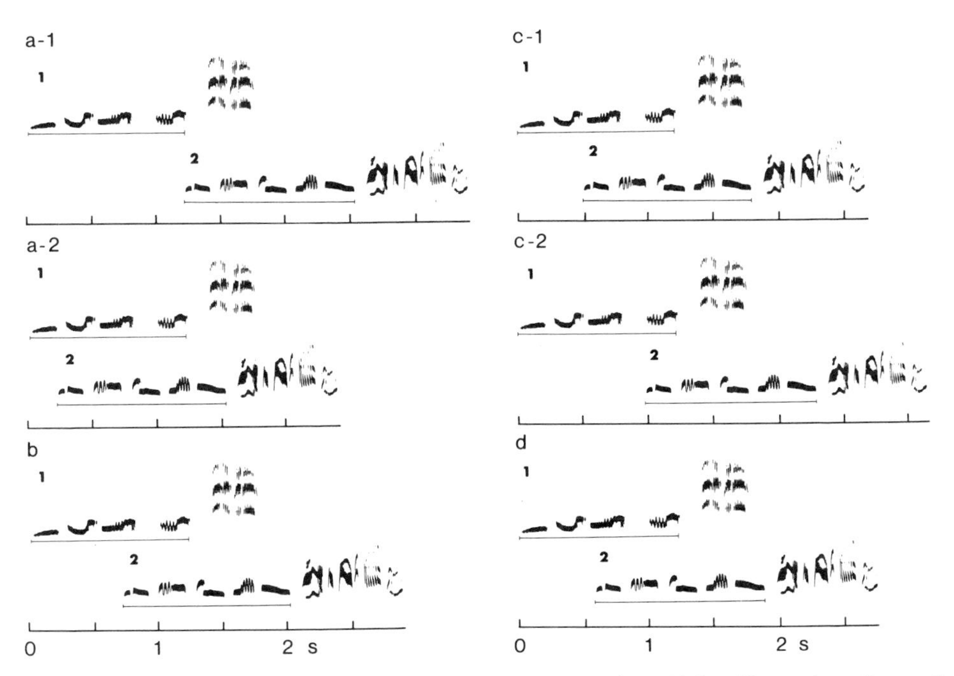

Figure 3. Sonagrams indicating how a tape recording of the songs of two birds will vary depending on the position of the microphone during the recording. The distance between the birds is 170m in this example and **Bird-1** emits a song (1) to which **Bird-2** responds with song **2**, 0.2s after receiving the first part of **Bird-1**'s song, i.e. in real time **Bird-2**'s song is delayed by 0.2s + (340m/s x 170m) = 0.7s. The microphone positions (**a-1**, **a-2**, **b**, **c-1**, **c-2** and **d**) are the same as shown in Fig. 2. The relative timing of arrival of the two songs (**1** and **2**) at each microphone position are shown as two sonagrams for clarity, rather than as overlapping songs on a single sonagram, which is how an actual recording would appear.

A possible solution to these problems is to record vocal interactions using a multichannel tape recorder and a fixed array of omnidirectional microphones. A specially developed computer program may then be used to extract the necessary time delays from the recordings and these delays used to calculate the location of each singer in three dimensional space (e.g. Watkins and Schevill 1972; Magyar et al. 1978; Clark et al. 1986). Therefore, the movements of birds can be tracked during the interactions and the data on distance apart could be used to back-calculate the timing of the interactions as heard at the locations of the different birds.

Interactive Playback Experiments: Tape Recorders *v.* Digital Sound Emitters

Having first determined the exact nature, including the timing, of vocal interactions, one may start designing experiments to test their biological function or signal value. One approach has been to use loop playback of single songs or series of different songs to test, for instance, which features of song cause a bird to match the playback song (e.g. Krebs et al. 1981; Falls et al. 1982; Wolffgram and Todt 1982) or which features cause a switch in song type (e.g. Kramer et al. 1985). Another approach is to examine the effects of different types of vocal interactions. This requires interactive field experiments in which playback simulates one of the parties of a singing interaction by means of finely tuned responses to a singing bird.

Until recently, such finely tuned natural responses were very difficult to achieve because of the inflexibility of the tape recorder, which was the only portable playback equipment easily available at that time. However sophisticated the use of tape recorders, the possibilities for interactive playback will always be limited in some way. For example, a multi-channel tape recorder or an array of tape recorders may be used to simulate song switching if each channel or tape recorder contains a different song type. However, an array of tape recorders will be an awkward *tool* to handle and there is a risk of scaring test birds with noise produced at some of the many frequent starts and stops of tape recorders. A multi-channel tape recorder involves the same risk plus a risk of beginning playback in the middle of a song. The latter risk is great if the songs and the inter-song pauses of the different channels vary in duration. The so-called *replicator*, consisting of two connected tape recorders which record songs on one machine and play them back from the second with an adjustable delay, has been an effective tool in interactive playback experiments (Todt 1970b). Amongst other things, it has been used to study the effect of matching a test bird's songs with an echo of its own songs with a variable delay (Todt 1981). Such studies may make important contributions to our understanding of song matching, although they are limited by the difficulties of interpreting responses since in a natural situation birds are never matched by delayed echoes of their own songs.

The development of computer technology has recently provided new portable equipment with unprecedented flexibility. The *digital sound emitter* (DSE) (Dabelsteen and Pedersen 1990, 1991) was one of the first pieces of playback equipment to use such computer technology. It was designed for use in interactive field playback experiments with songbirds. The DSE, which may be inserted into a carrying case (Fig. 4), consists of a signal synthesiser and a portable personal computer (PC) which is used to control the analogue output from the synthesiser. The synthesiser can be loaded with artificial sounds synthesised according to a mathematical model of the sound production of the test animal and/or with digitised natural sounds from high quality recordings. The technical specifications of the DSE (including a block diagram of the signal synthesizer and information about installation of sounds in the ROM-circuits of the synthesizer) and principles for the PC program are described elsewhere (Dabelsteen and Pedersen 1991). The program run by the PC may be tailored to support a vast number of different demands. It enables the human experimenter to start and stop the output from the DSE at any time and to choose freely between the available sounds and playback modes. The only limit on the use of the DSE for interactive experiments is the ability of the operator to perceive the sounds of the test animal and to operate the keyboard of the PC. Two examples of use of the DSE in playback experiments are given below and illustrate some of this equipment's possibilities.

Figure 4. The DSE ready for use with the signal synthesiser lowered into the carrying case and the PC placed on top of the synthesiser.

Playback Experiments with a Digital Sound Emitter

Experiment I. The Signal Value of Changing Song Intensity in Blackbirds

The different intensities of singing in blackbirds (MS, LI, HI, TS and SS) probably represent the steadily increasing arousal of the singer and a readiness to respond aggressively to males and sexually to females. SS should signal the highest arousal and should, all other things being equal, therefore release the strongest responses, but this has been impossible to demonstrate in males using normal loop playback (Dabelsteen and

Pedersen 1985). The details of the responses to loop playback suggested that the unnatural situation of loop playback (i.e. the speaker continued to emit SS without regard to the behaviour of the test bird) might be one of the causes of the low responses to SS. For example, the males sometimes responded strongly to SS at the start of playback of SS, and if the males were near the loudspeaker at a change from playback of full song (HI) to playback of SS their responses often increased for a short period. This seemed to be a situation in which interactive playback might increase the response to SS. An experiment using normal (i.e. fixed duration loop) and interactive playback of LI, HI and SS was therefore conducted. The synthesiser of the DSE was loaded with 171 representative blackbird sound elements originating from a two-minute recording of 30 different HI songs, and the PC was programmed to pick out the sound elements and combine them with their natural pauses to represent three *tape loops*. The first loop consisted of the HI songs originally recorded. The second loop had LI songs which were created by adding an extra motif part sound element to each HI motif part and by reducing each HI twitter part by one or two twitter sound elements and increasing each of the pauses between the 30 HI songs by 50% (Fig. 5). The third loop had SS made of the twitter parts from two or three HI songs (Fig. 5). The three "tape loops" could be started, stopped and alternated freely.

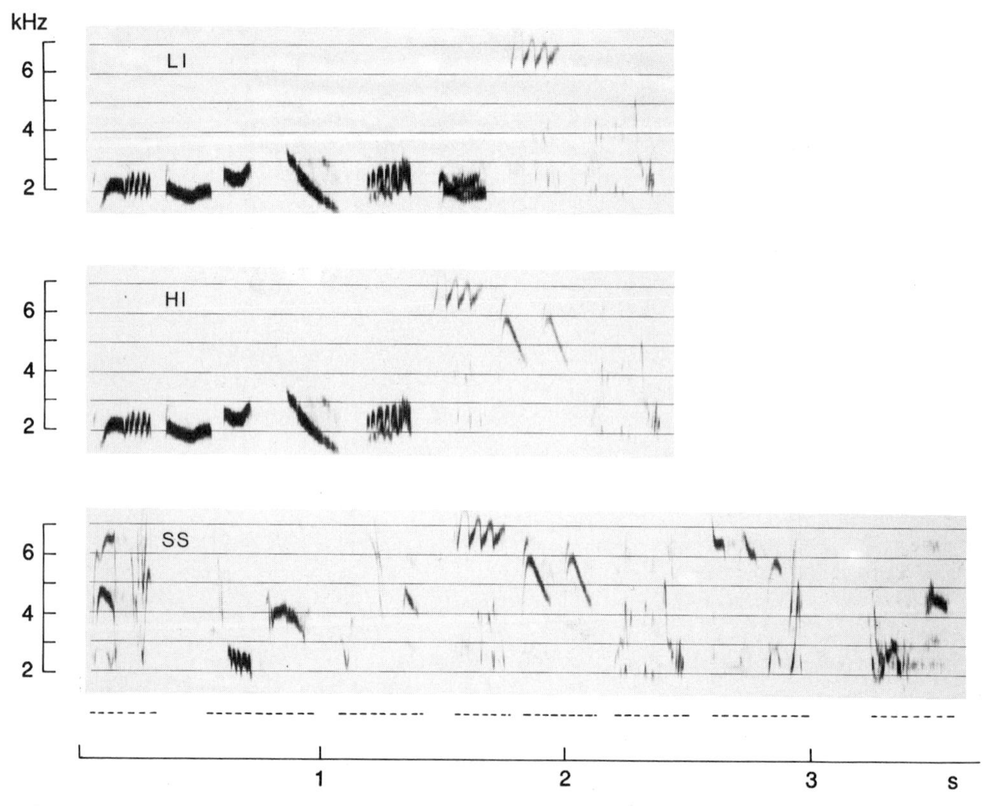

Figure 5. Sonagrams of three different intensities of blackbird song: low intensity song (**LI**), high intensity song (**HI**) and strangled song (**SS**). Sound elements are underlined; solid lines indicate motif part elements and dashed lines show twitter part elements. Reprinted with permission from Bioacoustics. Copyright 1991 ABA Publishers Ltd.

The responses of territorial males to LI, HI and SS were then compared using combinations of *normal playback* (continuous playback of a tape loop), *random "interactive" playback* (alternating between LI, HI and SS independently of the test bird but in fixed sequences derived from typical patterns of use found in previous experiments with interactive playback), and *interactive playback* (changing between LI, HI and SS in a natural way with the behaviour of the test bird). The speaker was placed in the middle of the territory to simulate an intruder. Interactive playback was usually started with LI. If the test bird responded we escalated the "intruder's" threat by changing playback to HI; and if the male came within about 10m of the speaker we changed to SS. If the male then retreated to > 10m from the speaker we changed back to HI, and if the male remained silent and out of sight for 1min we changed back to LI. If the male responded again we again followed the pattern of escalating to HI and SS (Dabelsteen and Pedersen 1990).

The results showed that LI elicited the weakest, and HI and SS the strongest aggressive responses when playback was a pattern of continuous playback of a tape loop, i.e. there was no difference in response to HI and SS with *normal* playback. However, interactive playback elicited stronger responses than normal playback of LI, HI and SS. *Interactive playback* elicited stronger responses than *random interactive* playback, indicating that the naturalness of the pattern of playback rather than the variation of playback *per se* elicited the strong responses to interactive playback. The details of the responses to interactive playback support the idea than SS represents the highest degree of aggressiveness because the escalations represented by the changes from HI to SS elicited similar escalated counter threats by the test birds, but only when the changes between HI and SS were interactive (Dabelsteen and Petersen 1990).

Experiment II. The Signal Value of Matched Singing in Great Tits

The male great tit sings in bouts consisting of a variable number of strophes, and each strophe is composed of a unit (the phrase) which is repeated in a stereotyped way a variable number of times (e.g. Gompertz 1961; McGregor and Krebs 1982) (Fig. 6). Males switch song types between two strophes by changing the repeated phrase. They sometimes match the strophes of a rival by singing strophes of the same song type or strophe duration or both (e.g. Krebs et al. 1981; Falls et al. 1982). During escalated encounters they can also overlap each others' strophes.

The signal value of matched alternating singing and matched overlapping singing has been examined in two different experiments using the DSE. In both experiments the synthesiser was loaded with 30 different great tit phrases (from 30 different individuals) representing 19 different song types, and the PC was programmed to pick out an individual phrase and to play it with the associated natural inter-phrase pauses inserted. A phrase could be played either as a 5s *loop* containing a 2s strophe and a 3s inter-strophe pause, or as single strophes consisting of a manually determined number of phrases. The playback loudspeaker was placed in the middle of great tit territories and the territory owners were stimulated to sing by means of *loop* playback.

In the first experiment, each test bird which had approached and started to sing was subsequently stimulated in one of four different ways and the responses to the different kinds of stimulation were compared. The first type of stimulation was *loop* playback of a phrase of a song type which was different from that sung by the test bird. In the three remaining types of stimulation, playback alternated with the test bird, that is, each of the strophes of the test bird was immediately followed by a playback strophe which matched that of the test bird in one of three different ways. Phrase type (song type) or number of phrases (strophe length) or both were matched. There were two main results. First, interactive playback elicited singing responses which were clearly different from those elicited by *loop* playback, but measures of approach response did not vary significantly between the two types of stimulation. Second, the changes in singing behaviour of the test birds supported the idea that matching may provide a precise indication of the intended receiver, especially when both the song type and the strophe length is matched (McGregor et al. *in press*).

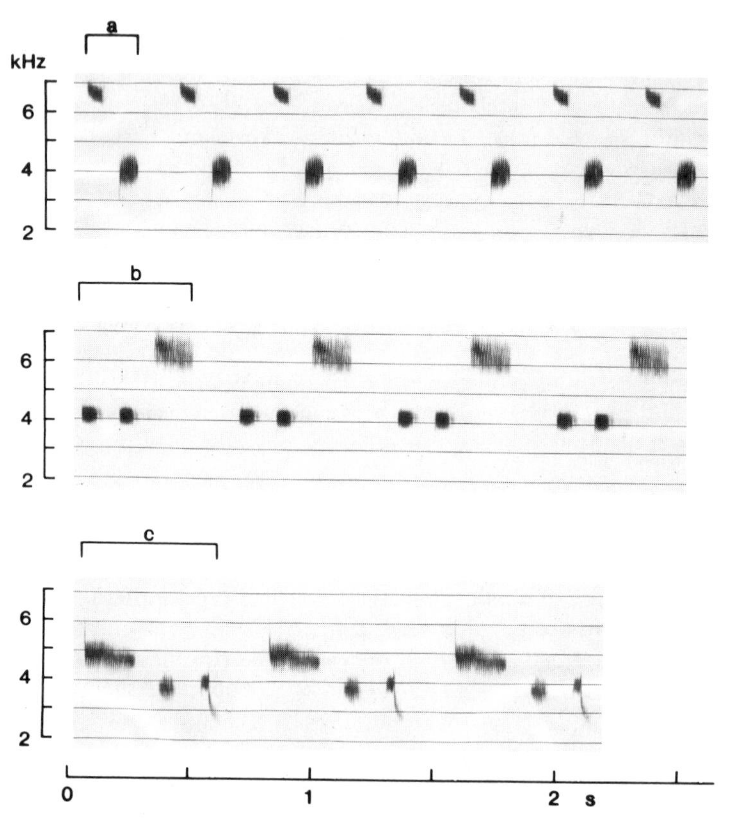

Figure 6. Sonagrams of a strophe of each of three different great tit song types. The top strophe has seven phrases, the middle has four phrases, and the lower has three phrases. Phrase lengths are shown for each song type (**a**, **b** and **c**).

The second experiment followed a similar design, it compared responses to two different types of song interaction. The first type of interaction was termed *alternating* because each of the strophes of the test bird were followed by a playback strophe which matched that of the test bird with respect to song type and strophe length (note that alternating matching was also used for the interactive playbacks in the first experiment). The second type of interaction (which also matched with respect to both song type and strophe length) was termed *overlapping* because playback was started immediately after the first phrase of the test bird's strophe, with the result that the test bird's strophe was overlapped by playback. The results of this experiment are still being analysed, however, the experimenters formed the strong impression that responses were highly variable between individuals. Specifically, some individuals seemed to try to avoid being overlapped by shortening their strophe length, others responded to overlapping with prolonged strophes, and some responded to alternating playback by trying to overlap the playback strophes (Dabelsteen et al. *in prep*).

Prospects for the Digital Playback Technique

It should be obvious from the previous section that none of these experiments would have been easy to do using conventional tape recorders. A further advantage of the DSE over tape recorders is that it automatically provides complete records of the experimental output. Nevertheless, there is potential for improving the DSE system in several respects. For example, the system could be extended beyond the limits imposed by the ability of the experimenter to identify and categorise sounds by incorporating a sound reception and signal recognition system which would allow direct response to the animal without the intervention of the experimenter. Such a system would require a powerful PC equipped with a suitable A/D converter and appropriate software. A more immediate direction for improvement is a reduction in the present size (37 x 30 x 26cm) and weight of the DSE when in its carrying case (10kg). This could be achieved by using a smaller synthesiser constructed from some of the current very small ROM-circuits with large memories together with a new-generation pocket-size PC. The powerful ROM-circuits now available will also make it possible to construct the signal synthesiser part of the DSE on a single circuit card and control the output from this synthesiser-card by means of a PC provided with an expansion slot in which the synthesiser-card is installed. Simple versions of such a synthesiser card are already available as storage units, but this means that the sounds must be loaded into the RAM memory of the computer before the start of each experiment. Such a solution has some of the same problems as another suggested solution, the installation of test sounds on the disk storage of a PC. Both the simple storage card and the disk storage solution add to the set-up time in the field since the signals would have to be loaded into the RAM before the experiment could start, and it may not be possible to use such solutions with the automatic *recogniser* mentioned above. The disk storage solution is also limited by the higher risk of disk reading errors and the higher overall power consumption.

In conclusion, digital signal synthesisers like the DSE or PCs equipped with extension cards containing the signal synthesiser part of the DSE system are much more flexible than conventional tape recorders. All of the acoustic interactions currently described in the literature can be simulated with these new digital playback systems. We may therefore expect to see a rapid increase in popularity of a variety of different types of interactive playback experiment in the near future. Such experiments may examine the function of different types of timing phenomena as well as vocalisation-specific, or type-specific, interactions. In birds, interactive techniques offer the prospect of studies of the many different calls which may also play an important, but currently overlooked, role in vocal interactions.

An important question remains: how far can we go in accurately mimicking acoustic signals and interactions in the absence of visual stimuli? This may not be a serious limitation for vocal interactions over the long distances used by many bird (and mammal) species in song duels, but such limitations may become important for some types of close range interactions. However, even these problems may be solved by another consequence of the increasing power and flexibility of computer technology. It is possible that the synergistic effect of sound and visual signals (and chemical signals in mammals) could be examined in experiments in which sound signal playback is co-ordinated with the movements of (and/or the secretion of chemicals from) a signaller in the form of a computer-controlled dummy or model. Such an approach has been used recently in experiments on the signal value of the honey bee (*Apis mellifera*) waggle dance (Michelsen et al. 1989).

Acknowledgements

I thank Peter McGregor for his valuable comments on the manuscript. The paper was written while holding a Nato Science Fellowship (11-8794-1).

References

Clark, C.W., Ellison, W.T. and Beeman, K. 1986. Acoustic tracking and distribution of migrating bowhead whales, *Baleana mysticetus*, off Point Barrow, Alaska in the spring of 1984. *Rep. int. Whal. Comm.*, **36**, 502.

Dabelsteen, T. 1984. An analysis of the full song of the blackbird *Turdus merula* with respect to message coding and adaptations for acoustic communication. *Ornis Scand.*, **15**, 227-239.

Dabelsteen, T. 1985. Messages and meanings of bird song with special reference to the blackbird (*Turdus merula*) and some methodology problems. *Biol. Skr. Dan. Vid. Selsk.*, **25**, 173-208.

Dabelsteen, T. 1988. The meaning of the full song of the blackbird *Turdus merula* to untreated and oestradiol-treated females. *Ornis Scand.*, **19**, 7-16.

Dabelsteen, T. and Pedersen, S.B. 1985. Correspondence between messages in the full song of the blackbird *Turdus merula* and meanings to territorial males, as inferred from responses to computerized modifications of natural song. *Z. Tierpsychol.*, **69**, 149-165.

Dabelsteen, T. and Pedersen, S.B. 1988. Song parts adapted to function both at long and short ranges may communicate information about the species to female blackbirds (*Turdus merula*). *Ornis Scand.*, **19**, 195-198.

Dabelsteen, T. and Pedersen, S.B. 1990. Song and information about aggressive responses of blackbirds, *Turdus merula*: evidence from interactive playback experiments with territory owners. *Anim. Behav.*, **40**, 1158-1168.

Dabelsteen, T. and Pedersen, S.B. 1991. A portable digital sound emitter for interactive playback of animal vocalizations. *Bioacoustics*, **3**, 193-206.

Dabelsteen, T. and Pedersen, S.B. *submitted.* Song features essential for species discrimination and behaviour assessment by male blackbirds (*Turdus merula*).

Dabelsteen, T., Larsen, O.N. and Pedersen, S.B. *submitted.* Habitat induced degradation of sound signals: quantifying the effect of communication sounds and bird location on blurring, excess attenuation and signal-to-noise ratio in blackbird song.

Dabelsteen, T., McGregor, P.K., Shepherd, M., Whittaker, X. and Pedersen, S.B. *in prep.* The signal value of overlapping versus alternating during matched singing in great tits (*Parus major*).

Falls, J.B., Krebs, J.R. and McGregor, P.K. 1982. Song matching in the Great tit (*Parus major*): the effect of similarity and familiarity. *Anim. Behav.*, **30**, 997-1009.

Gompertz, T. 1961. The vocabulary of the great tit. *Br. Birds*, **54**, 369-418.

Kramer, H.G., Lemon, R.E. and Morris, M.J. 1985. Song switching and agonistic stimulation in the song sparrow (*Melospiza melodia*): five tests. *Anim. Behav.*, **33**, 135-149.

Krebs, J.R., Ashcroft, R. and van Orsdol, K. 1981. Song matching in the great tit (*Parus major* L.). *Anim. Behav.*, **29**, 918-923.

Kroodsma, D.E. 1971. Song variations and singing behaviour in the rufous-sided towhee *Pipilio erythropthalmus*. *Condor*, **73**, 303-308.

Kroodsma, D.E. 1979. Vocal duelling among male marsh wrens: evidence for ritualized expression of dominance/subordinance. *Auk*, **96**, 506-515.

Lemon, R.E. 1968a. Coordinated singing by black-crested titmice. *Can. J. Zool.*, **46**, 1163-1167.

Lemon, R.E. 1968b. The relation between organization and function of song in cardinals. *Behaviour*, **32**, 158-178.

Magyar, I., Schleidt, W.M. and Miller, B. 1978. Localization of sound producing animals using the arrival time differences of their signals at an array of microphones. *Experientia*, **34**, 676-677.

McGregor, P.K. and Krebs, J.R. 1982. Song types in a population of great tits: their distribution, abundance and acquisition by individuals. *Behaviour*, **52,** 126-152.

McGregor, P.K. and Krebs, J.R. 1984. Sound degradation as a distance cue in great tit (*Parus major*) song. *Behav. Ecol. Sociobiol.*, **16,** 49-56.

McGregor, P.K., Dabelsteen, T., Shepherd, M. and Pedersen, S.B. *in press*. The signal value of matched singing in great tits: evidence from interactive playback experiments. *Anim. Behav.*

Michelsen, A., Andersen, B.B., Kirchner, W.H. and Lindauer, M. 1989. Honeybees can be recruited by a mechanical model of a dancing bee. *Naturwissenschaften*, **76**, 277-280.

Potts, W.K. 1984. The chorus-line hypothesis of manoeuvre coordination in avian flocks. *Nature*, **309**, 344-345.

Smith, D.G. and Norman, D.O. 1979. "Leader-follower" singing in red-winged blackbirds. *Condor*, **81**, 83-84.

Todt, D. 1968. Korrespondenz zwischen weckselsingenden Amseln. *Z. Naturforschg.*, **23b**, 1619.

Todt, D. 1970a. Gesang und gesangliche Korrespondenz der Amsel. *Naturwissenschaften*, **57**, 61-66.

Todt, D. 1970b. Gesangliche Reaktionen der Amsel (*Turdus merula* L.) auf ihren experimentell reproduzierten Eigengesang. *Z. vergl. Physiologie*, **66**, 294-317.

Todt, D. 1981. On functions of vocal matching: effect of counter-replies on song post choice and singing. *Z. Tierpsychol.*, **57**, 73-93.

Watkins, W.A. and Schevill, W.E. 1972. Sound source location by arrival-times on a three-dimensional hydrophone array. *Deep Sea Res.*, **19**, 691-706.

Wolffgram, J. and Todt, D. 1982. Pattern and time specificity in vocal responses of Blackbirds *Turdus merula* L. *Behaviour*, **81**, 264-286.

PLAYBACK AS A TOOL FOR STUDYING CONTESTS BETWEEN SOCIAL GROUPS

Karen McComb

Large Animal Research Group
Department of Zoology
University of Cambridge
Downing Street, Cambridge CB2 3EJ
U.K.

Introduction

When animals compete over resources, opponents are rarely equally matched (Parker 1974; Maynard Smith and Parker 1976; Parker and Rubenstein 1981; Maynard Smith 1982). Apart from differences in ownership of the resource at the outset of the contest (Davies 1978; Packer and Pusey 1982) and in the value of the resource to each opponent (Riechert 1979; Austad 1983), opponents differ in their ability to acquire / defend the resource - Resource Holding Potential (Parker 1974; Riechert 1978; Sigurjonsdottir and Parker 1981). Games theory models have shown that where fights endanger future survival and reproductive success, individuals should monitor the value of the resource and the Resource Holding Potential of their opponent, and withdraw without escalation if they would be likely to lose an ensuing fight (Parker 1974; Parker and Rubenstein 1981).

Playback experiments have shown that a variety of anurans, birds and mammals assess opponents through vocal displays (e.g. Arak 1983; Robertson 1986; Wagner 1989; Krebs et al. 1978; Clutton-Brock and Albon 1979). Until recently, only situations where single individuals compete in pair-wise contests had been examined in detail. The outcome of this type of contest is commonly determined by inter-individual differences in body size and stamina (Riechert 1978; Clutton-Brock et al. 1988; Robinson 1985), and the display characteristics used in assessment tend to reflect these individual differences. In some toads, for example, body size and croak pitch are negatively correlated and, when competing for females, males avoid opponents with low pitched croaks (Davies and Halliday 1978; Arak 1983; Robertson 1986). Red deer males (*Cervus elaphus*) that are successful in fights can sustain high roaring rates and individuals attempting to acquire harems avoid fights with opponents that they are unable to out-roar (Clutton-Brock and Albon 1979).

In practise, however, not all contests are between single individuals. Often groups of individuals compete as a unit for the resource under dispute. This is true of many social mammals including some primates and social carnivores, where individuals cooperate to defend shared resources such as territories and young. Where groups of individuals compete over resources, differences in the number of individuals per group might easily outweigh inter-individual differences in determining the outcome of the contest. Thus, while individual differences may still have underlying effects in group contests, individuals might be expected to adjust their agonistic behaviour primarily according to the relative number of individuals in their own and the opposing group. In this paper I discuss how playback experiments can be used to investigate contests between groups and to determine the factors that affect their outcome.

Social Mammals and Group Calling

Many social mammals engage in bouts of group calling where members of a group produce repeated, high intensity vocalisations in chorus (e.g. wolves *Canis lupus*, Harrington and Mech 1979; hyaenas *Crocuta crocuta*, East and Hofer 1991; lions *Panthera leo*, McComb et al. *in prep.*; howler monkeys *Alouatta seniculus*, Sekulic 1982; gibbons *Hylobates lar*, Raemaekers and Raemaekers 1984). One individual usually initiates but additional members of the group join in as the chorus progresses, often overlapping each other's calls. Calling is thought to co-ordinate the activities of widely spaced members of the group and to provide a mechanism for group spacing. (Schaller 1972; Waser 1975; Harrington and Mech 1983; Mitani 1985; Whitehead 1987). Some authors have suggested that competing groups assess one another on the basis of chorused vocalisations (Sekulic 1982; Harrington 1989), but until now this has not been tested systematically.

To demonstrate that assessment has occurred, it is necessary to show that individuals adjust their agonistic behaviour in a predictable way according to information that they receive about their opponents. For group chorusing to serve as an assessment cue it should contain information on some characteristic of the group that affects Resource Holding Potential. Whether opposing groups alter their response as this characteristic changes can be tested using playback experiments.

There is some evidence to suggest that larger groups tend to dominate smaller ones in inter-group encounters. When wolves are chasing intruders, the lead wolf sometimes pauses to let lagging members of the group catch up, or even goes back to find them before continuing with the chase (Harrington and Mech 1979). Sekulic (1982) found that troops of red howler monkeys with a larger number of males tended to approach groups with a smaller number of males, although there was no difference in the number of males of the troops that first left the encounter site. In African lions, where related adult females and their dependent offspring live in tight knit social groups known as prides and co-operate to defend a joint territory, inter-pride encounters that result in intense chases are more likely to be won by the larger group of females (Packer et al. 1990).

All of the above examples come from social mammals that use calling to mediate inter-group interactions (but see also Cheney 1986). If information on the number of individuals per group was broadcast in calling bouts then groups might be able to assess their opponents by listening to their calls. Harrington (1989) suggested that wolf packs use howling as a means of advertising group size, their choruses giving an inflated impression of the number of wolves present, but did not test this experimentally. Lion prides advertise territory ownership by scent marking and roaring (Schaller 1972). Recently it was demonstrated that groups of female lions presented with playbacks of unfamiliar females roaring within their territory assess the number of intruders that they expect to encounter if they approach, and the number of companions that they have to support them in a possible fight, and only approach if the odds of success are weighted in their favour (McComb et al. *in prep*). Using this study as an example, I now discuss a possible methodological framework for investigating assessment in group contests with playback experiments.

Experimental Design for Studies of Group Contests

If individuals are to predict the outcome of inter-group contests and adjust their behaviour accordingly, they must both assess the Resource Holding Potential of intruders calling from the loudspeaker and evaluate their own Resource Holding Potential, taking into account temporal changes in group composition that will affect it. By careful choice of sound recordings for playback, and of playback situations, two potentially important factors can be varied independently: number of intruders and number of individuals in the defending group.

Number of Intruders

McComb et al. (*in prep*) simulated two intruder group sizes, a single intruder and three intruders roaring in chorus. Intruder group size could, however, be set at potentially any value that is realistic for the species and within the bounds of its auditory processing system. Humans are limited to holding between five and nine items in their short term auditory memory (Miller 1956). Other mammals may have a similar capability although social species that make repeated use of chorused calls for inter-group communication could have developed superior capabilities.

To avoid confounding effects of individuals recognising opponents and associating them with some preconceived value of Resource Holding Potential (see below), the subjects should be unfamiliar with the intruders whose calls are used in playback. This may be achieved by using recordings of individuals from a distant part of the study area, individuals from an adjacent population, or individuals that have died subsequent to recording - provided that they are known to be unfamiliar to the present subjects.

Intruder chorus size need not necessarily, however, be limited by the existence of appropriate recordings of choruses containing the required number of participants. With some attention to the normal patterning of the calls of different individuals in the chorus, artificial choruses can be constructed using conventional sound mixing techniques. This method has some advantages in that the identities of the participants in the choruses can be tailored for particular comparisons. For example, if the question of interest is whether listeners respond differently to single individuals calling than to a number of individuals calling in chorus, playback tapes can be constructed so that the same individuals presented as single callers are also presented as part of the choruses. Here a difference in response to the two playback types is unlikely to be due to particular phenotypic characteristics of the individuals chosen as single callers. A similar effect can be achieved using natural choruses if the same individual can be recorded both when calling alone and when calling with several companions.

Calls can either be played back from a single loudspeaker, or from several speakers broadcasting in unison. For simulating group sizes of more than three it may be advisable to broadcast from two or more speakers as members of very large groups will often be spread out. When simulating a chorus size of five, for example, two individuals in the chorus might be played back from one speaker and the remaining three from another. The distance between the speakers should be representative of the normal spacing of group members. Different speakers could broadcast from different tracks on a single recorder, or from different recorders. If different tracks on the same recorder are used, the synchrony of the two sets of callers is fixed before the playback. Use of two recorders would necessitate careful control of the time at which each of the recorders is started during the playback session, if a realistic simulation of chorus synchrony is to be achieved.

Number in Defending Group

Number of individuals in the defending group can be varied systematically by playing back to subjects in groups of different sizes. In all cases external factors likely to affect response must be kept constant across playbacks. In territorial species, for example, playbacks could be given only when the group is in a central part of its territory and therefore likely to respond maximally to the simulated intruders (but see the "ceiling" effect of a maximal response to all stimuli discussed in the chapter by Falls in this volume).

Setting the Odds of Success in Group Contests

If social groups calculate the likelihood of success in a potential fight on the basis of number of individuals in their own and the opposing group then they should alter their response to playbacks of simulated intruders in a predictable way as these variables change. In particular they should be more cautious if their own group size is small and/or the number of intruders is large. This can be tested either by conducting tightly controlled experiments in which only one of the independent variables is allowed to change at a time and/or by using multivariate statistics to examine a data set in which more than one varies.

Figure 1. A typical approach: an adult female leads her group towards the loudspeaker after playback. (Photograph by Karen McComb.)

Female lions typically show unambiguous responses to playbacks of intruders and these can be broadly divided into approaches and non-approaches. In an approach, the defenders stand up and stalk towards the loudspeaker with their heads low (Fig. 1). When they reach the level of the loudspeaker they usually continue on past it for anything from a few to several hundred metres. If the distance between the loudspeaker and the subjects is held constant across tests, the time that it takes for the individuals in a group to reach the level of the loudspeaker provides a good measure of their readiness to approach the intruders. Individuals usually punctuate their approach with pauses during which they stand, or lie in alert posture, and look intently at the loudspeaker (Fig. 2). They sometimes turn their heads to look at companions while they pause, seemingly as a means of checking on the locations of other members of the group. Cautious approaches are accompanied by a greater number of pauses and looks at companions. Those groups that fail to approach either remain in their original position, looking more intently in the direction of the loudspeaker than prior to playback, or, more rarely, retreat in the opposite direction.

Figure 2. A group of three adult females with a year old male cub (3rd from right) pause during an approach. The females stand looking intently in the direction of the loudspeaker while the young male is less attentive in the midst of the group. (Photograph by Karen McComb.)

The response of female lions to playbacks of intruders roaring is a function both of the number of individuals in their own group and the number of intruders roaring from the loudspeaker (McComb et al. *in prep*). Females are less likely to approach playbacks of three intruders than of one intruder and if they do approach three intruders they take longer to reach the loudspeaker, they pause more often and they look at their companions more often. Controlling for number of intruders, the number of individuals in the defending group affects the likelihood of approach in the form of a logistic curve. Larger groups are more likely to approach than smaller groups or solitaries. The size of the defending group does not, however, affect time taken to reach the loudspeaker, number of pauses or number of looks at companions. It seems that once the decision to approach has been taken, individuals are always more cautious of approaching three intruders than one intruder, irrespective of their own group size. Odds of success, as calculated by the ratio of the number of individuals in the defending group to the number of intruders is the single best predictor of response in playback situations. In general females prefer to have odds of 1:1 before they will approach and most groups will only approach if the odds are at least 2:1 in their favour.

More Complex Forms of Inter-Group Assessment

So far I have considered the simplest possible situation where individuals are assessing opposing groups that they are unfamiliar with and where inter-individual differences in Resource Holding Potential amongst group members are assumed to have a negligible effect on the outcome of the contest. This will not always be the case.

In contests between groups of female lions inter-individual differences in Resource Holding Potential amongst group members might be expected to have minimal effects on overall fighting success, as there is little variation in body size amongst pride mates that have reached adulthood and dominance hierarchies are noticeably absent (Schaller 1972; Packer and Pusey 1985). However, adult male lions are larger and more powerful than adult females and when resident males are present with the pride their higher per capita Resource Holding Potentials could generate a significant advantage in contest situations. Sub-adults of both sexes would be expected to have lower per capita Resource Holding Potentials than adult females because of their smaller body size and lack of fighting experience. An assessor might, therefore, benefit by monitoring the age/sex classes of its opponents as well as the total number present. Female lions can differentiate between male and female roars when these roars are delivered in single-sex choruses (McComb, Packer, Pusey and Grinnell, *unpublished*). Whether they, or members of other social species, can assess the number of individuals of each age/sex class present in mixed age/sex choruses has not been tested.

In contrast to lions, some social mammals exhibit marked inter-individual differences in body size and dominance status even within single-sex groups of adults (e.g. wolves, Harrington and Mech 1979; hyaenas, East and Hofer 1991). In these cases, assessors might benefit by considering the body size and dominance status of each individual in the opposing group before deciding whether or not to participate in contests. The presence of the largest most dominant individual in a group of three, for example, may render it equivalent to a larger group that lacks this individual.

Assessment of individual differences in Resource Holding Potential through vocal means may be more difficult in group contests than in pair-wise contests. Chorusing of calls means that many of the characteristics of an individual's call usually available for assessment in pair-wise contests could be masked when individuals are overlapping each other's calls. More work on this, and on the inter-play of individual and group characteristics in contest situations is necessary.

If opposing groups are familiar with each other in contest situations the possibility of a more sophisticated method of assessing Resource Holding Potential must also be considered. On hearing a single familiar individual calling from a neighbouring group listeners may associate that individual with membership of a group of a particular size and respond to that individual according to the Resource Holding Potential of the group in which it is normally found rather than, or in addition to, its own Resource Holding Potential at the time when it is calling. Whether this sort of assessment is actually occurring can be tested by playing back to subjects recordings of single familiar individuals that belong to neighbouring groups of different sizes. If they associate each individual with a group of a different size, and group size is important in determining the outcome of contests, then they might be more likely to give way to an individual which belongs to a group that is larger than their own than to one that is from a smaller group. Possible confounding variables that should also be considered here are kinship relationships between neighbouring groups - particularly in species where new groups are formed as offshoots of already existing groups. Subjects might be expected to be more tolerant of neighbours to which they are closely related than to others with which they have few genes in common. Previous interactions that have occurred between the groups will also have a confounding effect - but this will usually be difficult to evaluate.

Group Contests and Assessment in Other Vertebrate Taxa

In this paper I have suggested a methodology for studying group contests in mammals, but the same principles could be applied to other vertebrates. Many communally breeding birds engage in joint defence of a group held territory (Brown 1987) and group chorusing can constitute part of this defence. In the Australian magpie (*Gymnorhina tibicen*), for example, group members sing together repeatedly as they face opponents in interactions at territory boundaries (Brown and Veltman 1987; Brown et al. 1988) and when rival groups of green woodhoopoes (*Phoeniculus purpureus*) confront each other a vigorous vocal dispute ensues (Ligon and Ligon 1982). Playback experiments could be used to examine the effect of group size on success in these territorial encounters and thus to define more clearly the benefits that individuals derive from living in groups. Male frogs and toads often deliver calls in chorus when displaying at mating sites but are not social and do not have a direct equivalent to the inter-group contests of mammals and birds discussed above. However, some species alter their calling behaviour according to the number of males present in the chorus and may use information on chorus size in deciding whether or not to join particular mating aggregations (Ryan et al. 1981; Green 1990). Again, this can be tested directly with playback experiments. Experiments of this sort in a wide variety of vertebrates would help us to understand an element of vocal perception that has hitherto been neglected - the ability of individuals to assess their opponents by counting them.

Acknowledgments

I thank: Tim Clutton-Brock, Sarah Durant, Craig Packer and the Editor for comments on the manuscript; Marion East and Heribert Hofer for proofs of their 1991 paper; John Fanshawe and Clare Veltman for advice on communally breeding birds; and fellow ARW participants for fruitful discussion. This was written while I was in receipt of a Research Fellowship from Newnham College, Cambridge and field research on lions was supported by NSF.

References

Arak, A. 1983. Sexual selection by male-male competition in natterjack toad choruses. *Nature*, **306**, 261-262.

Brown, J.L. 1987. *Helping and Communal Breeding in Birds*. Princeton University Press, Princeton, New Jersey.

Brown, E.D. and Veltman, C.J. 1987. Ethogram of the Australian magpie (*Gymnorhina tibicen*) in comparison to other Cracticidae and *Corvus* species. *Ethology*, **76**, 309-333.

Brown, E.D., Farabaugh, S.M. and Veltman, C.J. 1988. Song sharing in a group-living songbird, the Australian magpie, *Gymnorhina tibicen*. Part I: Vocal sharing within and among social groups. *Behaviour*, **104**, 1-28.

Cheney, D.L. 1986. Interactions and relationships between groups. In: *Primate Societies*. (Ed. by B.B. Smuts, D.L. Cheney, R.M. Seyfarth, R.W. Wrangham and T.T. Strhsaker), pp. 267-281. University of Chicago Press, Chicago.

Clutton-Brock, T.H. and Albon, S.D. 1979. The roaring of red deer and the evolution of honest advertisement. *Behaviour*, **69**, 145-169.

Clutton-Brock, T.H., Albon, S.D. and Guinness, F.E. 1988. Reproductive success in male and female red deer. In: *Reproductive Success: Studies of Individual Variation in Contrasting Breeding Systems*. (Ed. by T.H. Clutton-Brock), pp.325-343. University of Chicago Press, Chicago.

Davies, N.B. 1978. Territorial defence in the speckled wood butterfly *Pararge aegeria*: the resident always wins. *Anim. Behav.*, **26**, 138-147.

Davies, N.B. and Halliday, T.R. 1978. Deep croaks and fighting assessment in toads (*Bufo bufo*). *Nature*, **274**, 683-685.

East, M.L. and Hofer, H. 1991. Loud calling in a female dominated mammalian society II: Behavioural contexts and functions of whooping of spotted hyaenas, *Crocuta crocuta*. *Anim. Behav.*, **42**, 651-669.

Green, A.J. 1990. Determinants of chorus participation and the effects of size, weight and competition on advertisement calling in the tungara frog, *Physalaemus pustulosus* (Leptodactylidae). *Anim. Behav.*, **39**, 620-638.

Harrington, F.H. and Mech, L.D. 1979. Wolf howling and its role in territory maintenance. *Behaviour*, **68**, 207-249.

Harrington, F.H. and Mech, L.D. 1983. Wolf pack spacing : howling as a territory-independent spacing mechanism in a territorial population. *Behav. Ecol. Sociobiol.*, **12**, 161-168.

Harrington, F.H. 1989. Chorus howling by wolves: acoustic structure, pack size and the beau geste effect. *Bioacoustics*, **2**, 117-136.

Krebs, J.R., Ashcroft, R. and Webber, M. 1978. Song repertoires and territory defense in the great tit. *Nature*, **271**, 539-542.

Ligon, J.D. and Ligon, S.H. 1982. The cooperative breeding behavior of the Green woodhoopoe. *Scient. Am.*, **247**(1), 106-115.

Maynard Smith, J. and Parker, G.A. 1976. The logic of asymmetric contests. *Anim. Behav.*, **24**, 159-175.

Maynard Smith, J. 1982. *Evolution and the Theory of Games*. Cambridge University Press, Cambridge.

Miller, G.A. 1956. The magical number seven plus or minus two: some limits on our capacity for processing information. *Psych. Rev.*, **63**, 81-97.

Mitani, J.C. 1985. Gibbon duets and intergroup spacing. *Behaviour*, **92**, 59-96.

McComb, K., Packer, C. and Pusey, A. in prep. Roaring and assessment in African lions (*Panthera leo*).

Parker, G.A. 1974. Assessment strategy and the evolution of fighting behaviour. *J. Theor. Biol.*, **74**, 223-243.

Parker, G.A. and Rubenstein, D.I. 1981. Role assessment, reserve strategy and acquisition of information in asymmetric animal conflicts. *Anim. Behav.*, **29**, 221-240.

Packer, C. and Pusey, A. 1985. Asymmetric contests in social mammals: respect, manipulation and age-specific aspects. In: *Evolution : Essays in Honour of John Maynard Smith*. (Ed. by J.J. Greenwood & M.Slatkin), pp. 173-186. Cambridge University Press, Cambridge.

Packer, C., Scheel, D. and Pusey, A.E. 1990. Why lions form groups: food is not enough. *Am. Nat.*, **136**, 1-19.

Raemaekers, J.J. and Raemaekers, P.M. 1984. The Ooaa duet of the gibbon (*Hylobates lar*): a group call which triggers other groups to respond in kind. *Folia primatol.*, **42**, 209-215.

Riechert, S.E. 1978. Games spiders play: behavioral variability in territorial disputes. *Behav. Ecol. Sociobiol.*, **3**, 135-162.

Riechert, S.E. 1979. Games spiders play II: resource assessment strategies. *Behav. Ecol. Sociobiol.*, **6**, 121-128.

Robertson, J.G.M. 1986. Male territoriality, fighting and assessment of fighting ability in the Australian frog (*Uperoleia rugosa*). *Anim. Behav.*, **34**, 763-772.

Robinson, S.K. 1985. Fighting and assessment in the yellow-rumped cacique (*Cacicus cela*). *Behav. Ecol. Sociobiol.*, **18**, 39-44.

Ryan, M.J., Tuttle, M.D. and Taft, L.K. 1981. The costs and benefits of frog chorusing behaviour. *Behav. Ecol. Sociobiol.*, **8**, 273-278.

Schaller, G.B. 1972. *The Serengeti Lion: A Study of Predator-Prey Relations.* University of Chicago Press, Chicago.

Sekulic, R. 1982. The function of howling in red howler monkeys (*Alouatta seniculus*). *Behaviour*, **81**, 38-54.

Sigurjonsdottir, H. and Parker, G.A. 1981. Dung fly struggles: evidence for assessment strategy. *Behav. Ecol. Sociobiol.*, **8**, 219-230.

Wagner, W.E. 1989. Graded aggressive signals in Blanchard's cricket frog: vocal responses to opponent proximity and size. *Anim. Behav.*, **38**, 1025-1038.

Waser, P. 1975. Experimental playbacks show mediation of intergroup avoidance in a forest monkey. *Nature*, **225**, 56-58.

Whitehead, J.M. 1987. Vocally mediated reciprocity between neighbouring groups of mantled howler monkeys *Alouatta palliata palliata*. *Anim. Behav.*, **35**, 1615-1627.

SONG OVERPRODUCTION, SONG MATCHING AND SELECTIVE ATTRITION DURING DEVELOPMENT

Douglas A. Nelson

Animal Communication Laboratory
Department of Zoology
University of California
Davis, California 95616
U.S.A.

Introduction

The bewildering diversity among bird species in the number of songs that males sing and in the degree of geographic variation in song structure have long intrigued biologists (Krebs and Kroodsma 1980). This diversity in part is a consequence of song learning. At some time in their lives, male birds memorise adult song models that serve to guide the young male's own song efforts (Thorpe 1958; Konishi 1965; Marler 1970).

Two processes have been commonly thought to result in song sharing among neighbouring male song birds. In one, acquisition of song in the first few months of life, combined with limited pre-breeding dispersal would result in song sharing (Marler and Tamura 1962; Nottebohm 1970; Cunningham and Baker 1983). The second process assumes that song acquisition occurs after pre-breeding dispersal at the time males establish breeding territories (Kroodsma 1974; Payne 1981). Song memorisation early in life could also occur in the second process, but songs acquired then would not likely lead to song sharing if pre-breeding dispersal distances are large.

In this paper I will argue for the role of a third process, termed *action-based* learning (Marler 1990). In this instance, males acquire or invent a variety of songs which they produce in plastic song. They then select a subset of song(s) for retention in the repertoire. The selection process in some cases may be mediated by social interactions (Marler and Peters 1982a; West and King 1986; DeWolfe et al. 1989); in others endogenous factors may guide the attrition (Marler and Peters 1982a). The possibility of action-based learning means that one cannot assume that production of similar songs by neighbouring males has resulted from one male acquiring song(s) from the other. These three processes are not mutually exclusive, but their relative importance may vary within and among species.

Here I combine field and laboratory approaches to the study of song development. In Part I, field observations on the development of song repertoires and song matching in the field sparrow (*Spizella pusilla*) are described. In Part II, an hypothesis generated from

Playback and Studies of Animal Communication
Edited by P.K. McGregor, Plenum Press, New York, 1992

these observations is tested experimentally in a laboratory study of song development in the white-crowned sparrow (*Zonotrichia leucophrys*).

I. Observational Field Study of Song Development in the Field Sparrow

Male field sparrows sing two different kinds, or categories, of song (Nelson and Croner 1991). *Simple* song is used in long distance countersinging among territorial neighbours and in apparently *spontaneous* singing. It is the commonest song heard in mid-morning. Simple song structure differs significantly among individual males and males distinguish familiar and unfamiliar individuals based on their songs (Goldman 1973, Nelson 1989a). Banded males sing the same simple song throughout their lives (Nelson 1991). Most, if not all, males also possess a *complex* song that is used most frequently during the dawn chorus and in close chasing between territorial males (Nelson and Croner 1991). Here I describe variation in the size of repertoires of simple songs.

Territorial male field sparrows were observed from 1985 to 1989 at three sites in Dutchess County, New York. Field sparrows inhabit old fields that are distributed in a patchy mosaic of habitat islands amidst forests, farms and human settlements. An average of 98 territories were monitored each year. With the exceptions of 1985 and 1988, all but two to five birds were tape-recorded each year in the study area. Birds were identified on the basis of their songs, territory position and by colour bands.

Songs are composed of discrete *notes*, defined as continuous tracings on sonagrams. Notes are grouped into phrases of two kinds: sequences of repeated identical notes (*trills*), usually at the end of the song, or a sequence of different note types (*note complex*). All complete songs had at least two phrases. A song type is a consistently-produced song pattern that differs from other types in the repertoire of one male. Song types are distinguished by visual inspection of spectrograms (Nelson and Croner 1991).

In what follows, a *new* bird is one whose song differed from that of the territory's previous occupant. An *old* bird is one whose song was identical to the previous occupant's. Spectrograms were compared visually to assign birds into these groups. I employed a strict criterion in comparing songs, especially in frequency characters, which vary little within birds and can be reliably used to discriminate individuals (Nelson 1989a).

In 1988 and 1989 I documented the pattern of territory settlement and recorded repertoire sizes of males immediately upon arrival by visiting each field every other day beginning in early March before the first males returned. I listened to each male for at least five minutes to count the number of song types each sang (see below) and recorded a sample of song when possible.

Audiospectrograms of the different song types in the repertoires of four multiple-song-type males and eight neighbouring single-song-type males are presented in Figure 1. In 1988 and 1989 new birds were significantly more likely than old birds to sing multiple song types. Twenty three percent of new birds in 1988 and 29% in 1989 sang more than one song type. These percentages are likely to be underestimates, since the alternate-day sampling scheme could not always be followed. Indeed, five of the 13 (39%) new males found on their first day of territory residence in 1989 sang repertoires of multiple song types. The commonest multiple-song-type repertoire contained two types; a few birds each year sang three or four types.

In 1989 when I encountered males singing two or more song types I initiated a 15-minute focal watch in which I noted the sequence of songs and the numbers of songs sung by neighbours. Males with more than two song types were colour banded and followed for the remainder of the summer. I tape recorded each subject and its neighbours.

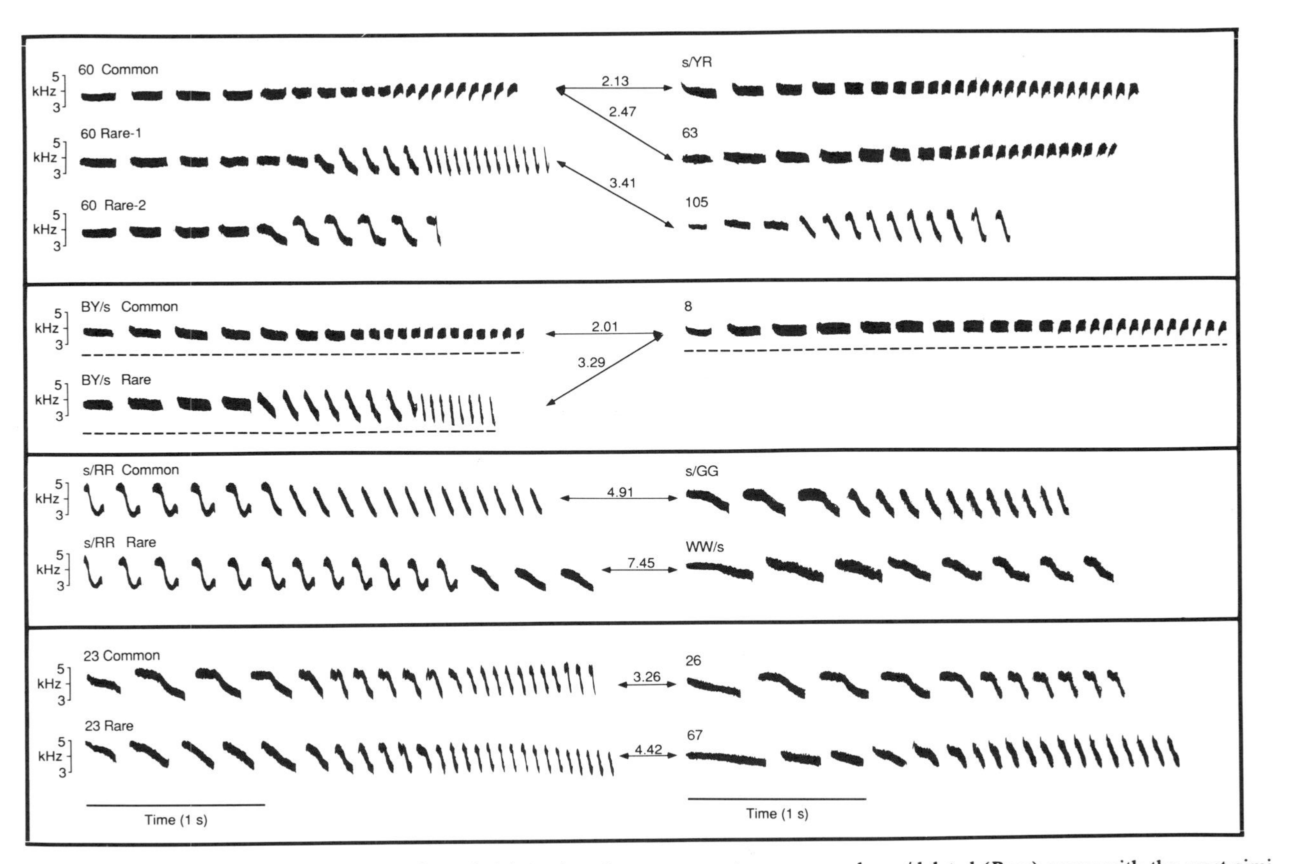

Figure 1. Spectrograms of simple song repertoires of eight males. Arrows connect common and rare/deleted (*Rare*) songs with the most similar neighbour's song. Numbers associated with arrows are acoustic distances, represented as Euclidian distances measured in a 14-variable space (see text). Dashed lines underneath songs denote the parts of songs that were used as stimuli in a playback experiment (see text).

In the first week on territory in 1989, 14 males with multiple-song repertoires sang each song type in their repertoires equally often (Fig. 2). Males sang with near-immediate variety, delivering each song type without immediate repetition, or repeating it a very few times before switching to another type (median = 2 repetitions). One song type came to predominate in each bird's repertoire over the ensuing weeks; this song was defined as the male's *common* song type; while the other song(s) were labelled *rare/deleted* types. In five males this attrition process occurred in a matter of days, as I heard their rare/deleted song(s) only on the first day I found them. Most males sang only their common song type after four weeks on territory; however, I heard one to three tokens of a rare/deleted type from three males after eight weeks of territory residence. From 1986 to 1989 four males (out of 33 encountered with > 2 song types) retained two song types in their repertoires over two or more years. Thus, on occasion rare/deleted types are not completely lost from a male's production repertoire.

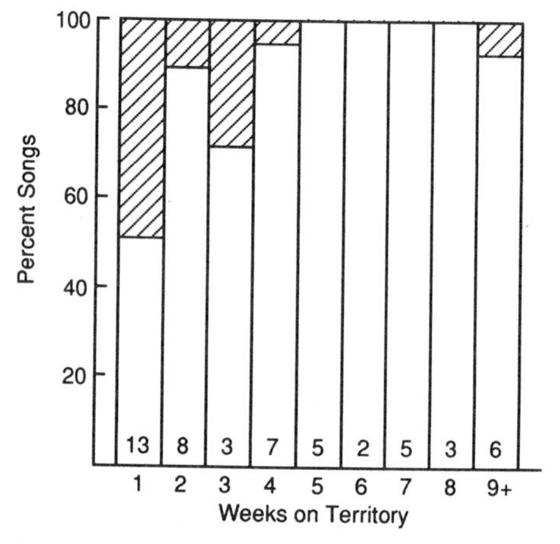

Figure 2. Percentages of common (open bars) and rare/deleted (hatched bars) song types sung in focal-animal samples as a function of time since initial territory occupancy. Number of males sampled each week is shown within each bar and declines with time since some males lost their territories. One male was not observed in week one.

The culling of songs from multiple-song repertoires was non-random. To compare the acoustic similarity of songs among neighbouring males in 1987 and 1989 I used Euclidian distances in an acoustic space formed by the 14 standardised variables measured on notes in the songs' first two phrases (Nelson 1989b). The first two phrases of common songs more closely resembled a neighbour's song than did rare/deleted songs (Figs. 1 & 3). Selection of the most similar song for retention in the repertoire resulted in song sharing among neighbours: common song types were more similar to a neighbour's song than to a song randomly chosen from the population. In contrast, rare/deleted songs were equally similar to a neighbour's song and to a randomly-chosen song. The first two phrases of common and rare/deleted song types within a bird's repertoire were no more similar to one another than were two songs randomly chosen from the population.

Song types were not selected on the basis of their absolute acoustic characteristics, but rather on their structure relative to the songs of neighbouring males. In a discriminant functions analysis, 31 common song types did not differ acoustically from the 38 rare/deleted song types sung by 31 multiple-song-type males.

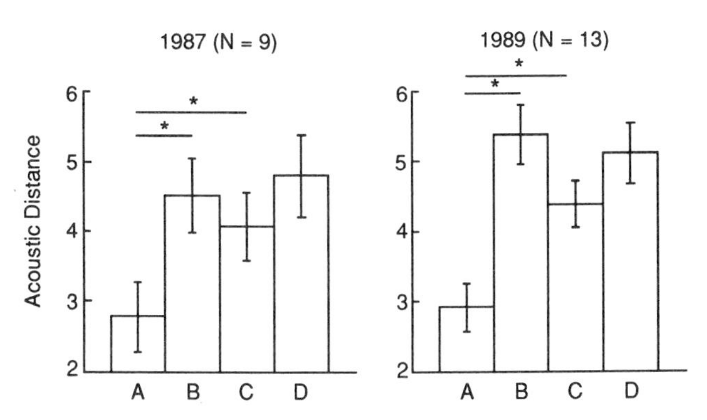

Figure 3. Acoustic distances (mean ± SE) between common and rare/deleted song types and the most similar neighbour and songs chosen at random from the entire local population: A) distance between a male's common song and his most similar neighbour's song; B) distance between common song and a randomly-chosen song; C) distance between rare/deleted song and the most similar neighbour's song; D) distance between rare/deleted song and randomly-chosen song. Horizontal bars with asterisks identify means that differ significantly.

If actual acquisition of common songs by new birds from neighbours was taking place during territory establishment, then we might expect that a neighbour's song would be the most similar song in the entire local population. This was true for only 27% of the 22 multiple-song-type birds in 1987 and 1989 (two in 1987, four in 1989). There was a median of 8.5 other songs (range 0 - 85) in the local population that were more similar to a bird's common song than was the most similar neighbour's song.

Possessing several song types in a repertoire did not necessarily enhance the probability of sharing a neighbour's song. Songs of new single-song-type birds also resembled neighbour's song more closely than songs chosen at random in 1987 and 1989.

New birds had an average of 2.5 neighbours (range = 0 - 5). Common song types of focal subject males were most similar to the neighbour that sang most frequently during the initial four weeks of territory occupancy in 1989. The neighbour with a song most similar to the subject's common type sang about twice as frequently as the average other neighbour (0.84 ± 0.14 versus 0.40 ± 0.10 songs per minute).

What are the perceptual consequences of a change of song status from common to rare/deleted? Males were presented with playback of four different simple songs: 1) *common self song* - the song retained and sung by the subject, 2) *rare/deleted self song* - a song either lost by attrition early in the male's occupancy of his territory, or a song sung much less frequently than common song, 3) *neighbour's song* and 4) *stranger's song*. The key predictions involved the relative responses made to common and rare/deleted self song. If rare/deleted songs are remembered, then they should elicit responses similar to common self song, and intermediate between neighbour and stranger song. If rare/deleted songs are forgotten, they should elicit the same response as a stranger's song.

Subjects were six male field sparrows that had been recorded immediately after their arrival on territories in 1987 (one male), 1988 (three), or 1989 (two). Only two subjects had sung their rare/deleted song type in the two months prior to testing. Different neighbour and stranger songs were used for each subject.

A male's song types sometimes shared either the introduction or trill phrases of the song. To prevent responses being influenced by such shared song phrases, only the phrase(s) that differed between songs for each male were used as experimental stimuli. Two subjects heard complete songs (BY/s in Fig. 1), two heard the first two phrases and two heard only the terminal trill. Incomplete songs consisting of the first one or two song phrases are commonly heard in nature; trills alone are rare.

Procedural details followed Nelson (1989a). Each treatment consisted of a 5 minute pre-playback period, 3 minute of stimulus playback and a 5 minute post-playback (quiet) period. Stimuli were broadcast from a speaker facing into the subject's territory located about 15m inside the subject's territory along the neighbour's boundary. Response strength was expressed as the first principal component derived from 12 variables: number of songs, flights, calls, time spent near the speaker in two distance categories and closest approach distance in both playback and quiet periods (see chapter by McGregor this volume; Nelson 1989a).

Males differed in their responses to the four stimuli (Fig. 4). Neighbour's song elicited significantly weaker responses than stranger's song. Common self song elicited weaker responses than stranger's song, but not significantly so. Rare/deleted self song elicited significantly weaker responses than stranger song. Common and rare/deleted songs elicited similar responses. Examination of the raw data revealed no obvious correlation between response levels and similarity to neighbour's song nor with recency of singing rare/deleted song.

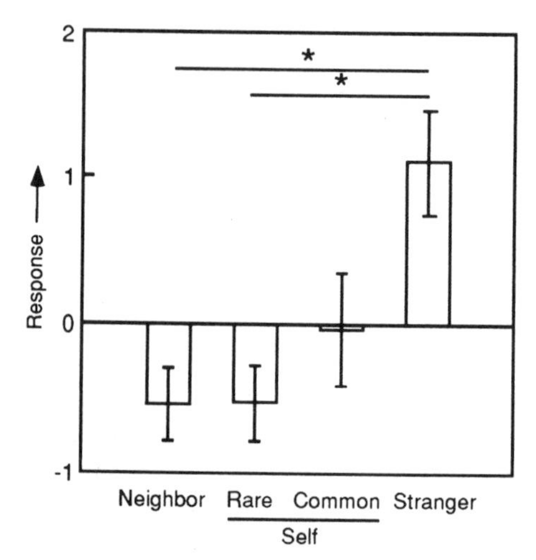

Figure 4. Responses (mean ± SE) of six males to playback of common and rare/deleted self song, neighbour's song and stranger's song. The ordinate is principal component one derived from 12 response variables (see text). The scale is arbitrary; positive values represent stronger responses, correlated with, a) close approach distances both during and after playback, b) increasing time spent in both distance categories during and after playback and c) increased number of flights during playback. Significant differences identified as in Figure 3.

Summary

Field sparrows settling on particular territories for the first time often sing two or more simple song types. All but one of these song types are subsequently lost from males' production repertoires. The retained song type is sung for the remainder of a male's life

(Nelson 1991). The attrition process appears to be guided by vocal experience with neighbouring males, with the result that males preferentially retain in their repertoires the simple song that best matches their most vocally active neighbour. As a consequence, local song groups are formed in which neighbours share songs. In a field playback experiment, rare/deleted self songs elicited responses that differed from those made to stranger's song. Evidently rare/deleted songs are not forgotten. Common self song elicited responses that were intermediate in strength.

II. The Effect of Song Matching during Plastic Song: Evidence from the White-Crowned Sparrow

The observational study of field sparrows suggested that matched counter-singing between territorial neighbours leads to the selective retention of the matching song type in a newly-arrived male's repertoire. Can this process be simulated in the laboratory by selectively playing back matching or non-matching songs to males in plastic song? Preliminary results from the white-crowned sparrow suggest that this is possible.

Singing behaviour of the white-crowned sparrow has been intensively studied (review in Kroodsma et al. 1985). White-crowned sparrows resemble field sparrows in that most males sing a single *advertising* song in their mature repertoires (nothing analogous to the complex songs of field sparrows have been described in white-crowned sparrows). However, two studies provided some evidence for overproduction in plastic song (Marler 1970, DeWolfe et al. 1989). In collaboration with Peter Marler, Luis Baptista and Martin Morton, I tried to shape the composition of male white-crowned sparrows' repertoires when they were in plastic song.

Nestling coastal (*Z. l. nuttalli*) and mountain (*Z. l. oriantha*) white-crowned sparrows were collected at five to eight days of age and brought into the laboratory where they were hand-reared to independence. Males were identified by laparotomy and were then housed individually in sound isolation chambers, where they remained until the end of the experiment.

Tutoring began at five to eight days of age and continued until the next spring. Birds were exposed to a changing roster of tape-recorded white-crowned sparrow songs. We selected 56 white-crowned sparrow songs (28 *nuttalli*, 28 *oriantha*) from different areas. By examining sonagrams, we selected songs that appeared distinctive, so that if subjects learned them we would be able to clearly assign the imitations to the source tutor song. Birds heard ten minutes of one *nuttalli* song and ten minutes of one *oriantha* song each morning for ten consecutive days before changing to a different pair of tutor songs. Each pair of songs was broadcast for ten days except in winter when there was one block of 21 days and three blocks of 30 days.

Birds were tape-recorded twice a month in the fall and early winter. When evidence of imitations of tutor songs appeared (in February for the earliest *nuttalli*) recordings were made about twice weekly. Recordings were analysed by real-time spectrography (Hopkins et al. 1974) to identify stable song patterns. Based on analysis of their plastic song repertoires, the 25 males were classified into four groups: 1) ten males imitated one tutor song; 2) three birds produced two or more minor variants based on a single tutor song; 3) nine birds imitated two or more tutor songs; and 4) three birds sang songs that did not closely resemble any tutor song. With the possible exception of one bird, all males copied tutor songs that they were exposed to in the first two months of life. Although most sang subsong during the winter, clear evidence of imitated syllables and phrases did not appear until late February in *nuttalli* or April in most *oriantha*.

The nine males that imitated two tutor songs and the three that developed two or more variants derived from a single tutor song were subjects for further experimentation. In a design to mimic matched counter-singing between territory neighbours, one song type in their plastic song repertoires was chosen at random and the tutor song that most closely resembled it was played back to the bird for 20 minutes each day for ten consecutive days. Half the birds were treated in this way and half served as controls. Controls were subjected to a novel song for ten days. Subjects were recorded every other day during playback and every three days after that.

Results for three males in the experimental group are shown in Figure 5. PM24 crystallised his song type that was most similar to the tutor song within a week after the beginning of playback. PM36 and PM56 both sang their matching type about 50% more frequently after playback than they sang it prior to playback.

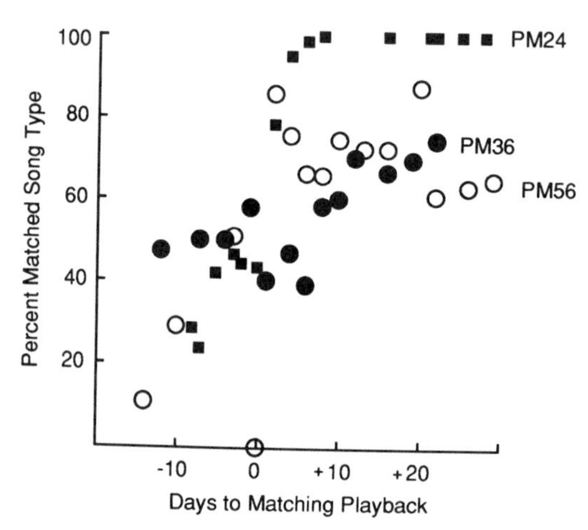

Figure 5. Results of matching song playback to three male white-crowned sparrows in plastic song. Ordinate is the percentage of subjects' songs that matched playback.

None of the control birds acquired the novel song type they were exposed to, despite having twice as much experience with it as the songs they memorised six to eight months earlier. There is a suggestion that the rate of song crystallisation may be lower than in birds exposed to matching playback (cf. Figs. 5 & 6).

Two males that sang two or more song variants on a single *theme* responded quite differently to playback matched to one of their variants than did the males that imitated two distinct tutor songs. Both varied the terminal trill of their song while keeping the introductory notes constant (Fig. 7). Playback of a song resembling one of the trills did not increase the singing frequency of that variant. A third, control, bird resembled the other controls and did not incorporate the novel song into his repertoire.

Though the sample is small, there is a hint that matching song playback has differing effects depending on the source and/or acoustic dissimilarity of the song material upon which selection acts. Songs imitated from different tutor song types differed more acoustically than were improvised variants that differed only in the last song phrase. Male white-crowned sparrows appear to classify song variants together, at least when frequency of production is used as the dependent variable.

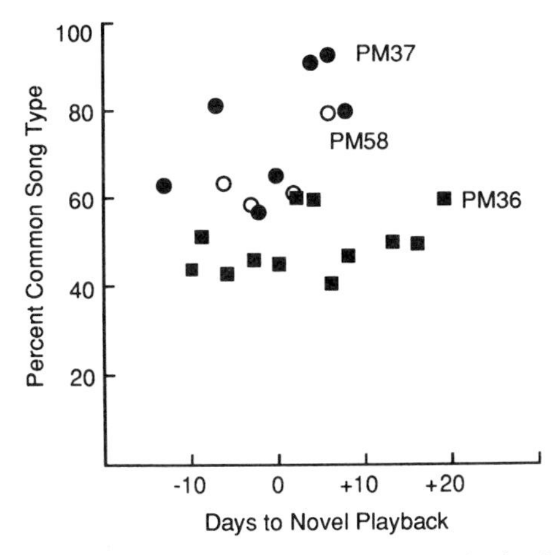

Figure 6. Results of novel song playback to three males in plastic song. Each male sang two different song types learned before 60 days of age.

General Discussion

These observational and experimental data indicate that song overproduction and selective attrition during plastic song may play important roles in the development of song repertoires in individual male song birds and in the formation of local song groups or *neighbourhoods*. In common with other species (nightingale, *Luscinia megarhynchos*, Todt et al. 1979; indigo bunting, *Passerina cyanea*, Payne 1981, 1983; white-crowned sparrow, Baptista and Petrinovich 1984; zebra finch, *Taeniopygia guttata*, Clayton 1987; Slater et al. 1988, Williams 1990), the nature of a male sparrow's social environment appears to influence the course of song development in its first year of life.

Two hypotheses concerning the development of song sharing have been proposed to apply to migratory songbirds in which pre-breeding dispersal distances are large. In comparing these it is important to distinguish between the time of song acquisition (memorisation) and the time of song production (performance). The hypotheses differ in the time of life when males acquire songs from adult models. In the *late acquisition* hypothesis, the sensitive phase for acquisition extends until (or possibly begins at) the time that young males disperse to their first breeding territory. Males then acquire songs from the local established males. The selective attrition hypothesis assumes that song acquisition is restricted to early life and that song sharing results from the selective attrition of songs from the repertoire.

The selective attrition hypothesis can explain several findings from the field sparrow study more readily than can the late acquisition model. Firstly, males with multiple song-type repertoires were often the first males to return to their respective fields in the spring. Since no adults were present at that time to serve as models, this suggests that these birds were singing songs they had acquired earlier. Second, rare/deleted songs did not resemble neighbour's song more than would be expected by chance. These songs must have been either acquired from other sources or they could be invented or improvised. Third, if new males were acquiring songs from their neighbours, we would expect neighbours' songs to be the most similar song in the population. They are not in most instances, indicating that the fidelity of song sharing among neighbours may be constrained by the number and structure of songs acquired early in life.

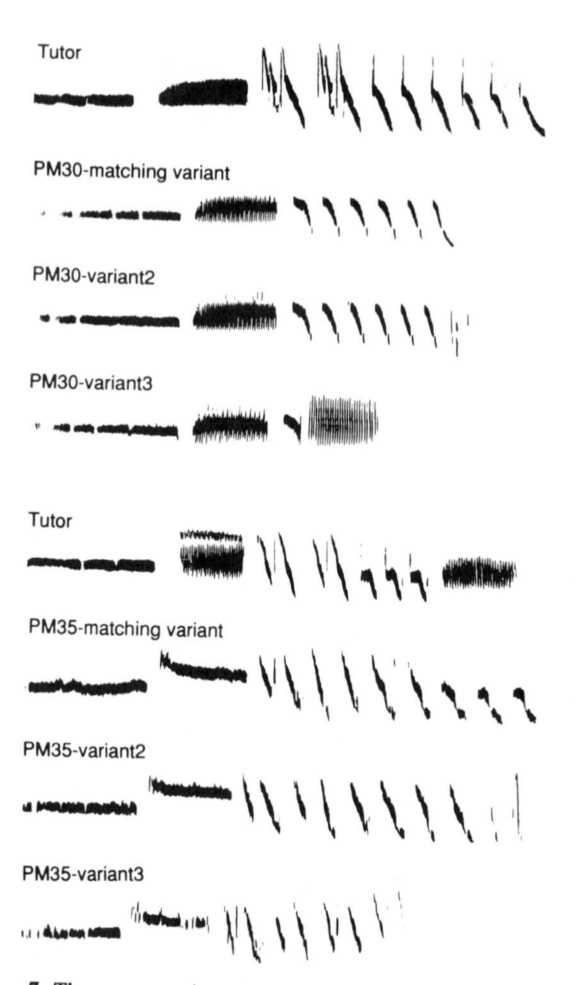

Figure 7. Three song variants improvised from one tutor song in the plastic song repertoires of two males. The top song in each group of four is the tutor song that most closely resembled the song variant immediately below it.

Late acquisition may play a role also, since field sparrows with single songs in their plastic song repertoires also shared songs with neighbours. It is likely that some of these birds may have actually sang more than one song type before I found them. More intensive sampling is needed to address this issue. However, in the white-crowned sparrow experiment, no male acquired novel songs after exposure in their first spring (about 300 days of age): all acquisition ended at least five months previously. Rather, matching playback facilitated the retention of song types that already existed in plastic song repertoires. Interactive playback, in the form of a live tutor (Baptista and Petrinovich 1984) or computer (Dabelsteen and Pedersen 1991) may affect the time course of acquisition and/or selection.

The selection hypothesis identifies a constraint that may explain why not all males share their song with a neighbour. In field sparrows, there is great variability in the degree of song similarity among neighbouring males. Some males sing songs that are nearly identical to a neighbour's, while others are not particularly similar (Fig. 1). If the ability to acquire songs for production is restricted to early in life and if a male had not acquired a certain song type then, he would not be able to match a neighbour singing that song type during territory establishment. Random differences in the sample of songs learned early in life (e.g. Baptista 1974) may constrain a bird's options for song matching later on (Marler and Peters 1982b).

In field sparrows, song sharing appears to be guided in the first few weeks on territory by vocal interaction with neighbouring males. Males retained in their repertoires the song type that most closely resembled the song of their most actively singing neighbour. Payne (1983) also reported that indigo buntings counter-sang more with neighbours that sang a matching song than with other males; but most young buntings shared a song with a neighbour when first found (Payne 1981). The playback experiment with white-crowned sparrows supports the prediction generated from these field studies: specific auditory input during plastic song affects the composition of a male's repertoire by selecting song types from a pre-existing repertoire.

The results of the field playback experiment provide evidence that common and rare/deleted self songs differ in their perceptual significance to male field sparrows. The significantly weaker response to neighbour song relative to stranger song agrees with previous experiments on field sparrows (Goldman 1973; Nelson 1989a) and on other species (reviewed in Falls 1982). In other species as well, self song (presumably common) elicits a response intermediate to, though not always statistically distinguishable from, that made to neighbour's and stranger's songs (Weeden and Falls 1959; Brooks and Falls 1975; Searcy et al. 1981). Brooks and Falls (1975) suggested that the intermediate response white-throated sparrow (*Zonotrichia albicollis*) males made to self song was the net result of habituation from hearing one's own song repeatedly and novelty from hearing self song broadcast at a distance.

To my knowledge, Hinde's (1958) study on caged male chaffinches (*Fringilla coelebs*) is the only other study to contrast responses to self songs that differ in their frequency of performance. Chaffinches sang more often to and matched song types with, the common song in their repertoires as opposed to rarer song types. Hinde also noted that this occurred even if the common song had an atypical structure. It seems unlikely that the lower response to rare self song than to common self song in either field sparrows or chaffinches can be explained in terms of different levels of stimulus-specific habituation and sensitisation. We would not expect birds to habituate more to a less frequently occurring stimulus (rare/deleted song). Rather, it seems that different songs within the repertoire have different *values* or *weights* associated with them and that common songs are more heavily weighted than rare/deleted song types.

That male song birds store representations of more songs than they commonly sing has implications for modelling processes of song recognition in birds (Nelson and Marler 1990). If males learn a greater variety of conspecific songs than they produce in their mature repertoires, this could help explain how the level of responsiveness to experimentally-altered songs correlates with the normal limits of song variation within the local population of field sparrow songs (Nelson 1988, 1989b). The store of songs learned early in life when combined with the songs of neighbours that are learned but not sung (Falls 1982), would provide a large sample, and hence more accurate estimate, of the normal limits of acoustic variation present in conspecific song. The strength of a male's response may be a statistical average of the comparisons between a stimulus and a number of songs stored in memory. Some comparisons may be weighted more heavily than others, as suggested by evidence presented here and from other studies (Falls et al. 1982; McGregor and Krebs 1984; Kreutzer 1987).

Acknowledgements

Support was provided by NRSA F32 NS0629, a fellowship from the Charles H. Revson Foundation, and Bio-medical research support grant GRSG S07 RR07065 to the Rockefeller University. I thank Peter Marler for his encouragement and comments on the manuscript, and Luis Baptista and Martin Morton for their generous help.

References

Baptista, L.F. 1974. The effects of songs of wintering white-crowned sparrows on song development in sedentary populations of the species. *Z. Tierpsychol.*, **34**, 147-171.

Baptista, L.F., Petrinovich, L. 1984. Social interactions, sensitive phases and the song template hypothesis in the white-crowned sparrow. *Anim. Behav.*, **32**, 172-181.

Brooks, R.J. and Falls, J.B. 1975. Individual recognition by song in white-throated sparrows. I. Discrimination of songs of neighbors and strangers. *Can. J. Zool.*, **53**, 879-888.

Clayton, N.S. 1987. Song tutor choice in zebra finches. *Anim. Behav.*, **35**, 714-721.

Cunningham, M.A. and Baker, M.C. 1983. Vocal learning in white-crowned sparrows: sensitive phase and song dialects. *Behav. Ecol. Sociobiol.*, **13**, 259-269.

Dabelsteen, T. and Pedersen, S.B. 1991. Song and information about aggressive responses of blackbirds, *Turdus merula*: evidence from interactive playback experiments with territory owners. *Anim. Behav.*, **40**, 1158-1168.

DeWolfe, B.B., Baptista, L.F. and Petrinovich, L. 1989. Song development and territory establishment in Nuttall's white-crowned sparrows. *Condor*, 91, 397-407.

Falls, J.B. 1982. Individual recognition by sounds. In: *Evolution and Ecology of Acoustic Communication in Birds. Vol.II.* (Ed. by D.E. Kroodsma, E.H. Miller & H. Ouellet), pp. 237-278. Academic Press, New York.

Falls, J.B., Krebs, J.R. and McGregor, P.K. 1982. Song matching in the great tit (*Parus major*): the effect of similarity and familiarity. *Anim. Behav.*, **30**, 997-1009.

Goldman, P. 1973. Song recognition by field sparrows. *Auk*, **90**, 106-113.

Hinde, R.A. 1958. Alternative motor patterns in chaffinch song. *Anim. Behav.*, **6**, 211-218.

Hopkins, C.D., Rossetto, M. and Lutjen, A. 1974. A continuous sound spectrum analyzer of animal sounds. *Z. Tierpsychol.*, **34**, 313-320.

Konishi, M. 1965. The role of auditory feedback in the control of vocalization in the white-crowned sparrow. *Z. Tierpsychol.*, **22**, 770-783.

Krebs, J.R. and Kroodsma, D.E. 1980. Repertoires and geographical variation in bird song. *Adv. Study Behav.*, **11**, 143-177.

Kreutzer, M. 1987. Reactions of cirl buntings (*Emberiza cirlus*) to playback of an atypical song: the use of own and neighbors' repertoires for song recognition. *J. Comp. Psychol.*, **101**, 382-386.

Kroodsma, D.E. 1974. Song learning, dialects, and dispersal in the Bewick's wren. *Z. Tierpsychol.*, **35**, 352-380.

Kroodsma, D.E., Baker, M.C., Baptista, L.F. and Petrinovich, L. 1985. Vocal "dialects" in white-crowned sparrows. In: *Current Ornithology. Vol.II.* (Ed. by R.F. Johnston), pp. 103-133. Plenum Press, New York.

McGregor, P.K., Krebs, J.R. 1984. Sound degradation as a distance cue in great tit (*Parus major*) song. *Behav. Ecol. Sociobiol.*, **16**, 49-56.

Marler, P. 1970. A comparative approach to vocal learning. *J. Comp. Physiol., Psychol. Monogr.*, **71**, 1-25.

Marler, P. 1990. Song learning: the interface between behaviour and neuroethology. *Phil. Trans. R. Soc. Lond. B.*, **329**, 109-114.

Marler, P. and Peters, S. 1982a. Developmental overproduction and selective attrition: new processes in the epigenesis of birdsong. *Develop. Psychobiol.*, **15**, 369-378.

Marler, P. and Peters, S. 1982b. Subsong and plastic song: their role in the vocal learning process. In: *Evolution and Ecology of Acoustic Communication in Birds. Vol.II.* (Ed. by D.E. Kroodsma, E.H. Miller & H. Ouellet), pp. 25-50. Academic Press, New York.

Marler, P. and Tamura, M. 1962. Song "dialects" in three populations of white-crowned sparrows. *Condor*, **64**, 368-377.

Nelson, D.A. 1988. Feature weighting in species song recognition by the field sparrow (*Spizella pusilla*). *Behaviour*, **106**, 158-182.

Nelson, D.A. 1989a. Song frequency as a cue for recognition of species and individuals in the field sparrow (*Spizella pusilla*). *J. Comp. Psychol.*, **103**, 171-176.

Nelson, DA. 1989b. The importance of invariant and distinctive features in species recognition of bird song. *Condor*, **91**, 120-130.

Nelson, D.A. 1991. Song overproduction and selective attrition during song development in the field sparrow (*Spizella pusilla*). *Behav. Ecol. Sociobiol.* (*in review*)

Nelson, D.A. and Croner, L.J. 1991. Song categories and their functions in the field sparrow (*Spizella puzilla*). *Auk*, **108**, 42-52.

Nelson, D.A and Marler, P. 1990. The perception of birdsong and an ecological concept of signal space. In: *Comparative perception. Vol II.* (Ed. by W.C. Stebbins & M.A. Berkley), pp. 443-478. John Wiley and Sons, New York.

Nottebohm, F. 1970. Ontogeny of bird song. *Science*, **167**, 950-956.

Payne, R.B. 1981. Song learning and social interaction in indigo buntings. *Anim. Behav.*, **29**, 688-697.

Payne, R.B. 1983. The social context of song mimicry: song-matching dialects in indigo buntings (*Passerina cyanea*). *Anim. Behav.*, **31**, 788-805.

Searcy, W.A, McArthur, P.D., Peters. S.S. and Marler, P. 1981. Response of male song and swamp sparrows to neighbour, stranger, and self songs. *Behaviour*, **77**, 152-263.

Slater, P.J.B., Eales, L.A. and Clayton, N.S. 1988. Song learning in zebra finches (*Taeniopygia guttata*): progress and prospects. *Adv. Study Behav.*, **18**, 1-34.

Thorpe, W.H. 1958. The learning of song patterns by birds, with especial reference to the song of the chaffinch, *Fringilla coelebs*. *Ibis*, **100**, 535-570.

Todt, D., Hultsch, H. and Heike, H. 1979. Conditions affecting song acquisition in nightingales (*Luscinia megarhynchos* L.). *Z. Tierpsychol.*, **51**, 23-35.

Weeden, J.S. and Falls, J.B. 1959. Differential responses of male ovenbirds to recorded songs of neighboring and more distant individuals. *Auk*, **76**, 343-351.

West, M.J. and King, A.P. 1986. Song repertoire development in male cowbirds (*Molothrus ater*): its relation to female assessment of song potency. *J. Comp. Psychol.*, **100**, 296-303.

Williams, H. 1990. Models for song learning in the zebra finch: fathers or others? *Anim. Behav.*, **39**, 745-757.

MALE QUALITY AND PLAYBACK IN THE GREAT TIT

Marcel M. Lambrechts

Department of Biology
University of Antwerp, UIA
Universiteitsplein 1, B-2610 Wilrijk
BELGIUM
and
Department of Zoology
University of Wisconsin-Madison
Madison, WI 53706
U.S.A.

Introduction

Although it is widely accepted that male songbirds sing to repel males or to attract or stimulate females, it is not well understood why the structure of bird song is so complex. In most songbirds, individuals sing different versions of the species-specific song (i.e. song or syllable types) to form a song repertoire. The composition of repertoires (the song types that constitute the repertoire), the size of repertoires (number of song types), the rate of song type switching and the way single song types are performed (song rate, percentage performance time, song duration), all differ considerably within and among species (reviews in Kroodsma and Miller 1982; Searcy and Andersson 1986; Kroodsma and Byers 1991).

Song variation contains cues for species recognition and individual recognition, but also contains information on motivation and intention, on geographic origin (e.g. song dialects, habitat structure and population density) and on individual characteristics, such as age or territory quality (reviews in Falls 1982; Becker 1982; Searcy and Andersson 1986; Kroodsma and Byers 1991).

In songbirds, different song characteristics (e.g. song repertoire size, song type structure, song rate, song output and song versatility) are considered to be indicative of male quality (e.g. reproductive experience, parental care, health, physical condition, working capacity, survival ability, fighting ability and overall genetic fitness) and it has been suggested that advertisement of male quality could increase success in male-male competition or in female attraction (e.g. McGregor et al. 1981; West et al. 1981; Yasukawa 1981a, 1981b; Greig-Smith 1982; Jarvi 1983; Searcy and Yasukawa 1983; Baker et al. 1986; Lambrechts and Dhondt 1986, 1987; Gottlander 1987; Radesater et al. 1987; Reid 1987; Rothstein and Fleischer 1987; Bijnens 1988; Hiebert et al. 1989; Eens et al. 1991; Moller, 1991). Indeed,song features that reflect male quality may influence mate choice, because females could gain genetic benefits such as high quality offspring, or nongenetic benefits such as resources, by mating with a high quality male, that is, a male with more success in resource defence. Song may also reflect vigour or fighting ability and advertisement of these male characteristics could therefore be a means used by high quality males to scare away potential rivals during territorial defence.

The aim of this paper is to discuss and evaluate playback techniques that have been used, and which could be used, to examine the following three questions:

1) Are song features that are correlated with measures of male quality used as reliable acoustic signals of male quality?
2) Do birds perceive variation in song features that reflect male quality?
3) Are song characteristics associated with high male quality more successful in male repulsion or in female attraction than song characteristics associated with low male quality?

I will focus on the singing performance of the great tit *Parus major*, because it is one of the few songbird species that has been studied in detail and in which relationships between song characteristics and possible measures of male quality have been reported.

Great Tit Singing Performance: a Reliable Indicator of Male Quality?

Descriptive Studies of Individual Differences in Singing Ability and Male Quality

Great tits have a simple song that they repeat in a stereotyped fashion. The smallest unit of great tit song is a group of one to more than five different notes (or elements) that is called a phrase (see Fig. 1). One to more than 20 phrases are rapidly repeated in a burst of song that is called a strophe. A strophe is normally between one and five seconds in duration. After the production of a strophe, great tits pause for several seconds before a new strophe is produced. Great tits normally sing a bout of 20 to more than 100 strophes of a single phrase type before they introduce a new song bout of another phrase type. Each male can sing between one and eight distinct phrase types (also termed song types) to form a song repertoire. Phrase types can be distinguished by the number of notes per phrase and by the structure of notes that make up the phrase (McGregor and Krebs 1982; Lambrechts and Dhondt 1988a).

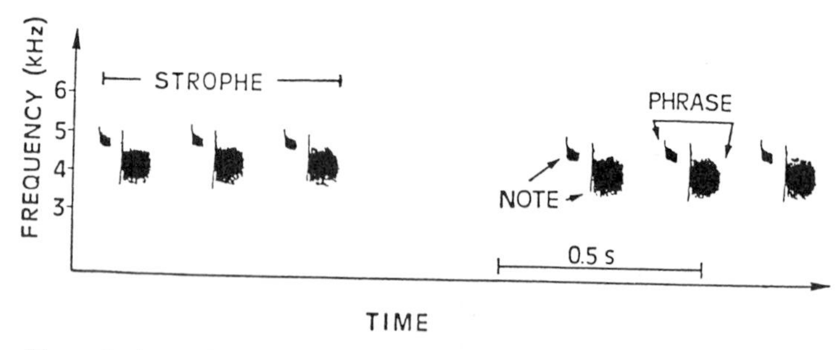

Figure 1. A sound spectrogram of a great tit strophe with six two-note phrases.

Lambrechts and Dhondt (1986, 1987, 1988a) proposed that three measures of great tit singing performance give information about singing ability:

1) the average number of *phrases per strophe* (i.e. a measure of strophe length);
2) average *drift* (drift is the increase in the time intervals between notes and between phrases within strophes);
3) the average *percentage performance time*, or *PPT*, within song bouts; PPT is calculated as the ratio of strophe duration to strophe duration plus the subsequent pause (see also Weary et al. 1988, 1991; Lambrechts 1988).

Males vary considerably in average number of phrases per strophe, strophe duration (in seconds), average PPT and drift within strophes (Lambrechts and Dhondt 1986, 1987, 1988a; Weary et al. 1990; McGregor and Horn 1991). Males singing more phrases per strophe on average generally show less drift within strophes and produce more sound during song bouts (Lambrechts and Dhondt 1987, 1988a).

Drift happens over a time span of a few seconds, it appears in consecutive strophes and it is more pronounced when a male produces strophes that are longer than its average strophe length. These factors led Lambrechts and Dhondt to suggest that drift within strophes is probably not caused by short-term changes in motivation of the singer, but could be the result of physiological or respiratory problems for the male. Also, the strophe rate (number of strophes per minute) can be very low when males sing extremely long strophes, for example, some males sing more than 50 phrases per strophe (lasting more than 15s) and in these situations the number of phrases per strophe will decrease rapidly and the intervals between strophes will increase throughout a song bout. To explain such observations, Lambrechts and Dhondt (1988a) proposed that a decrease in the song output over the course of a song bout could be caused by neuro-muscular exhaustion (see Wells and Taigen (1986) for a similar argument in frogs). However, Weary et al. (1988, 1991) showed experimentally that motivation could also be involved.

In a number of species of songbird the average song output increases when extra food is provided on the territory, suggesting that singing is energy-limited (Searcy 1979; Morton 1982; Davies and Lundberg 1984; Gottlander 1987; Reid 1987; Cuthill and Macdonald 1990). Moller (1991) found that parasite load has a negative effect on song output in swallows (*Hirundo rustica*) and proposed that song output may reflect working capacity. These results suggest that the average singing performance of great tits might also be influenced by food or parasites and therefore give further support to the idea that singing performance may reflect physical condition or health.

Although the exact causes of variation in great tit singing performance have yet to be established, descriptive studies have shown consistent inter-individual differences in the average number of phrases per strophe and in the average drift within strophes, both throughout the territorial period within a year and in consecutive years (Lambrechts and Dhondt 1986, 1987). Furthermore, the average strophe length (measured as average number of phrases per strophe or as average strophe duration) and the average drift within strophes are correlated with social dominance at a winter feeder, age, survival and individual lifetime reproductive success (Lambrechts and Dhondt 1986, *unpublished data*). These findings suggest that individual male great tits differ consistently in singing performance and that the average singing performance of a male could be considered to be a reliable measure of male quality.

Playback Experiments to Test the Reliability of Singing Performance

Changes in Average Singing Performance in Response to Playback

If within-year differences between males in average singing performance were maintained in a variety of contexts (e.g. at different periods during the breeding season, in conflict situations and during spontaneous song; cf. Lambrechts and Dhondt 1987), then average singing performance would be a more reliable measure of male quality. However, it is known that singing performance can be influenced by the motivation of the singer (see Becker 1982, p. 240) and average singing performance would be a less reliable measure of male quality if it was mainly influenced by motivation (Weary et al. 1988).

One of the contexts in which the reliability of an acoustic signal could be examined is a playback experiment. Territorial males presented with a speaker broadcasting conspecific song commonly increase song output (references in Becker 1982). Therefore, playback experiments could be used to see whether males significantly change their average singing performance in response to a playback stimulus. If average singing performance is a reliable signal, then males should differ in average singing performance in response to playback and an individual's average singing performance during spontaneous song and during playback should not differ significantly,

Table I summarises the results of field studies that looked at the effect of playback on the average singing performance of territorial great tits (measured as averages of the number of phrases per strophe, strophe duration and PPT). Three different playback paradigms were used. First, the average singing performance during spontaneous song was compared with that during playback (Lambrechts and Dhondt 1987; Lambrechts 1988). Second, the singing performance just before playback of two strophes was compared with that just after playback, to look for short-term changes in song output (Weary et al. 1988, 1991). Finally, the average singing performance during playback of very long strophes (of 17s duration) was compared with that during playback of short strophes (of 1.7s duration) (McGregor and Horn 1991).

None of the studies found a significant overall effect of playback on average strophe length (measured as average number of phrases per strophe or as average strophe duration) of the subjects (Lambrechts and Dhondt 1987; Weary et al. 1988; McGregor and Horn 1991). Two studies found a significant short-term increase in PPT in response to a short playback stimulus of two strophes (Weary et al. 1988; 1991). One study did not find a significant difference between the average PPT during spontaneous song and the average PPT produced during a playback trial, although the subjects started to match the song type of the playback stimulus (Lambrechts 1988, Table I). In five out of six studies that looked for inter-individual differences in average singing performance during song playback, the subjects differed significantly in the average number of phrases per strophe, in the average strophe duration or in the average PPT (Table I). One study did not find significant inter-individual differences in the average PPT during a playback experiment when the motivation of the singers was relatively low (PPT lower than 40%), but this study did not examine whether the subjects differed in the average PPT during spontaneous song (Weary et al. 1988). Another study found significant differences in the average PPT among males when the motivation of the singers was relatively high (Weary, Lambrechts and Krebs *unpublished data*). Finally, one study found significant inter-individual differences in the average PPT during spontaneous song and during playback when the motivation of the singers was relatively high (PPT greater than 40%) and the average PPT during spontaneous song and during playback did not differ significantly within individuals (Lambrechts 1988).

Table I. The effect of playback on various measures of singing performance of territorial male great tits. The measures of singing performance are described in detail in the text. The playback paradigms used were: spontaneous song *v.* song during playback; song before playback *v.* song after; and song during playback of long strophes *v.* during playback of short strophes. The **Altered** column notes whether the measure of singing performance was significantly changed by playback, the **Individual Diffs** column notes if there were significant between-individuals differences in response to playback.

Performance Measure	Paradigm	Altered?	Individual Diffs?	Reference
Phrases/strophe	spont. *v.* during	no	yes	Lambrechts & Dhondt 1987 Table II, this chapter
Strophe duration	before *v.* after	no	yes	Weary et al. 1988
Strophe duration	long *v.* short	no	yes	McGregor & Horn 1991
PPT	spont. *v.* during	no	yes	Lambrechts 1988
PPT	before *v.* after	yes	no	Weary et al. 1988
PPT	before *v.* after	yes	yes	Weary et al. 1991 Weary et al. *unpubl*

Table II shows the average number of phrases per strophe of nine great tits before they were presented with a playback stimulus and during playback of 12 phrases per strophe with an average PPT of 50%. The correlation between the spontaneous average number of phrases per strophe and the average number of phrases per strophe during playback is significant ($r = 0.97$, $N = 9$, $P < 0.01$; re-analysis of data from Lambrechts and Dhondt 1987). In comparison with spontaneous song, some individuals produced one or two more phrases during playback, however most differences were small, and any increase in average number of phrases per strophe caused by playback were relatively small compared with the inter-individual variation in average number of phrases per strophe found in the study population. For example, some individuals sing an average of four phrases per strophe, whereas others sing more than 12 phrases per strophe (e.g. Lambrechts and Dhondt 1987).

We cannot reject the hypothesis that the average singing performance of an individual gives reliable information about male quality on the basis of the results of these playback experiments. As Table II shows, the change in average singing performance of a male caused by playback was relatively small in comparison to the amount of inter-individual variation in singing performance within a population. However, we cannot be certain that the playback subjects were motivated to respond at their maximum ability. Therefore, other information is needed to determine the cause of inter-individual differences in great tit singing performance and to test the reliability of these acoustic signals.

Table II. The number of phrases per strophe sung by nine male great tits before and during playback. Values are means ± sd (sample size), t and P values are for two-tailed t-tests. See text for details of playback stimulus and Lambrechts and Dhondt 1987 for experimental design details.

Male	Before	During	t	P
A	7.17 ± 0.75 (6)	8.95 ± 1.52 (22)	2.76	< 0.02
B	6.41 ± 0.94 (17)	7.44 ± 1.82 (18)	2.09	< 0.05
C	5.60 ± 0.84 (10)	5.77 ± 0.90 (17)	0.49	n.s.
D	5.58 ± 1.66 (13)	5.25 ± 0.62 (12)	0.64	n.s.
E	5.00 ± 2.19 (11)	4.16 ± 0.97 (48)	1.99	≈ 0.05
F	4.82 ± 1.31 (14)	4.70 ± 1.92 (5)	0.16	n.s.
G	4.50 ± 0.58 (5)	4.50 ± 0.61 (17)	0.00	n.s.
H	4.25 ± 0.50 (4)	3.83 ± 0.75 (6)	0.96	n.s.
I	3.53 ± 0.96 (16)	3.12 ± 0.52 (39)	2.08	< 0.05

Individual Differences in Response Strength

Individual males may differ in the way they respond to a speaker broadcasting song. The strength of response to playback has been considered as a measure of "aggression" or a willingness to fight (e.g. Lewis 1986; McGregor 1988; McGregor and Horn 1991) and could be considered as a measure of male quality.

Observational studies suggest that the average singing performance of great tits could reflect vigour or fighting ability, because the average number of phrases per strophe and average drift within strophes are correlated with social dominance at a winter feeder (Lambrechts and Dhondt 1986). Therefore, we would expect to find a correlation between measures of average song performance and strength of response during playback.

In a sample of nine great tits, Lambrechts and Dhondt (1987, see Table II) found a relationship between the average number of phrases per strophe of a male before playback started and the change in response strength during playback. The two males with the highest average number of phrases per strophe significantly increased their average strophe length in response to the playback stimulus, whereas the male with the shortest average number of phrases per strophe significantly decreased its average strophe length during playback. In another great tit population, McGregor and Horn (1991) found positive correlations between the average number of phrases per strophe and eight measures of strength of response. Both these results support the hypothesis that average number of phrases per strophe gives information about a male's willingness to attack and therefore could reflect vigour or fighting ability.

Perception of Variation in Measures of Singing Performance

Determining whether males perceive variation in measures of singing performance, such as strophe duration, strophe rate and drift, can help to understand if, how and when quality advertisement based on song is important in natural populations. If, for instance, laboratory experiments showed that males were unable to perceive variation in such factors, we could conclude that these factors are not used to assess the quality of conspecifics. (But see important distinction between obtaining a lack of response and inferring inability to perceive which is discussed below and in the chapters by Ratcliffe and Weisman, Horn, and Weary in this volume). A number of playback experiments in the field and in the laboratory have looked at whether males can perceive variation in measures of singing performance.

Field Experiments

The difficulties of interpretation in studies of perception are fully discussed elsewhere in this volume (chapters by Ratcliffe and Weisman, Horn, and Weary), but briefly the problem hinges on equating a lack of a difference in response to stimuli with an inability to perceive the difference between stimuli. If territorial males respond differently to playback of a high quality male than to playback of a low quality male, we could conclude that the subjects have been able to perceive the differences between the two playback stimuli and that quality advertisement may have a functional meaning in territorial defence. However, if the males respond in a similar way to the two playback stimuli, we will never be sure whether the males were unable to perceive the differences between the playback stimuli, or whether they perceived the differences, but such differences were unimportant in the context in which the playback experiment was carried out.

McGregor and Horn (1991) found that territorial great tits do not respond significantly differently to long strophes produced at a high rate, short strophes produced at a high rate, and long strophes produced at a low rate. There are at least three explanations of these findings. First, the great tits did not perceive variation in strophe rate and strophe length, meaning that they do not use these song measures to assess the quality of conspecifics. Second, the subjects did perceive the differences in strophe length or strophe rate, but variation in such factors does not function in territorial defence signals. A third possibility is that males do use strophe length or strophe rate in male-male competition, but that the experimental protocol did not allow a test of this hypothesis, for example, male quality assessment could be important during territory establishment, but not in the later stages of the territorial period when the experiment was carried out.

Operant Procedures in Laboratory Conditions

Recently, operant procedures have been developed in which the strength of response to song characteristics is not influenced by the biological meaning of these song characteristics for the species (references in chapter by Weary in this volume).

In one type of operant procedure, a male would be trained to respond (e.g. by approaching a feeder) *only* if it hears a sound with very specific characteristics - the *training song*, which may be a single note or a single strophe. The next stage in the procedure is to present the male with a training song or a *test song* (which differs from the training song in the sound characteristics of interest) in a random order. If the male is able to distinguish between training and test songs then this will be shown by a difference in response (in this example, feeder approach) (e.g. Weary 1990, 1991, chapter in this volume).

Weary (1991) used operant procedures to show that great tits do not respond differently to long and short strophes, nor do they respond differently to strophes with large and small amounts of drift, when the training song and the test song is a single strophe. He concluded that great tits are unable to perceive the natural variation in drift and strophe duration when they are only played a single strophe.

Weary did not rule out the importance of redundancy in communication and suggested that great tits may be sensitive to variation in drift and/or in strophe duration when they hear a number of repetitions of a strophe, as they would under natural conditions (because they normally repeat a strophe a number of times in a rather stereotyped way).

Future operant procedures could examine the importance of repetition of strophes (redundancy) for the perception of variation in measures of singing performance.

Singing Performance, Male Quality and Territorial Defence

Information Exchange about Male Quality in Conflicts between Intruders and Territory Owners and Effects of Site-related Dominance

Animals can use vocalisations to signal their fighting ability when the chance of escalated conflict is high, thus avoiding unnecessary fights between individuals that differ greatly in quality (e.g. Davies and Halliday 1978; Clutton-Brock and Albon 1979). The advantage to low quality individuals is that they would avoid injuries, high quality individuals would avoid wasting time and energy in a fight they would have won anyway.

Lambrechts and Dhondt (1987) and McGregor (1988) pointed out two factors that complicate the role of number of phrases per strophe in advertising male quality. Many songbirds have a song repertoire of different song types and song type structure could effect the average number of phrases per strophe (Lambrechts and Dhondt 1987; Weary et al. 1990). Lambrechts and Dhondt (1987) therefore proposed that if two males sang the same song type (i.e. song type matching or matched countersinging, see also Krebs et al. 1978) it would be easier to assess the relative quality of opponents. This could explain why matched countersinging appears in contexts where the chance of escalated conflicts is high in great tits (Krebs et al. 1981) and why individuals sometimes match song types at close distance (< 5m, *pers. obs.*).

Lambrechts and Dhondt's hypothesis on the function of song type matching can generate predictions about the pattern of song duels. First, a male may be more willing to reply with the same song type if the strophe length of the opponent is shorter than its own, because the male with the longer strophes has a higher chance of winning the conflict. Furthermore, the duration of song type matching will be relatively short, because the male with longer strophes will chase the other male rather than sing at it. Second, a male may avoid song type matching with a male that produces very long strophes, because the male with shorter strophes has a higher chance of losing the contest. Third, the duration of song type matching is longest when opponents have approximately equal strophe lengths, because the males have more difficulty deciding who is stronger and the chance of a costly fight is higher. These predictions could be investigated with a playback design similar to that used by McGregor and Horn (1991); by playing long and short strophes of a song type in the subject's repertoire, the probability of song type matching could be related to the relative difference in strophe length between subject and playback.

Information exchange about male quality will be most important in situations where the outcome of a fight is largely influenced by male quality. However, field studies in the great tit suggest that prior-occupancy, rather than intrinsic male quality, may have a greater effect on the outcome of territory owner *v.* intruder conflicts (Krebs 1982;

Lambrechts and Dhondt 1986: p.58, Lambrechts and Dhondt 1988b: p.599) and the same may be true of other territorial songbirds. The most extreme case of site-related dominance would be that the territory owner wins all fights with other males, whatever the quality of the male involved. The relative importance of male quality and site-related dominance in male-male conflicts could be examined with playback.

One of the more popular playback techniques to study the function of song variation in territorial defence is to present a territory owner with a single speaker broadcasting song and to record the response strength of the territory owner before, during and after the playback period. The same male could be presented with song of a high quality male in one trial and song of a low quality male in another trial (cf. McGregor and Horn 1991). An alternative experimental design would be a choice experiment, in which the song of a high and a low quality male are played alternately from two loudspeakers that are some distance apart and in which a territory owner is forced to respond more strongly to one of the two speakers (cf. Searcy et al. 1981; Kroodsma 1986).

Table III. Summary of results of playback experiments in the literature in relation to the predictions of the male quality (**Quality**) and site-related dominance (**Site Dom.**) hypotheses (see text for details). PB = Playback, SL = Strophe length.

		Consistent with		
Effect	**Occurs?**	**Quality**	**Site Dom.**	**Reference**
Response occurs before 1st PB strophe ends	Yes	No	Yes	*unpubl* cited in Weary 1991
Increase in SL during long strophe PB	Yes	No	Yes	Lambrechts & Dhondt 1987 (individual level)
Decrease in SL during short strophe PB	Yes	Yes	No	Lambrechts & Dhondt 1987 (individual level)
Response $>$ to long than to short PB strophes	Yes?	No	Yes	McGregor & Horn 1991 (population level)
Response $>$ to short than to long PB strophes	No?	Yes	No	McGregor & Horn 1991 (population level)
Response $\approx$ to long and short PB strophes	Yes	No	Yes	McGregor & Horn 1991 (population level)

Table III summarises the results of playback experiments in the great tit that are available in the literature and indicates whether or not each result supports predictions of the *male quality* or *site-related dominance* hypothesis. The prediction of the male quality hypothesis have been discussed above. In contrast to the male quality hypothesis, the site-related dominance hypothesis predicts that males should respond in a similar way to long and short strophes because outcomes of contests are decided by site dominance, not male quality. However, if we assume that average strophe length gives information about the motivation of the singer (Weary et al. 1988) and that territory owners have greater difficulty chasing away highly motivated intruders (i.e males with relatively long strophes), then a stronger response to a playback stimulus with relatively longer strophes is predicted.

Most of the results of playback experiments are inconsistent with the above predictions of the male quality hypothesis. However some experiments support the modified prediction of the site-related dominance hypothesis (stronger response to long strophes). For example McGregor and Horn (1991) found a weak tendency to respond more strongly to long than to short strophes (see also Lambrechts and Dhondt 1987; Table II above). Confusingly, some experiments give support to the opposite interpretation; for example, Lambrechts and Dhondt (1987, see also Table II above) found that males with short strophes may decrease their strophe length in response to playback of long strophes. These results are consistent with the male quality hypothesis, showing an inhibitory effect of long strophes.

McGregor and Horn (1991) proposed that short-term changes in strophe length of the individual, rather than average strophe length, could be important in resolving conflicts between males. This is because short-term changes in strophe length could give information about fighting intentions; for example, males increase their own strophe length to show willingness to start a fight. This hypothesis would work in a situation in which site-related dominance is important *and* in which the territory owner wants to avoid energy-demanding conflicts with potential intruders.

To date, experimenters have done playback experiments on well-established territorial males with well-known territory boundaries in order to avoid loudspeaker location effects on strength of response (e.g. edge-centre effect; Dhondt 1966; Searcy et al. 1981; Giraldeau and Ydenberg 1987). However, to study the importance of male quality in male-male competition, playback experiments should also be carried out in situations where males are less familiar with a place or territory, for example, before or during the period when territorial borders are established, or when non-territorial males try to take over a territory just after the former territory owner has died or been removed (cf. Krebs 1982).

In some circumstances males may not change overt behaviour in response to song, but show physiological responses. For example, in laboratory conditions Diehl and Helb (1986) demonstrated that blackbirds (*Turdus merula*) increase their heart rate when presented with playback of conspecific song. Similar experiments could be carried out to investigate how heart rate changes in response to the singing performance of high and low quality males.

Song Performance, Male Quality and Intrusion Rate

Song is as a long distance signal (Wiley and Richards 1982); high quality males could reduce intrusion rates and avoid energy-demanding conflicts by long-range advertising of fighting ability. Potential intruders would avoid territories with very long strophes or a high PPT because they would have a higher chance of being attacked in these territories (see above).

To date, no playback experiments have been carried out to test more directly whether singing performance, as measured by averages in number of phrases per strophe, strophe duration, drift or PPT, increases the effectiveness of male repulsion. Speaker-replacement experiments (in which territorial males are replaced by loudspeakers) would be the obvious way to look at this; for example, by examining whether unfamiliar individuals intrude more often in territories with song playback of a low quality male than in territories with song playback of a high quality male. The speaker-replacement experiments that have been done have shown that re-occupation of territories is slowest when speakers broadcast large song repertoires (great tits, Krebs et al. 1978). In red-winged blackbirds (*Agelaius phoeniceus*), the intrusion rate of unfamiliar individuals is highest in territories broadcasting small song repertoires (Yasukawa 1981a). The conclusion of these experiments was that a large song repertoire may increase the success in territorial defence and one reason could be that song repertoire size reflects male quality (Yasukawa 1981a; Baker et al. 1986).

Male Quality and Female Choice

In many species of songbird characteristics of song are correlated with different measures of breeding success; for example, pairing date (Howard 1974; Catchpole 1980; Yasukawa et al. 1980; Jarvi 1983; Radesater et al. 1987; Eens et al. 1991) and reproductive success (McGregor et al. 1981, Cosens and Seely 1986). One explanations of such relationships could be that females are attracted to these song characteristics because they signal male quality.

The relationship between male song characteristics and female choice is complicated by a number of factors. First, relationships have been reported between song characteristics and measures of territory quality (reviews in Leonard and Picman 1988; Lambrechts and Dhondt 1988b), therefore a correlation between song features and breeding success may not be the result of a female preference for these song features, but because the female chooses on the basis of territory quality (e.g. Alatalo et al. 1986). Second, extra-pair copulations are a common feature of songbird reproductive behaviour including great tits (Hinde 1952: pp. 54, 91; Norris and Blakey 1989). Therefore, song characteristics may be related to extra-pair copulations rather than intra-pair copulations (e.g. Searcy 1984; Eens et al. 1991; Kroodsma and Byers 1991).

Playback experiments present the opportunity to investigate female preference for song characteristics without the influence of such confounding variables. The chapter by Searcy in this volume reviews the techniques available for such studies and the chapter by Catchpole (this volume) details a specific example of this approach.

Playback Experiments in the Field

Eriksson and Wallin (1986) found that unmated female pied and collared flycatchers (*Ficedula hypoleuca* and *F. albicollis*) arriving from migration were caught more frequently at sites with a male dummy and a speaker broadcasting conspecific song, than at sites with a *silent* male dummy. This is one of the few field experiments demonstrating that song attracts females. The same experimental design could also test whether female songbirds are attracted more to song of high quality males, than to song of low quality males.

Such a playback design is probably less useful in a resident species with high site-fidelity and high mate fidelity (such as the great tit), because both males and females can influence the result of the experiment. Females of territory owners will influence the intrusion rate of *strange* females if there is competition for males or territories. Males will sing in response to the speaker broadcasting song, therefore it will be difficult to tell whether female intruders were attracted to the speaker or to singing territorial male. A modification of the design would be to replace the territory owner by two speakers alternately broadcasting song of a high and a low quality male. Assuming that singing performance is used in female choice, we would predict that the *widowed* female would be more attracted to the singing performance of the high quality male. This experiment could be carried out in any species that defends territories, it would simulate those occasions when males disappeared during the territorial period and when the female has to choose between available males (e.g. in great tits, Hinde 1952: pp. 54, 91).

Oestradiol Experiments in the Laboratory

One technique regularly used to investigate female preferences for song features is to use *copulation solicitation* displays as an assay of preference; such displays are given by captive female songbirds that are treated with oestradiol and held on a long-day photoperiod (*oestradiol experiments* are reviewed in the chapter by Searcy in this volume).

In common with many other species (see Table I in the chapter by Searcy in this volume) captive female great tits that are treated with oestradiol show more copulation solicitation displays when presented with more song types and when played local, rather than distant, songs (e.g. Baker et al. 1986, 1987). Investigators have therefore suggested that female songbirds prefer larger song repertoires or song types of the local population (but see Searcy 1984; Chilton et al. 1990; Kroodsma 1990). Below, I report an experiment to investigate the *sexual response* (copulation solicitation displays) of female great tits to variation in song characteristics that reflect male quality.

Sexual response of female great tits to the singing performance of high and low quality males Both strophe length and strophe rate could influence the sexual response of female great tits, therefore a good experimental design should control for strophe rate when female response to strophe length is examined. However, when male great tits are highly motivated, strophe length is positively correlated with the % performance time within song bouts (PPT, see above) and high quality males with very long strophes and high PPT can produce strophes at a very low rate (of a few strophes per minute) (see above) (Lambrechts and Dhondt 1988a; *pers. obs.*).

Singing performance is a more accurate measure of male quality when the motivation of the singer is high (Lambrechts 1988) and assuming that females use measures of singing performance to assess the quality of potential mates (see above), I would predict a stronger sexual response to very long strophes than to very short strophes, even if the strophe rate is somewhat lower in the stimulus with very long strophes, as is the case in natural singing performances (*pers. obs.*). Furthermore, I would always expect to find a stronger response to the singing performance of a high quality male whatever song type that is used.

I tested these predictions with female great tits that were captured at the Antwerp University campus in September 1988. Females were treated with 17ß-oestradiol and held on a long-day photoperiod, a silent male was also present outside the experimental cage (cf. Baker et al. 1986; more details in Table IV caption).

Each female received the two stimuli of the same song type common to the local area; the song type used differed between the subjects. One stimulus represented a high quality male: strophe length was 12 phrases per strophe, lasting between 5.2s and 6.7s depending on the song type used; PPT was 60%; the strophe rate was 6 strophes per minute, this varied between 5 and 7 strophes per minute depending on the song type used. The low quality male stimulus had a strophe length of 6 phrases per strophe, lasting between 2.7s and 3.3s; PPT was 40%; the strophe rate was 8 strophes per minute, this varied between 7 and 9 depending on the song type used.

In contrast to the results of Baker et al. (1986; 1987), the subjects started to respond to the silent male before playback started. This was not unexpected because females in the field can display vigorously in the presence of a male that does not sing (Hinde 1952, p.90-91; Hailman et al. m/s; *pers. obs.*). I controlled for the effect of the male on the sexual response of the subjects in the following ways. First, females were tested in the same order on different days. Second, the same male was presented in all trials. Third, I considered the duration of response during playback. Fourth, the duration of response before playback started was subtracted from the duration of response during playback.

The preliminary results are presented in Table IV. In contrast to the prediction, females responded more strongly to the singing performance of the low quality male, but the result was not statistically significant ($P > 0.05$). There were no important differences between song types and there was no significant difference between responses on day 1 and day 2. If anything, the response was slightly higher on day 2, showing no evidence for habituation during the second trial.

Table IV. The duration of wing quivering display of 11 females to a male before song started and to the same male during playback of high quality and of low quality male song. Three different song types were used; A had 2 notes/phrase, B had 4 notes/phrase and C had 3 notes/phrase. Females were tested in the order 1-11; six females had high quality song on day 1 followed by low quality on day 2 (Order 1) and five had the reverse (Order 2). **Before** = period between entering cage and start of playback (3min), **During** = period during song playback (one 3min song bout followed by 2min silence, repeated 3 times, i.e. 9min of song, 6min of silence). Photoperiod December - March was 8L:16D and March 4th - April 23rd was 17L:7D. Females were habituated to the test room from February 1st onwards (cf. Baker et al. 1986) and implanted with 8-10mm of 1.95mm outside and 1.47mm inside diameter Silastic tubing on April 10th. During testing (April 22-23rd) a male was positioned 1.5m from the test cage and a cassette recorder used for playback was 2.2m from the cage. Sound pressure level at the test cage was 79dB (Bruel and Kjaer 2209), background noise was 52dB. Song phrases were recorded in the field and stored in a computer to produce strophes and bouts of song. The females were used in other experiments April 17-21.

Female	Song type	Order	High Quality Before	High Quality During	Low Quality Before	Low Quality During
1	B	1	69.9	4.7	62.7	8.4
2	B	1	0.0	0.0	0.0	3.3
3	B	2	12.5	0.0	32.9	1.1
4	A	1	27.7	17.6	34.8	65.9
5	B	2	9.3	0.0	11.6	0.0
6	C	2	4.0	12.9	7.8	119.8
7	C	1	0.0	0.0	0.0	0.0
8	B	2	0.0	22.5	0.0	10.9
9	C	2	0.0	0.0	0.0	0.0
10	B	1	74.4	66.3	33.4	133.4
11	A	1	82.5	97.2	71.4	108.0

Even assuming that the female display given in oestradiol experiments is a *copulation solicitation* display (but see below), these preliminary results suggest that there is no simple relationship between male quality as indicated by song and the sexual response of female great tits.

How should oestradiol experiments in the great tit be interpreted? When captive female great tits implanted with oestradiol are presented with song they vibrate their wings rapidly in combination with the production of a soft, high-pitched vocalisation. This *wing quivering* display has been described as a copulation solicitation display (detailed description in Baker et al. 1986; see also Hinde 1952: p. 90). However, a recent review of the literature shows that in the great tit or its relatives, the wing quivering display has been observed in more than 15 different contexts and that in most contexts the display has been observed after pair formation and after the fertile period of the female. The display has been described as a solicitation display, an appeasement display or a mutual display to strengthen the pair-bond between mates (Hailman et al. m/s; *pers. obs.*). This suggests that the wing quivering display described in the laboratory conditions could be more than a copulation solicitation display, making interpretation of oestradiol experiments in the great tit and its relatives more difficult, as the arguments below illustrate.

Table IV shows that females tend to respond more strongly to short strophes produced at a somewhat higher rate than to long strophes produced at a lower rate. This result could indicate that strophe rate rather than strophe duration influences female choice. The argument that strophe rate gives a reliable indication of physical condition has appeared in the literature quite commonly (Gottlander 1987; Radesater et al. 1987; Reid 1987). However, an alternative explanation is that strophe length reflects a willingness to attack (see above) and that female great tits are more sexually stimulated by less aggressive signals (i.e. shorter strophes produced with a lower PPT). This hypothesis is supported by the observation that before copulation male great tits do not approach the female while singing vigorously, rather they may wing quiver in response to the female and utter soft vocalisations (Hinde 1952: p.90; *pers. obs.*; Hailman et al. m/s). In other words, I am suggesting that the vocal characteristics used to attract a female (loud, aggressive song which may indicate male quality) are different from those used to sexually stimulate a female (soft, non-aggressive sounds that may not indicate male quality). Clearly a better understanding of the biological meaning of the wing quiver display is needed to aid the interpretation of oestradiol experiments in the great tit.

Concluding Remarks

Although I have concentrated almost exclusively on singing performance in the great tit, some similar playback experiments have been carried out or developed in other species. Furthermore, the problems of interpretation in the great tit are likely to be equally important in other species of songbird. One important conclusion is that detailed descriptive studies in the field are essential to determine if, when and how a playback experiment should be carried out. For instance, field studies in the great tit showed that prior-occupancy rather than male quality may determine the outcome of a fight between a territory owner and an intruder. A better understanding of the phenomenon of site-related dominance could help to determine if and when male quality could be important in male-male competition and details of playback experiment execution (centre or edge of territory? early or late in the season? etc.)

As pointed out in the discussion of oestradiol experiments, a correct interpretation of playback experiments in laboratory conditions is only possible if we know how birds behave in natural situations. Field experience with the species involved also helps in the design of a more realistic playback experiment. A nice example is the study of female choice in European starlings (*Sturnus vulgaris*). In natural situations male starlings start to sing inside or near the nest hole when an unmated female approaches (Eens et al. 1990). Mountjoy and Lemon (1991) showed that females are more attracted to nestboxes with a speaker broadcasting song than to boxes without song. Since male starlings with larger song repertoires and longer songs (song bouts) have more success in female attraction (Eens et al. 1991), the same experimental design could test if females are more attracted to boxes broadcasting song of a high quality male.

Finally, a correct interpretation of the results of playback experiments is made more difficult by a lack of knowledge about causes of inter-individual differences in song characteristics that reflect male quality. Song repertoires provide a good example of this. Song repertoire size could be influenced by brain morphology for instance, limiting the number of song types a male can learn (e.g. Nottebohm et al. 1981). Large song repertoires may increase success in territorial defence or in female attraction for reasons other than male quality (e.g. the anti-habituation hypothesis or Beau Geste hypothesis, Krebs 1976, Krebs 1977). However, some individuals may have acquired larger song repertoires because they have been involved in more conflicts which give more learning opportunities. If this is the case song repertoire size could give reliable information about vigour or fighting ability and such an advertisement could be used in male-male or male-female assessment (Baker et al. 1986; Baker 1988). Unfortunately, little is known about the causes of variation in song features that reflect male quality. Further experimental research could establish whether song characteristics that reflect male quality, such as singing performance, are influenced by morphological or physiological limitations (e.g. oxygen consumption, testosterone levels, muscle glycogen reserves), by the genetic quality of the male, or by environmental factors during stages of song development or song production.

Acknowledgments

I am grateful to Torben Dabelsteen, Bruce Falls, Don Kroodsma, Peter McGregor and Bill Searcy for discussions during the ARW conference; Jack Hailman for comments on an early version of the manuscript; and Andre Dhondt and Marcel Eens for discussions prior to the oestradiol experiment. I am grateful to the following people who helped with the oestradiol experiment: Marcel Eens helped with the experimental set up; Jenny De Laet implanted oestradiol; Karel Verbeeck built the test cage; Fonny De Wulf helped to make the playback tapes; Andre Dhondt (FKFO project no. 2.0044,80; Krediet aan Navorsers) and R. F. Verheyen provided materials. Great tits were held in captivity under a licence of the Ministerie van de Vlaamse Gemeenschap. At Antwerp University I was financially supported by the Belgian National Fund for Scientific Research (Senior Research Assistant). I wrote this paper while I was a Guyer Postdoctoral Research Fellow at the University of Wisconsin-Madison (1990-1991).

References

Alatalo, R.Y., Lundberg,A. and Glynn, C. 1986. Female pied flycatchers choose territory quality and not male characteristics. *Nature*, **323**, 152-153.

Baker, M.C. 1988. Sexual selection and size of repertoire in songbirds. *Acta XIX Cong.Int. Orn.*, 1358-1365.

Baker, M.C., Bjerke, T.K., Lampe, H. and Espmark, Y. 1986. Sexual response of female great tits to variation in sizes of males' song repertoires. *Am. Nat.*, **128**, 491-498.

Baker, M.C., McGregor, P.K. and Krebs, J.R. 1987. Sexual response of female great tits to local and distant songs. *Ornis. Scand.*, **18**, 186-188.

Becker, P.H. 1982. The coding of species-specific characteristics in bird sounds. In: *Acoustic Communication in Birds*. (Ed. by D.E. Kroodsma, E.H. Miller & H. Ouellet), pp. 213-252. Academic Press; New York.

Bijnens, L. 1988. Blue tit *Parus caerulus* song in relation to survival, reproduction and biometry. *Bird Study*, **35**, 61-67.

Catchpole, C.K. 1980. Sexual selection and the evolution of complex songs among European warblers of the genus *Acrocephalus*. *Behaviour*, **74**, 149-166.

Chilton, G., Lein, M.R. and Baptista, L.F. 1990. Mate choice by female white-crowned sparrows in a mixed-dialect population. *Behav. Ecol. Sociobiol.*, **27**, 223-227.

Clutton-Brock, T.H. and Albon, S.D. 1979. The roaring of red deer and the evolution of honest advertisement. *Behaviour*, **69**, 145-169.

Cosens, S.E. and Seely, S.G. 1986. Age-related variation in song repertoire size and repertoire sharing of yellow warblers *Dendroica petechia*. *Can. J. Zool.*, **64**, 1926-1929.

Cuthill, I.C. and MacDonald, W.A. 1990. Experimental manipulation of the dawn and dusk chorus in the blackbird (*Turdus merula*). *Behav. Ecol. Sociobiol.*, **26**, 209-216.

Davies, N.B. and Halliday, T.R. 1978. Deep croaks and fighting assessment in toads, *Bufo bufo*. *Nature*, **274**, 683-685.

Davies, N.B. and Lundberg, A. 1984. Food distribution and a variable mating system in the dunnock *Prunella modularis*. *Ibis*, **127**, 100-110.

Dhondt, A.A. 1966. A method to establish the boundaries of bird territories. *Gerfault*, **56**, 404-408.

Diehl, P., and Helb, H.-W. 1986. Radiotelemetric monitoring of heart-rate responses to song playback in blackbirds (*Turdus merula*). *Behav. Ecol. Sociobiol.*, **18**, 213-219.

Eens, M., Pinxten, R. and Verheyen, R.F. 1990. On the function of singing and wing waving in the European starling *Sturnus vulgaris*. *Bird Study*, **37**, 48-52.

Eens, M., Pinxten, R. and Verheyen, R.F. 1991. Male song as a cue for mate choice in the European starling. *Behaviour*, **116**, 210-238.

Eriksson, D. and Wallin, L. 1986. Male bird song attracts females - a field experiment. *Behav. Ecol. Sociobiol.*, **19**, 297-299.

Falls, J.B. 1982. Individual recognition by sound in birds. In: *Evolution and Ecology of Acoustic Communication in Birds. Vol.II.* (Ed. by D.E. Kroodsma, E.H. Miller & H. Ouellet), pp. 237-273. Academic Press, New York.

Giraldeau, L.A. and Ydenberg, R.C. 1987. The centre-edge effect: the result of a war of attrition between territorial residents? *Auk*, **104**, 535-538.

Gottlander, K. 1987. Variation in the song rate of the male pied flycatcher *Ficedula hypoleuca*: causes and consequences. *Anim. Behav.*, **35**, 1037-1043.

Greig-Smith, P.W. 1982. Song-rates and parental care by individual male stonechats (*Saxicolla torquata*). *Anim. Behav.*, **30**, 245-252.

Hailman, J.P., Lambrechts, M.M., Clemmons, J.R., Hafthorn, S., Hailman, E.D., Kempernaers, B. and Rost, R. *m/s*. A display of many contexts: wing quivering in tits (*Parus*).

Hiebert, S.M., Stoddard, P.K. and Arcese, P. 1989. Repertoire size, territory acquisition and reproductive success in the song sparrow. *Anim. Behav.*, **37**, 266-273.

Hinde, R.A. 1952. The behaviour of the great tit (*Parus major*) and some other related species. *Behav. Suppl.*, **2**, 1-201.

Howard, R.D. 1974. The influence of sexual selection and interspecific competition on mockingbird song (*Mimus ployglottos*). *Evolution*, **28**, 428-438.

Jarvi, T. 1983. The evolution of song versatility in the willow warbler *Phylloscopus trochilus*: a case of evolution by intersexual selection explained by the "female choice of best mate". *Ornis Scand.*, **14**, 123-128.

Krebs, J.R. 1976. Habituation and song repertoires in the great tit. *Behav. Ecol. Sociobiol.*, **1**, 215-227.

Krebs, J.R. 1977. The significance of song repertoires, the Beau Geste hypothesis. *Anim. Behav.*, **25**, 475-478.

Krebs, J.R. 1982. Territorial defence in the great tit: do residents always win? *Behav. Ecol. Sociobiol.*, **11**, 185-194.

Krebs, J.R., Ashcroft, R., and van Orsdol, K. 1981. Song matching in the great tit (*Parus major* L.). *Anim. Behav.*, **29**, 918-923.

Krebs, J.R., Ashcroft, R. and Webber, M.I. 1978. Song repertoires and territory defence in the great tit. *Nature*, **271**, 539-542.

Kroodsma, D.E. 1986. Design of song playback experiments. *Auk*, **103**, 640-642.

Kroodsma, D.E. 1990. Using appropriate experimental designs for intended hypotheses in 'song' playbacks, with examples for testing effects of song repertoire size. *Anim. Behav.*, **40**, 1138-1150.

Kroodsma, D.E. and Byers, B.E. 1991. The function(s) of bird song. *Amer. Zool.*, **31**, 318-328.

Kroodsma, D.E. and Miller, E.H. (Eds.). 1982. *Acoustic Communication in Birds. Vols I & II.* Academic Press; New York.

Lambrechts, M.M. 1988. Great tit song output is determined both by motivation and by constraints in singing ability: a reply to Weary et al. *Anim. Behav.*, **36**, 1244-1246.

Lambrechts, M.M. and Dhondt, A.A. 1986. Male quality, reproduction, and survival in the great tit (*Parus major*). *Behav. Ecol. Sociobiol.*, **19**, 57-63.

Lambrechts, M.M. and Dhondt, A.A. 1987. Differences in singing performance between male great tits. *Ardea*, **75**, 43-52.

Lambrechts, M.M. and Dhondt, A.A. 1988a. The anti-exhaustion hypothesis: a new hypothesis to explain song performance and song switching in the great tit. *Anim. Behav.*, **36**, 327-334.

Lambrechts, M.M. and Dhondt, A.A. 1988b. Male quality and territory quality in the great tit *Parus major*. *Anim. Behav.*, **36**, 596-601.

Leonard, M.L. and Picman, J. 1988. Mate choice by marsh wrens: the influence of male and territory quality. *Anim. Behav.*, **36**, 517-528.

Lewis, R.A. 1986. Aggressiveness, incidence of singing, and territory quality of male blue grouse. *Can. J. Zool.*, **64**, 1426-1429.

McGregor, P.K. 1988. Song length and "male quality" in the chiffchaff. *Anim. Behav.*, **36**, 606-608.

McGregor, P.K. and Horn, A.G. 1991. Strophe length and response to playback in great tits. *Anim. Behav.*, *in press*.

McGregor, P.K. and Krebs, J.R. 1982. Song types in a population of great tits (*Parus major*): their distribution, abundance, and acquisition by individuals. *Behaviour*, **79**, 126-152.

McGregor, P.K., Krebs, J.R. and Perrins, C.M. 1981. Song repertoires and lifetime reproductive success in the great tit (*Parus major*). *Amer. Nat.*, **118**, 149-159.

Moller, A.P. 1991. Parasite load reduces song output in a passerine bird. *Anim. Behav.* , **41**, 723-730.

Morton, E.S. 1982. Grading, discreteness, redundancy, and motivation-structural rules. In: *Evolution and Ecology of Acoustic Communication in Birds. Vol.I.* (Ed. by D.E. Kroodsma, E.H. Miller & H. Ouellet), pp. 183-212. Academic Press; New York.

Mountjoy, D.J. and Lemon, R.E. 1991. Song as an attractant for male and female European starlings and the influence of song complexity on their response. *Behav. Ecol. Sociobiol.*, **28**, 97-100.

Norris, K.J. and Blakey, J.K. 1989. Evidence for cuckoldry in the great tit *Parus major*. *Ibis*, **131**, 436-441.

Nottebohm, F.S., Kasparian, S. and Pandazis, C. 1981. Brain space for a learned task. *Brain Res.*, **213**, 99-109.

Radesater, T., Jakobsson, S., Andbjer, N., Bylin, A. and Nystrom, K. 1987. Song rate and pair formation in the willow warbler, *Phylloscopus trochilus*. *Anim. Behav.*, **35**, 1645-1651.

Reid, M.L. 1987. Costliness and reliability in the singing vigour of Ipswich sparrows. *Anim. Behav.*, **35**, 1735-1743.

Rothstein, S.I. and Fleischer, R.C. 1987. Vocal dialects and their possible relation to honest signalling in the brown-headed cowbird. *Condor*, **89**, 1-23.

Searcy, W.A. 1979. Sexual selection and body size in male red-winged blackbirds. *Evolution*, **33**, 649-661.

Searcy, W.A. 1984. Song repertoire size and female preferences in song sparrows. *Behav. Ecol. Sociobiol.*, **14**, 281-286.

Searcy, W.A. and Andersson, M. 1986. Sexual selection and the evolution of song. *Ann. Rev. Ecol. Syst.*, **17**, 507-533.

Searcy, W.A., McArthur, P.D., Peters, S.S. and Marler, P. 1981. Response of male song and swamp sparrows to neighbor, stranger and self songs. *Behaviour*, **77**, 152-163.

Searcy, W.A. and Yasukawa, K. 1983. Sexual selection and red-winged blackbirds. *Amer. Sci.*, **71**, 166-174.

Weary, D.M. 1990. Categorization of song notes in great tits: which acoustic features are used and why? *Anim. Behav.*, **39**, 450-457.

Weary, D.M. 1991. How great tits use song-note and whole-song features to categorize their songs. *Auk*, **108**,187-189.

Weary, D.M., Krebs J.R., Eddyshaw, R., McGregor, P.K. and Horn, A. 1988. Decline in song output by great tits: exhaustion or motivation? *Anim. Behav.*, **36**, 1242-1244.

Weary, D.M., Lambrechts, M.M. and Krebs, J.R. 1991. Does singing exhaust male great tits? *Anim. Behav.*, **41**, 540-542.

Weary, D.M., Norris, K.J. and Falls, J.B. 1990. Song features birds use to identify individuals. *Auk*, **107**, 623-625.

Wells, K.D. and Taigen, T.L. 1986. The effect of social interaction on calling energetics in the gray treefrog *Hyla versicolor. Behav. Ecol. Sociobiol.*, **19**, 9-18.

West, M.J., King, A.P. and Eastzer, D.H. 1981. Validating the female bioassay of cowbird song. Relating differences in song potency to mating success. *Anim. Behav.*, **29**, 490-501.

Wiley, R.H. and Richards, D.G. 1982. Adaptations for acoustic communication in birds: Sound transmission and signal detection. In: *Evolution and Ecology of Acoustic Communication in Birds. Vol.I.* (Ed. by D.E. Kroodsma, E.H. Miller & H. Ouellet), pp. 131-181. Academic Press, New York.

Yasukawa, K. 1981a. Song repertoires in the red-winged blackbird (*Agelaius phoeniceus*): a test of the Beau Geste hypothesis. *Anim. Behav.*, **29**, 114-125.

Yasukawa, K. 1981b. Male quality and female choice of mate in the red-winged blackbird (*Agelaius phoeniceus*). *Ecology*, **62**, 922-929.

Yasukawa, K., Blank, J.L. and Patterson, C.B. 1980. Song repertoires and sexual selection in the red-winged blackbird. *Behav. Ecol. Sociobiol.*, **7**, 233-238.

MECHANISMS AND FUNCTION OF CALL-TIMING IN MALE-MALE INTERACTIONS IN FROGS

Georg M. Klump [1] and H. Carl Gerhardt [2]

[1] Institut für Zoologie
Technische Universität München
Lichtenbergstrasse 4
W-8046 Garching
GERMANY

[2] Division of Biological Sciences
University of Missouri
Columbia, Missouri 65211
U.S.A.

Introduction

In many species of frogs females choose between males on the basis of their advertisement calling (e.g. see reviews by Wells 1977, 1988). Acoustic signals are also important in the context of territorial male-male interactions (e.g. see reviews by Wells 1977, 1988). For maximising the number of intended receivers, calling males have evolved signals that have a very high sound pressure level (SPL) in the range of 90 to 120dB SPL (measured at 0.5m distance; see Loftus-Hills and Littlejohn 1971; Gerhardt 1975; Passmore 1981; Narins and Hurley 1982). In all species of anurans studied, acoustic advertisement was found to be the most energetically expensive behaviour observed (e.g. Taigen and Wells 1984; Prestwich et al. 1989). Thus, strategies that improve a frog's ability to transmit acoustic signals more efficiently are likely to be selectively advantageous.

Because frogs seldom call in isolation, but typically assemble in more or less dense choruses, the signals of the males surrounding an individual caller are one of the most important factors in limiting the range over which an individual's calls can be perceived and analysed by the intended receiver (see Gerhardt and Klump 1988a). Thus, calling frogs have evolved patterns of acoustic interaction which reduce the jamming of a caller's signal by the other calling males around it (e.g. Narins and Zelick 1988).

The timing of signal production by a male frog with respect to the calls of nearby males is the most important mechanism to avoid the jamming of its signal. The main aim of this paper is to provide methods to demonstrate and quantify interaction between males in the timing of their calls and thereby to infer the mechanisms involved. Furthermore, we will review the evidence on how female choice is influenced by the patterns of call

Playback and Studies of Animal Communication
Edited by P.K. McGregor, Plenum Press, New York, 1992

timing in male-male interactions. Here, we will concentrate on the timing of individual calls. The timing of call bouts has been demonstrated to effect female choice (Whitney and Krebs 1975), but little is known about the mechanisms involved in calling in bouts by frogs (e.g. see Schwartz 1991).

Anecdotal evidence for non-random timing of calls by male frogs has long been known. For example, duos, trios and even larger groups of interacting frogs have been described by a number of researchers (e.g. see Goin 1949; Hardy 1959; Duellman 1967; Foster 1967; Schneider 1967; Lemon 1971; Rosen and Lemon 1974; Schneider et al. 1988; reviews of the genera and species in which "duetting" is known were given by Wickler and Seibt 1974, and Wells 1977). In frogs, non-random calling usually enables males to avoid or minimise overlap with other callers. In some species this leads to a relatively regular alternating in the production of calls by the two individuals (e.g. in *Hyla arborea*, see below). In only a few frog species observations suggest non-random overlapping of calls. This pattern of call timing may lead to a reduced risk of predation (e.g. in *Smilisca sila*, Tuttle and Ryan 1982) or increase the distance over which the chorus sounds are transmitted (e.g. in *Bombina bombina*, Lörcher 1969, or in *Bufo americanus*, see Wells 1977).

Measuring Acoustic Male-Male Interactions

Two methods have been applied to the study of acoustic male-male interactions. First, natural interactions have been recorded for a time period during which two or more recorded individuals produced a large number of calls. Second, playbacks using tape recorders, special analogue call-synthesisers, or digital techniques, have been directed toward a calling male in the field. Moreover, in some of these experiments the frog's own calls were used to trigger the output of the playback-device (e.g. Narins 1982; Schwartz 1987b, 1991). The latter technique not only simulates natural interactions but also allows more freedom in experimental manipulation. For example, the playback device can be programmed to output signals that overlap the calls of the target male, or occur with some delay afterwards. Moreover, the parameters of the acoustic stimuli and the timing of the playback relative to the target male's calls can be varied beyond the range shown by conspecifics in natural interactions. The stimuli that have been used in playback experiments have ranged from analogue recordings of natural exemplars of the advertisement or aggressive calls (e.g. Wells and Schwartz 1984) to synthetic signals that were simple (e.g. Narins and Capranica 1978) or complex (e.g. Schwartz 1991) in acoustic structure.

There are three main goals of measuring patterns of acoustic interaction. The first goal is to establish whether the two frogs actually interact in a natural setting, and, if so, over what distances or range of SPL. The second goal is to learn about the mechanisms involved in the acoustic interactions, i.e. discovering how a frog's call is timed relative to its own last call and the occurrence of the call of its interacting neighbour. The third goal is to understand how a male's timing of calls may influence its reproductive success.

Quantitative Analysis of Natural Interactions

The primary goal in the analysis of natural interactions has been to demonstrate that the individual frogs adjust the timing of their own calls in relation to those of neighbouring frogs, i.e. testing whether their timing is non-random relative to the timing of the neighbour's call. With the exception of the study of Brush and Narins in *Eleutherodactylus coqui* (1989), quantitative analyses of the interaction patterns of neighbouring frogs have been limited to only two individuals at a time (*Kassina senegalensis*, Wickler and Seibt 1974; *Bufo regularis*, Wickler and Seibt 1974, *Hyla cinerea*, Klump and Gerhardt *unpublished*; *H. arborea*, Klump *unpublished*).

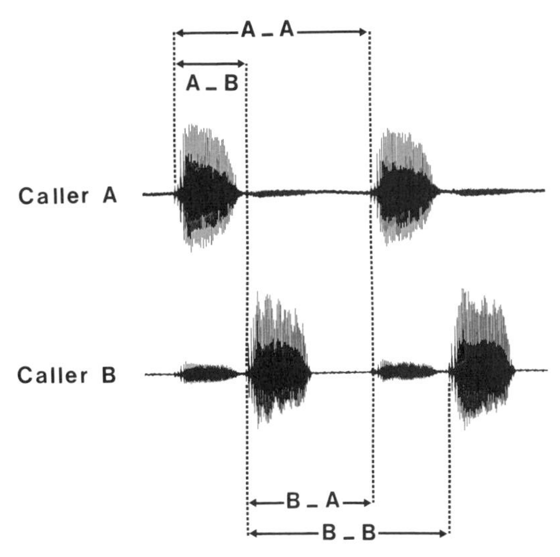

Figure 1. Oscilloscope trace of a recording of two individual *Hyla cinerea* males: Caller A leads the calling, the calls of B are lagging. Intervals A_A and B_B are the individual's inter-call intervals, intervals A_B and B_A are the between-individual call latencies.

Non-random vocal responses between individuals can be characterised by a number of different measures. A straightforward way is to measure the time delay between an individuals own calls and the preceding calls of the neighbour (intervals A_B and B_A in Fig. 1). In Figure 2 we show distributions of call latencies of six males of *H. cinerea* males with reference of the calls of their nearest neighbour. The patterns are all very similar. A few calls, which are about 150-180ms in duration, may be given during the first 50ms after the onset of the neighbour's call. These may be calls that have been initiated and were not suppressed by the perceived signal. In the time interval up to the end of the neighbour's call, calling is nearly completely suppressed. With the exception of the pattern shown in Figure 2d, the suppression period is followed by a short time period during which many more calls are given than expected from a random, uniform distribution of the call latencies relative to the calls of a neighbour. In the male whose pattern is summarised in Figure 2d, most calls cluster around a time period representing a much longer latency than in the other individuals shown. In this male, however, there is also a small number of calls elicited at the end of its neighbour's calls, but the period of suppression extends much longer than in the other subjects shown.

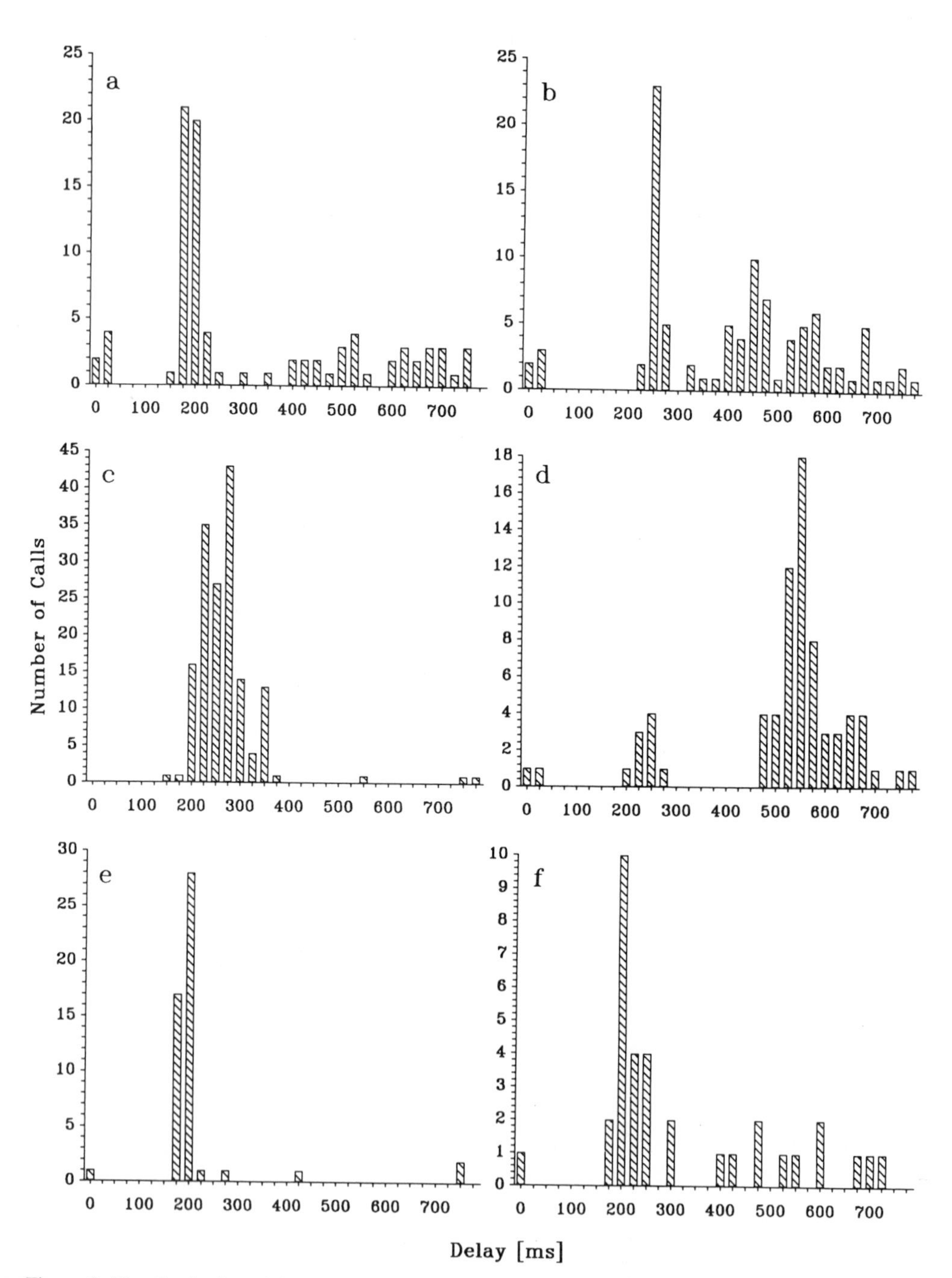

Figure 2. The distribution of time intervals between the beginning of a neighbour's call and the beginning of a frog's own call in six different males of *Hyla cinerea* (a to f).

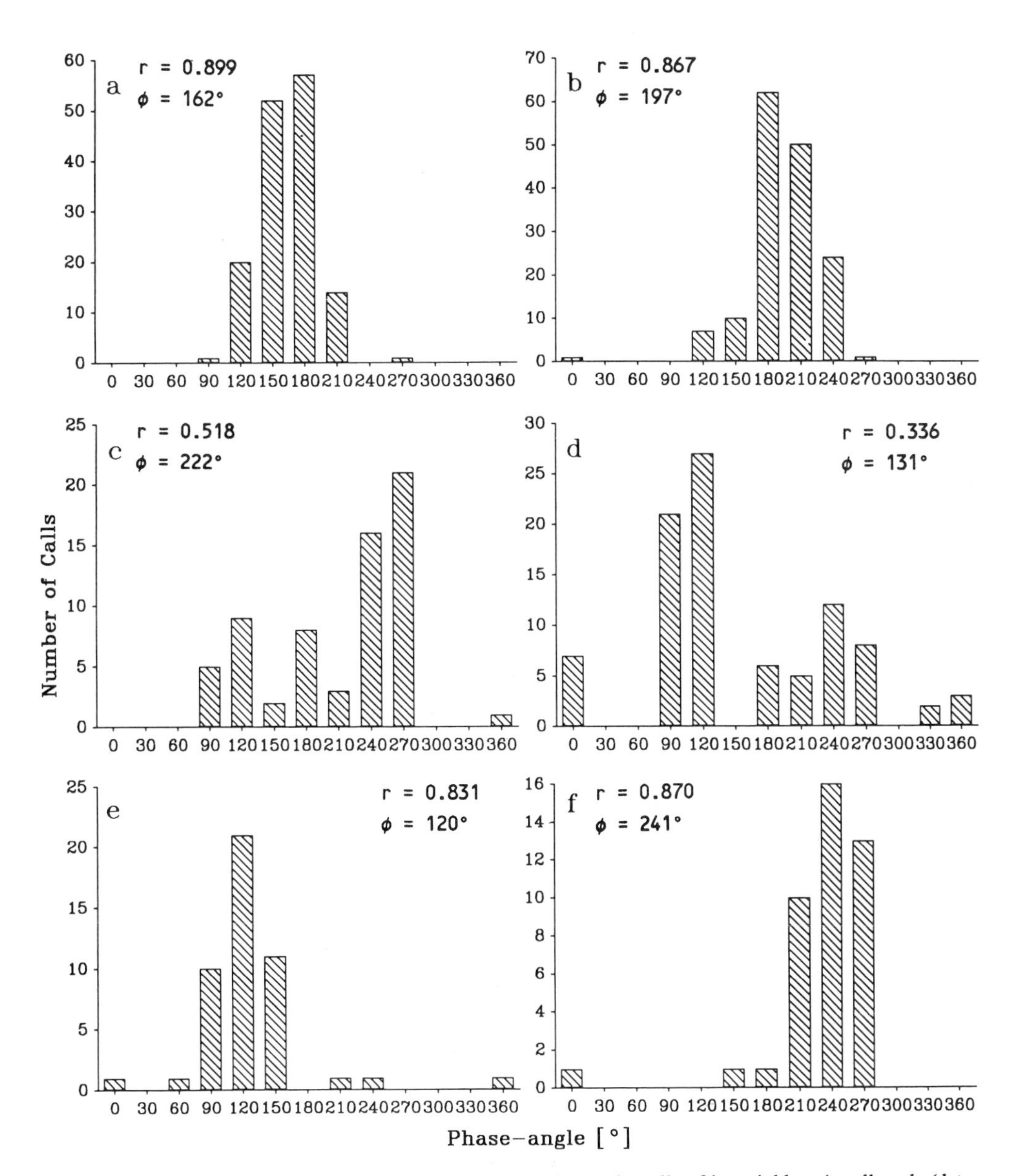

Figure 3. The relative phase of an individual's call in relation to the calls of its neighbour's call cycle (data from the six males of *Hyla cinerea*; males in Figures a and b, c and d, e and f were calling next to each other, respectively).

Because the distribution of inter-call intervals in spontaneously calling single frogs in different species showed that rhythmic calling may be based on an internal oscillator (e.g. see Loftus-Hills 1974; Lemon and Struger 1980; Zelick and Narins 1985), the next step is to demonstrate that a constant time delay between the neighbour's call and the frog's own call is not the result of a random but constant phase difference between the two uncoupled periodic-call oscillators. Wickler and Seibt (1974) have demonstrated this within a single recording by making the following comparison. First, they determined the distribution of delays between the signals from simultaneous records of the timing of the two individual frogs' calls (frogs 1 and 2). Then, for frog 1 they used a second time record of its calls that was not simultaneous to the first record of frog 2, but was obtained a few minutes later (i.e. the frogs do not directly interact in these two time records). In their second analysis of the distribution of delays, they treated this non-simultaneous time record as if it were obtained simultaneously with that of frog 2. They compared this new distribution with the one obtained from the simultaneous time record. If this analysis resulted in a narrow distribution of delays, this would indicate that the observed distribution in the simultaneous time record may have resulted by chance. This is because the two oscillators governing the timing of calls in the two individuals were so similar in their frequency and constant in their relative phase that no direct interaction is needed to give a constant delay between the calls. Usually, the analysis of the two non-simultaneous time records results in a uniform distribution of call latencies, which suggests that the observed distribution for the simultaneous time records is due to a real interaction. There are two other reasons why the results in Figure 2 support the hypothesis of male-male interaction. First, most of the subjects show the same behaviour with reference to their neighbour's call, which is not to be expected if their calling is independent. Second, the variation in the call period of individual males even over a short time is often large so that the two calling rhythms would drift out of phase if they were uncoupled (i.e. the underlying oscillator is "noisy", see Zelick and Narins 1985). Finally, if the response latency in different individuals is random but constant with respect to the neighbour's calls, then we also should expect to find different delays in different individuals and a uniform distribution of delays in a sample of recordings of a series of individuals. Bimodal distributions of response latencies can occur if different callers trigger their delay period on either the onset or the offset of the neighbour's call (e.g. see Loftus-Hills 1974), or if one caller leads the other and the lead changes (for the discussion of leading, see below).

Rather than measuring the time delay between the stimulus and the caller's response, one can also characterise the timing of one individual's calls relative to a neighbour's calls by plotting the relative phase of the individual's calls with respect to the neighbour's call period. A phase angle of 180° would indicate that both individuals alternate their calls, a phase angle of 0° would indicate that both call simultaneously, and intermediate phase angles (180° to 0° or 180° to 360°) would indicate that one individual leads the call cycle and the other follows. The strength of the coupling between the timing of two individuals can be expressed by the length of the mean vector in the phase diagram (vector strength r) which also can be tested for significance using a Rayleigh-test (see Batschelet 1981). The mean phase angle (ϕ) characterises the amount of lead or lag in the call cycle (for examples see Fig. 3). A frequency distribution of phase angles in the population may reveal the overall pattern that is characteristic of a species. In *H. arborea*, for example, the distribution of phase angles clusters around 180°, and the distribution shown in Figure 4 can well be approximated by a normal distribution with a mean of 180°.

The relative-phase plot does not reveal, however, whether both callers adjust their call cycle in a natural interaction, or whether only one caller adjusts its call cycle to the other individual's call period. Wickler and Seibt (1974), in their study of call-interactions in *K. senegalensis* and *B. regularis*, correlated the delay of one individual's calls (A) with the corresponding call period of the other caller (B). That is, they analysed all call periods between the two calls of B during which an intermittent call was given by A. If caller B uses the calls of A as a trigger for its own calls, i.e. its calls are driven by A, then the call period of B will increase if the call delay of A is increased, because B "waits" for the calling of A to occur.

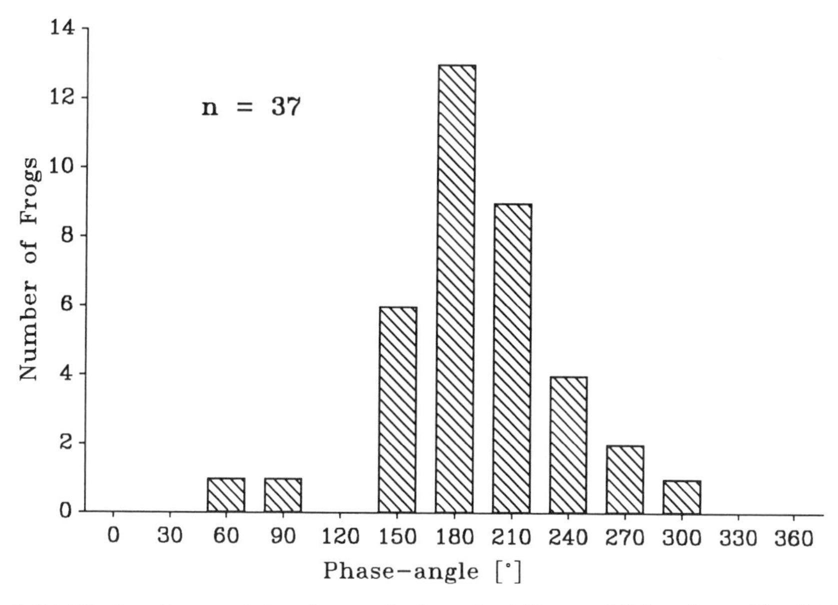

Figure 4. Distribution of mean relative phase angles in next neighbours of *Hyla arborea*. The distribution may be well approximated by a normal distribution (Kolmogorov-Smirnov test, $p > 0.6$).

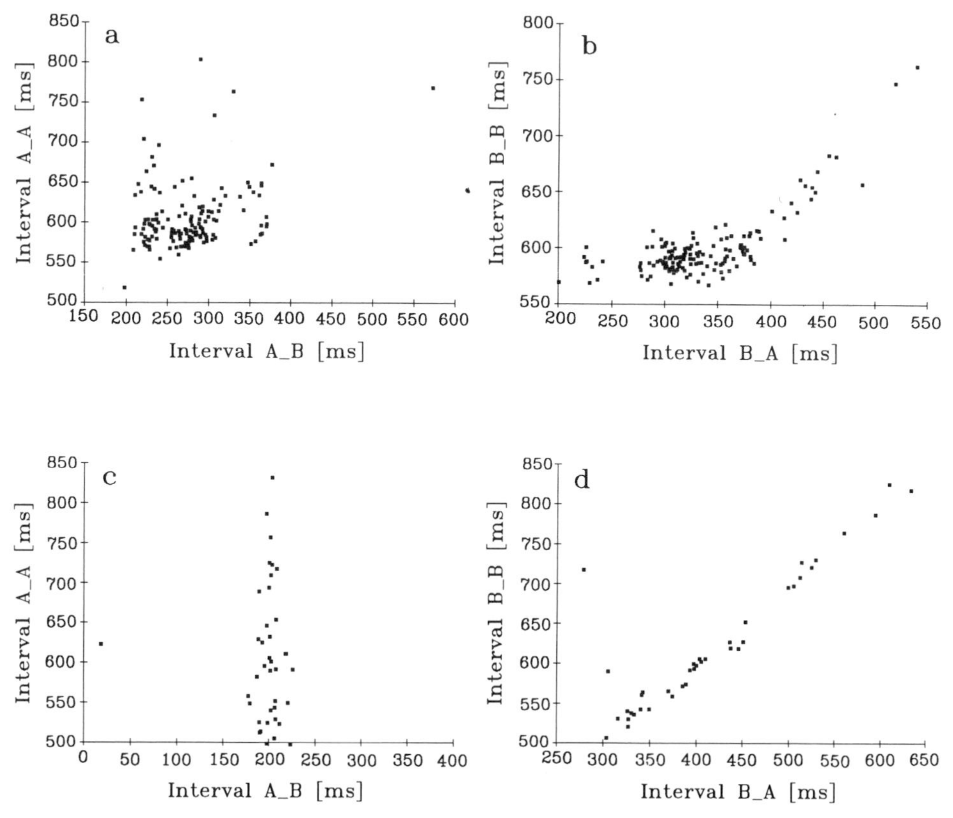

Figure 5. Duration of the call period of male *Hyla cinerea* in relation to the time interval between the individual's own call and the call of the neighbour (males in Figures a and b, and in Figures c and d called next to each other).

In Figure 5 we show examples from interacting males of *H. cinerea*. The scatter diagrams show the call period of one individual plotted in relation to the delay of calls of the other individual. The analysis can be performed in reciprocal pairs such that the influence of A on B and vice versa can be analysed. The left and right graphs in this figure are from neighbouring *H. cinerea* in a chorus. The two animals, whose patterns are shown in the top graphs, are only weakly correlated in the timing of their calls. In the bottom graphs, notice that only individual B adjusts its call timing to that of A (Fig. 5d); individual A varies its call period independent of B, apparently ignoring the calls of B (Fig. 5c). The strength of the coupling can be estimated by the magnitude of the correlation coefficient between the duration of the call period of one frog relative to the response latency of the other (r is 0.268 *, 0.763**, 0.013, and 0.964** for Figs. 5a, 5b, 5c, and 5d, respectively; * $p_2 < 0.01$, ** $p_2 < 0.001$). However, in *H. cinerea* at least an individual caller may change its strategy over relatively short observation times (e.g. the leader-follower relationship may change). Thus, it is necessary to establish first that the samples of intervals are drawn from the same distribution, which is a prerequisite for this type of correlation analysis. In case of changing durations of call cycles, i.e. in the process of speeding up calls etc., the measured durations can be normalised with reference to the animal's call period.

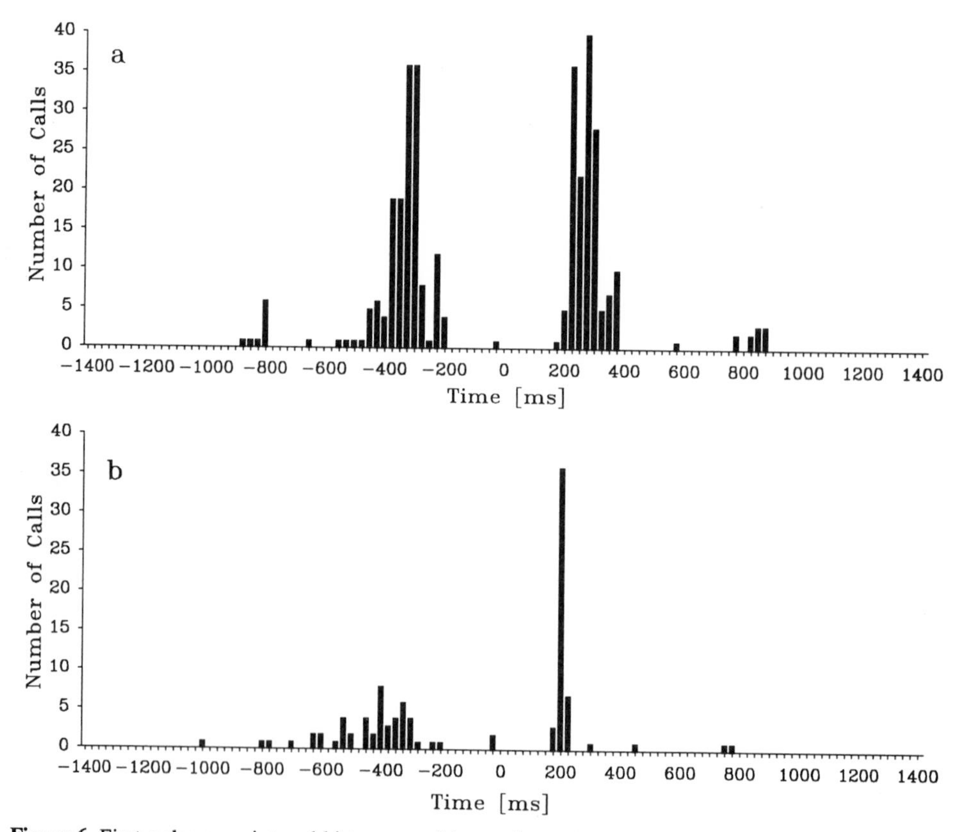

Figure 6. First order cross-interval histograms of interacting males of *Hyla cinerea*: (a) males that alternate their calls and adjust the call intervals to each others calls; (b) one male adjusting its calling times to the calls of its neighbour without reciprocal adjustment.

The last method to be described here for the analysis of natural interactions is the use of cross-correlation techniques. Wickler and Seibt (1974) used the cross-correlation technique developed by Perkel et al. (1967) for the analysis of neural responses. In the first order cross-interval histogram shown in Figure 6, a frequency distribution of the time intervals between an individual's own call and the preceding and following call of its neighbour is plotted. The data in the top and in the bottom diagram show the same *H. cinerea* as in the top and bottom diagrams in Figure 5. Positive time intervals indicate that the neighbour's call is delayed relative to the calls of the individual under study; negative time intervals indicate that the neighbour's calls precede those of the individual under study. The individuals shown in the top graph of Figure 6 are only weakly coupled in the timing of their calls. Although the intervals cluster around a mean, call timing in both the positive and the negative time interval shows considerable variance. In the bottom part of the figure, it is evident that the two individuals behave very differently: the distributions of positive and negative time intervals show considerable differences. The variance in the response latency of one individual relative to the other is very small (right distribution), but not so for the reciprocal interaction (left distribution). As in the previous example, a comparison between the two time records of periodic but independent calling (i.e. separated by a few minutes recording time) could be used as a control.

In the cross-correlation technique used by Brush and Narins (1989) diagrams were constructed in which the Pearson product-moment correlation coefficient r for the two records of the calls of the interacting frogs on the ordinate was plotted against the time shift between the two records on the abscissa. In the case of an autocorrelation function, both records of calls were the same. If one time record of the calls was shifted by the inter-call interval, the correlation coefficient r reached another maximum and the inter-call interval could be determined from the autocorrelation function. If the two time records were from different frogs, the cross-correlation function revealed the strength of the coupling of call timing and also the relative delay of one frog's calls against those of the other frog. The correlation coefficient r can be tested for significance using a t-test (Sokal & Rohlf 1969).

Playback Techniques with Periodic-Call Stimuli

An early demonstration of the entraining of a treefrog's call to a periodic stimulus was given by Busnel and Dumortier (1955) who showed that the calling rate of a male of *H. arborea* could be modified by the beats of a metronome with which the male alternated his calls. Loftus-Hills (1974) in his study of Strecker's chorus frog, *Pseudacris streckeri*, presented tape-playbacks to individual males calling singly in a laboratory setup. The repetition rate of the stimuli was varied over a range of 25% to 214% of the average spontaneous calling rates of the frogs. Loftus-Hills (1974) found that the calling of males could be entrained only over a limited range of calling rates relative to their own rate. Above and below this range, the subjects reverted to their own spontaneous calling rate and did not adjust their rate to that of the playback. Moore et al. (1989) in their study of call-timing in the white-lipped frog, *Leptodactylus albilabris*, found that males of this species entrained over a wide range of playback rates, including rates well beyond the natural calling rate. If the stimulus rate in the playback was too high, the frog skipped a few cycles before giving one of his own calls synchronised with the playback. Zelick and Narins (1985) found that in *E. coqui* the rate at which a male switched from a 1:1 to a 1:2 interleaf of its calls was related to its spontaneous call period.

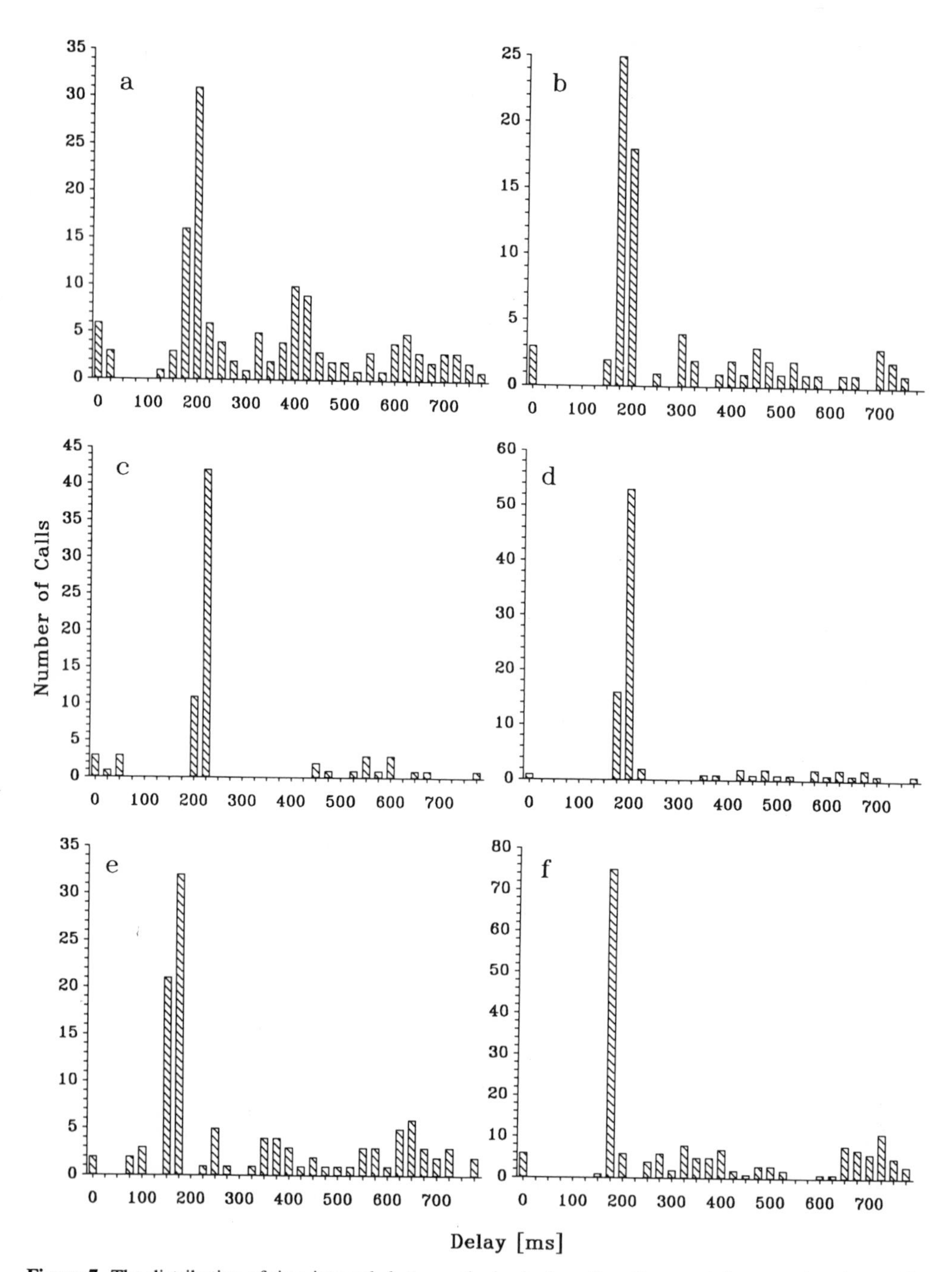

Figure 7. The distribution of time intervals between the beginning of a call presented at a random inter-call interval and the beginning of a frog's own call in six different males of *Hyla cinerea* (a to f).

Playback Techniques with Random-Call Stimuli

Theoretically, the use of periodic playback stimuli may result in apparent coupling if both the playback and the caller have the same calling rates such that for extended periods of time (e.g. during the recording time of a few minutes) there will be a constant time delay between the individuals calls and the playback stimuli from the tape (see also Popp 1989). Although this problem is usually insignificant because the variance in the inter-call interval is large enough so that the caller automatically drifts in the timing of its calls relative to the playback stimuli if it is unaffected by the playback, one way to avoid this problem completely is the use of playbacks with random call intervals. Lemon and Struger (1980) presented 30 randomly-timed single calls to spring peepers (*Pseudacris crucifer* = *Hyla crucifer*) and observed how much the expected inter-call interval of their experimental subjects was changed by the intermittent stimulus. The results of the observations bear directly on the mechanisms involved in call timing during interactions between males. In our studies of *H. cinerea*, we conducted playbacks to animals that were calling from locations that were relatively isolated from the rest of the chorus; the stimulus had an average inter-call interval equal to that of a typical male of *H. cinerea*, but the inter-call intervals varied randomly between the shortest interval that can permit non-overlapping calls of 180ms to intervals of 1.6s that were about two times the average inter-call interval. If the experimental subjects failed to react to the playback stimuli, then no apparent difference from chance performance is expected because of the large variance of the inter-call intervals in the playback tape. As shown in Figure 7, the pattern of interaction of individuals of *H. cinerea* was very similar in relation to the randomised stimuli as to that observed in natural interactions with another male (Fig. 2).

Playback Techniques with Temporal-Gap Stimuli

In the natural environment a male frog often vocalises in a dense chorus in which large numbers of nearby conspecific and heterospecific males are also calling. Thus, rather than alternating calls with the loudest neighbour to avoid overlap, it may be more efficient to wait for a male to vocalise during a low-noise gap in the noisy background. To determine if individuals are employing this strategy, callers may be presented with stimuli having temporal gaps, in which the acoustic energy is omitted or lowered. Zelick and Narins (1982, 1983, 1985) used a background sound with spectral energy in the frequency region of the calls of their experimental species (*E. coqui*) that had regular or pseudo-randomly placed temporal gaps. Moore et al. (1989) presented male white-lipped frogs (*L. albilabris*) with series of periodic stimuli in which some stimuli were omitted so that a temporal gap occurred. Males of *E. coqui* clustered their calls in the gaps, i.e. calls were synchronised with the reduction in sound level of the masking background. Loftus-Hill (1974) found a similar response in the frog *P. streckeri*, in which most individuals called with a fixed delay after the offset of periodically presented playback stimuli of differing duration; only one of the five individuals called with a fixed latency after the stimulus onset. In *L. albilabris* males apparently lack a mechanism to make use of temporal gaps. Rather, males in this species showed a delayed suppression of their own calls with the suppression period triggered by the onset of an acoustic stimulus (Moore et al. 1989).

Schwartz (1991) found that males of *Hyla microcephala* did not confine calling to periods of silence ("inter-bout" intervals) between computer-generated playbacks of stimuli that simulated choruses of conspecifics. Although the male was sometimes stimulated to call more often by these playbacks, follow-up experiments using digitised natural calls showed that the target male did not call randomly with respect to calls that simulated two interacting males. In fact, male calling was triggered by brief gaps between the digitised calls. This ability to respond rapidly to brief gaps in background noise level, similar to that found in *E. coqui*, may explain how males of *H. microcephala* are able to alternate precisely call notes in one-on-one interactions (Schwartz, *pers. comm.*).

Playback Techniques with Interacting-Call Stimuli

In interactive experiments, the frog's calls serve as reference points for the timing of playbacks. In a typical interactive experiment, the subject's call triggers a call synthesiser or a computer with a digital-to-analogue interface board (Narins 1982; Schwartz 1987b, 1991; Schwartz and Rand 1991; Moore et al 1989). Typically the investigator varies not only the properties of the acoustic stimulus (e.g. its duration), but also the delay between the production of the frog's call (trigger) and the stimulus. Narins (1982) studied the amount of entrainment (i.e. the percentage of short-latency responses in relation to the total number of calls during playback) as a function of the delay of the playback in two species of frogs. In both species there was a species-specific time period after an individual's call, during which the animal failed to respond in a time-locked fashion. Narins (1982) called this the absolute behavioural refractory period, which was 210ms and 1133ms in *Hyla ebraccata* and *E. coqui*, respectively. At the end of the absolute refractory period, males of both species showed increasing entrainment as the delay of the playback was increased until a maximum response was reached after about 0.5s. Narins (1982) termed this period (between the end of the absolute refractory period and the delay time at which the maximum entrainment was observed) the relative behavioural refractory period. The differences between the refractory periods of the two species cannot be explained by temperature effects that may influence neural timing loops (Narins 1982). However, in *H. ebraccata* there may have been a strong selective pressure to rapidly echo the calls of neighbours that overlap a frog's own calls (see below section on female choice).

Mechanisms Involved in Call-Timing

Moore et al. (1989) suggested a general mechanism that may underlie the timing of calls in various frog species. Their model contains a number of different components which, in combination, are sufficient (at least for a first-order approximation) to explain the interactions observed between frogs or between a caller and playback. The first component is an absolute behavioural refractory period that follows an individual's own call (see Fig. 8 for a schematic representation of the different components).

Figure 8. A schematic representation of the phases forming the call-cycle in frogs (absolute behavioural refractory period, relative behavioural refractory period, C.A. = call-activation phase, call-emission phase): a) Internally triggered call generation. The period of the call-cycle is thought to be determined by an internal oscillator. When calling spontaneously (i.e. calls are not triggered by external events), the internal call oscillator determines the timing of the individual's next call. b) Externally triggered call generation. Hearing other calls will suppress calling after a delay (S.D. = suppression delay, presumably used for processing the perceived call and halting the calling mechanism) for a limited time (call supp. = call suppression period). The suppression of calling is followed by a call excitation period (E.P) in which initiating a call is very likely. Calling cannot be suppressed when the call activation phase is entered, and other calls are less likely to be perceived during the call emission.

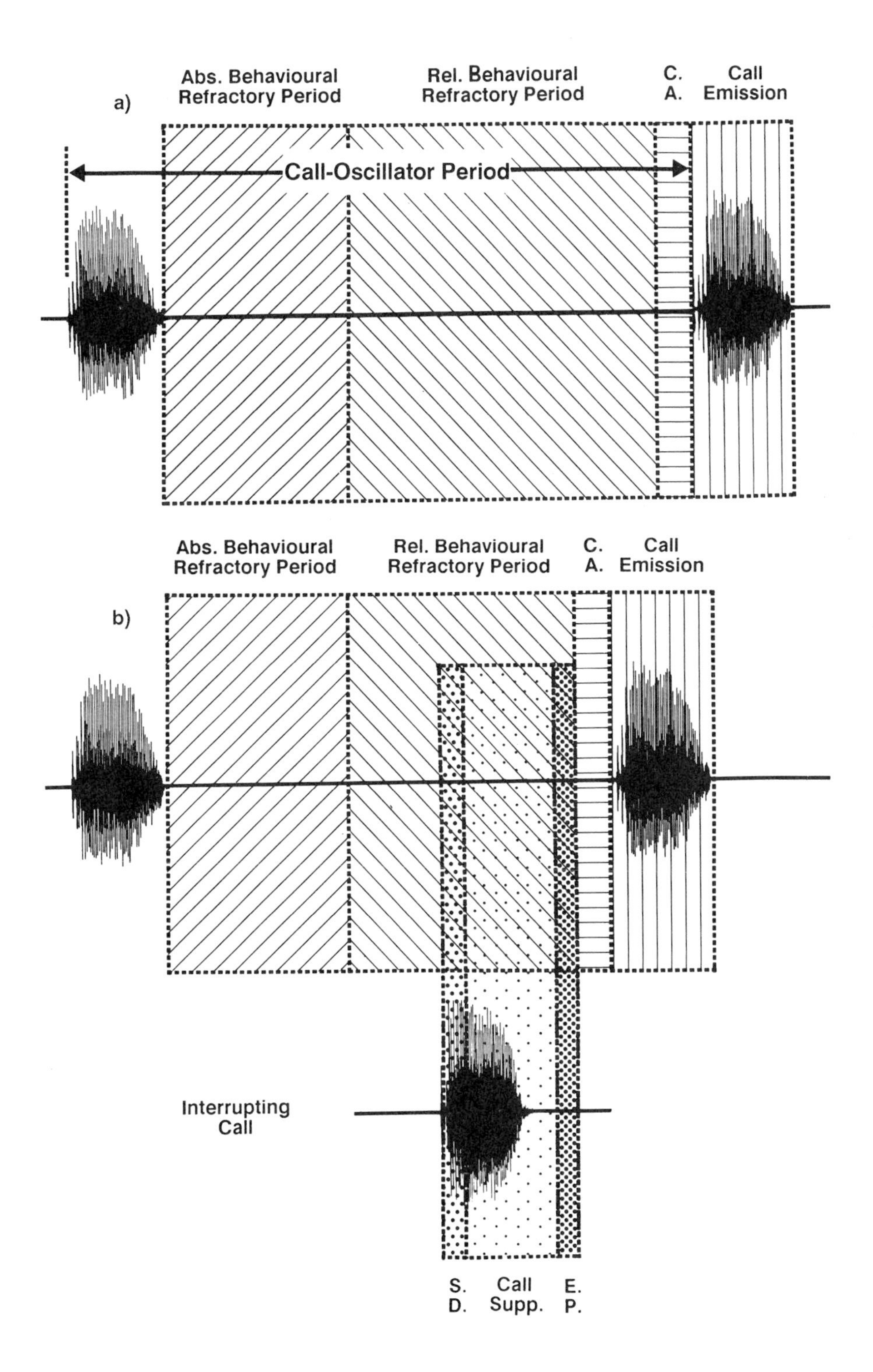
a)
Abs. Behavioural Refractory Period
Rel. Behavioural Refractory Period
C. A.
Call Emission
Call-Oscillator Period
b)
Abs. Behavioural Refractory Period
Rel. Behavioural Refractory Period
C. A.
Call Emission
Interrupting Call
S. D.
Call Supp.
E. P.

During this period an external stimulus cannot elicit a call with a short response latency (relative to the call-cycle period; e.g. Narins 1982). Although there is no short latency response, the duration of the call-cycle period (i.e. the time from the onset of one call to the next) may well be influenced within a certain range (e.g. Lemon and Struger 1980). The detection of a change in the period of the call-cycle, however, may be impeded if additional calls are presented directly after the end of the refractory period, or if no calls are presented during the subject's refractory period. This occurs when the frog has entrained its calls with a short latency response to the periodic playback stimulus and the stimulus interval is too large to allow additional calls to be presented to the animal during the refractory period. A relative behavioural refractory period (Narins 1982) follows the absolute refractory period. During this period, a call can be elicited with a short-latency response. If no stimulus is perceived, the frog terminates the relative refractory period with a spontaneously generated own call. The call timing is defined by an internal oscillator.

The second important component of the model suggested by Moore et al. (1989) is a period of call suppression or inhibition, during which fewer calls are given than expected from a random placement of calls with reference to the acoustic stimulus. In some species, the period of suppression starts shortly after the onset of a perceived stimulus and extends throughout the stimulus and for a short time past its end (see the schematic example in Fig. 8b; also *E. coqui*, Zelick and Narins 1982 and *H. cinerea*, Klump and Gerhardt, this study). In *L. albilabris* the period of suppression takes the form of a period of delayed inhibition that is triggered by the onset of the perceived stimulus (Moore et al. 1989). This is probably a reflection of the brief duration of the (playback) chirp in this species. The period of suppression may be followed by a short period of excitation, during which the spontaneous generation of a call is very likely (e.g. in *E. coqui*, Narins 1982; *H. cinerea*, Klump and Gerhardt, this study).

The call itself is preceded by a short activation phase of constant duration, during which other calls will have no influence on when the next call is given (i.e. this is the time duration from the triggering of the neural circuitry that generates the call until the call is sounded). Although Moore et al. (1989) point out that the refractory period in combination with the delayed inhibition alone are sufficient to explain the observed entraining of the vocal response in *L. albilabris*, they suggest that an additional *sharpening* mechanism is involved in the timing of the calls leading to a much smaller scatter of the data points in the interval-versus-phase diagrams than expected. Such a sharpening of the distribution of times of delayed calling could also be achieved in this species by a short excitatory period, with a high probability of calling, that follows the delayed inhibition.

The mechanism hypothesised here can explain both the entrainment of calling found in natural interactions between calling frogs and the increase in an individual's calling rate that is often found when comparing the calling rates of single callers with that of two or more interacting callers. The minimum calling period that is possible will be the refractory period plus the times for the continuing suppression of the response by the perceived call and the duration of a frog's excitation plus activation phases. The mechanism suggested by Moore et al. (1989) can also account for the 2:1 and higher interleaf factors found in the response to stimuli presented periodically at rates much higher than the target male's spontaneous calling rate.

What Characterises an Effective Playback Technique for the Evaluation of the Mechanism?

In an efficient design a single playback experiment will provide estimates of many or all of the parameters that can be used to determine the mechanism involved in acoustic interactions. In our view, this requirement is met by using random time intervals between the presented stimuli rather than a periodic presentation of the playback signals. Random timing of the playback reduces the probability that the timing relationship of the experimental subjects relative to the stimulus is observed merely because the call oscillator and the periodic playback have a similar call period. If single playback signals are not sufficient to elicit a response, bouts of stimuli presented at a constant rate could be used, but the time intervals between the bouts should be random. Random time intervals between stimuli or bouts should be long enough to minimise interactions between stimulation events. The minimum range of times over which the random interval should vary should equal the duration of a few call cycles of the target male. This ensures that random variation between the timing of the stimuli and the subject's responses would be detected if they were uncoupled. Widely spaced single stimuli should be used preferentially, because only such stimuli allow the investigator to study the effect of single stimuli on the call period of the subject.

Besides revealing adjustments of the call period of the oscillator of the experimental subject, the use of randomly timed stimuli will also allow the assessment of a subject's absolute and relative behavioural refractory periods. Furthermore, this type of playback will reveal any periods of call suppression or inhibition that may exist, if, in addition to random intervals, at least two stimulus durations are presented at random. This procedure can be used to determine the trigger points to which the subjects responds, i.e. whether the frog adjusts its call timing to the onset or the offset of the signals presented. Finally, the suggested design will also give information about periods of excitation in relation to the playback stimulus and about the duration of the call activation phases.

In the initial evaluation of the mechanism underlying the calling behaviour interactive playback techniques are not a prerequisite. If the design incorporates the suggestions above, the parameters of the playback would, however, be limited to time periods of stimulation similar to those at which the frogs themselves call. The playback period should be long enough to allow a sufficient number of responses to map out inhibitory periods, refractory periods, adjustments of the period of the call oscillator, etc. A minimum of 30 playback stimuli may be sufficient (e.g. Lemon and Struger 1980), but a larger number will result in a more precise estimation of the parameters of the mechanism involved in the subject's call timing. The observation of the experimental subject during the playback should be combined with a sufficiently long observation prior to the playback. This pre-playback observation period then allows one to measure the spontaneous calling behaviour, e.g. the spontaneous call period of the individual's call oscillator, which may be related to other features of the mechanism revealed by the playback (e.g. the duration of refractory periods).

The design suggested here is by no means the only way to execute playback experiments to unravel the mechanism involved in call timing. Rather, it is a synopsis of the successful approaches used in previous studies that are reviewed above.

Computer Simulation of Acoustic Male-Male Interactions

Although the structure of most frog calls is simple and there is a relatively detailed knowledge of the features of the mechanisms involved in the triggering of call production, only a few attempts have been made to simulate male-male interactions with computer models representing the caller's behaviour patterns. Fox and Wilczynski (1986) developed computer models to study the effect of chorus formation on a male's individual ability to attract female frogs over larger distances (this was described as the per-capita active space). The boundary of an individual caller's active space was defined by the distance from the caller at which the sound-pressure level (either peak or time-weighted level) would reach the perceptual threshold. The authors did not specify whether they meant auditory threshold or threshold for triggering a phonotactic approach. The model lacks so much realism in the basic assumptions, however, that the importance of the conclusions for frogs in the "real world" is difficult to assess. Most importantly, the effects of masking on the active space within a chorus are ignored (see for example Gerhardt and Klump 1988a, for a study of the effects of masking on female choice in treefrogs). Thus, maximising the duty cycle of the signal (one of the optimal strategies of the model) may not pay off for a frog in a natural chorus, because large duty cycles mean large amounts of masking by neighbouring callers. Furthermore, in the only treefrog species (*Hyla gratiosa*) in which the active space in long range attraction of females has been measured (Gerhardt and Klump 1988b), optimal inter-individual spacing as predicted by the model would be so large (in the range of 50 to 200 m) that few frogs would fit in a pond of the size which forms the normal breeding site of this species.

The computer simulation of interacting Puerto Rican treefrogs (*E. coqui*) presented by Brush and Narins (1989) uses a model developed for the transmission of packages of information between computer terminals on a network using the carrier sense multiple access protocol (the exact version of the model was the P-persistent carrier sense multiple access model; for details see Brush and Narins 1989). Performance criteria of the frog assembly that was modelled were the absolute number of non-overlapped calls of the group, the relative number of non-overlapped calls of the group (compared to the group's best performance) and the relative number of non-overlapped calls per frog. The frog's behaviour was simulated as being composed of five states: silence (state 1: In this state the animal will initiate a calling process spontaneously with probability P), the call activation phase (state 2: in this state call emission is prepared and will follow immediately), the call emission phase (state 3), the refractory period (state 4), and a random delay until state 1 is achieved (state 5). This last state is only entered if a call of another frog is perceived during state 2 or state 4 (presumably the frog has difficulties detecting other frogs while its own call is sounding, i.e., in state 3). The durations of the states 2 to 4 were chosen on the basis of experimental data, and the duration of state 5 was uniformly distributed between 0 and 800ms. The probability P and the number of interacting frogs in the chorus were the model parameters which were varied in the simulation.

For small group sizes (<10) the probability P of initiating a call could be varied by at least a factor of 10 ($1.0 \geq P \geq 0.1$) without any strong effect on the group performance (i.e. the number of non-overlapped calls per group per min). Optimal group size for calling probabilities $P > 0.1$ was four to six animals. The individual's optimal group size for transmitting its own call is one, i.e. it is performing suboptimally when it shares signalling time with others. Brush and Narins (1989) suggest that the number of frogs found interacting in the field corresponds with the group size at which the functions of the relative proportion of clear calls per group and per individual intersect. This group size is three over a wide range of variation in the calling probability P.

Although this computer simulation of the calling frogs' behaviour incorporates some knowledge of the mechanism involved in the triggering of calls, it also lacks at least one essential component on the side of the perception of the signal. The basic assumption is made that all frogs "connected to the network" perceive each other equally well. This assumption may be violated, since in choruses of frogs, masking of single calls by the chorus background may severely limit the perception of a particular individual's calls (Gerhardt and Klump 1988a). The data presented by Zelick and Narins (1983) suggest that the change in intensity needed for the detection of the ending of a neighbour's call in *E. coqui* (i.e. a 4 to 6dB decrement) is similar to what a female of *H. cinerea* needs for detecting individual callers in a chorus background reliably (i.e. about a 3dB increment). Thus, the conclusion in the paper by Gerhardt and Klump (1988a) that a female *H. cinerea* often can only perceive the neighbouring males around it (about three) may also apply to interacting males of *E. coqui*. Masking in itself may be a sufficient explanation why only immediate neighbours interact, thus forming a group of about three males that interact vocally.

In conclusion, more realistic models are needed that not only incorporate the mechanisms involved in call production, but also mechanisms involved in call perception. Only then can we more reliably assess whether the proximate mechanisms underlying calling behaviour and call perception provide a sufficient basis for what we observe in nature. Furthermore, as will be demonstrated in the next section, female choice experiments are necessary to identify parameters of calling behaviour that should be optimised in order to accomplish one of the major functions of calling: the attraction of mating partners in a noisy environment.

Female Choice in Relation to Call-Timing in Male-Male Interactions

The evolutionary significance of call-timing mechanisms can best be answered by studying how females respond to the various strategies of individual callers. In general, the basic assumption is that non-random timing of calls allows a male to maximise the number of unmasked advertisement calls, which are thus broadcast clearly. Overlapping of an individual's calls by a neighbour's calls may reduce its attractiveness to females, because features of the male's calls that are important for female choice may be masked or obscured. For example, by overlapping calls the fine-temporal structure of the call's envelope may deteriorate and information about pulse rate, an important character for species recognition in some species, may be less discernible to the female. By presenting choices between two simulated callers with calls overlapping in a way that degraded the signal envelope and a simulation of two callers producing non-overlapping calls, Schwartz (1987b) showed that females preferred non-overlapping signals in two species in which the calls are composed of rapid pulse trains (*Hyla versicolor*, *H. microcephala*). In a third species, the spring peeper (*P. crucifer*), which has a call with an unmodulated envelope, females did not prefer non-overlapping signals. Additionally, Forester and Harrison (1987) have demonstrated that in *P. crucifer* females neither prefer leading or lagging callers, nor do overlapped calls impede the females' preference for average frequency calls. In the other two species studied by Schwartz (1987b), *H. versicolor* and *H. microcephala*, there was no preference for the non-overlapping calls if the overlapping calls were in phase so that the structure of the envelope was preserved, and thus pulse-rate information was discernible to the female. Gerhardt (1978; see also Gerhardt and Doherty 1988) has shown that pulse rate is one of the most important parameters used by female *H. versicolor* in mate recognition. In experiments in which synthetic calls partially overlapped, females of this species did not prefer the signal that was first heard and which was then overlapped by a second stimulus played back from another speaker; nor did females prefer the stimulus that partially interrupted the first call (Table I). Thus, overlapping the calls of a neighbour does not confer an advantage to an individual in this species (but see below). Similarly, pulse rate seems to be an important character in *H. microcephala*, since females prefer calls with the pulse rate typical of males of this species (Schwartz 1987a).

There are frog species, however, in which at least some males may benefit by overlapping the calls of other males. In *S. sila*, Tuttle and Ryan (1982) suggest that overlapping calls may reduce a male's probability of being detected by a frog-eating bat. A more recent study shows, however, that the precision of synchronisation in this species may have been over-estimated (Ibanez 1991). In two other species of hylids, only one of two males experiences an advantage. In *H. ebraccata*, Wells and Schwartz (1984) found that females prefer the lagging call of two overlapping calls. This species has an advertisement call with two note types. The long primary note is followed by one or more short secondary notes, which make the call especially attractive to the female (see Wells and Schwartz 1984). The lagging caller masks the short secondary notes of the leading caller with the long primary note of its own call and the secondary short notes of the lagging frog's call are unobstructed. In *H. ebraccata*, males respond to playbacks of a periodically repeated call with a delay of 140 to 200ms after stimulus onset (Wells and Schwartz 1984). This response latency leads to a maximal masking effect of the secondary notes by the primary note. In response to overlapping calls, males may increase the number of secondary notes such that they may extend their calls past the end of the calls of the interfering male.

Table I. The relationship between the timing of calls of two acoustically interacting males and female choice in the gray treefrog, *Hyla versicolor* (call duration 875ms, 85dB SPL, 1 call/4s). *Delay* is the delay of the lagging call (ms). *Preference* is preference for the leader (%).

	Number of females			
delay	**tested**	**responding**	**preference**	**P**
420	16	13	46	n.s.
1000	9	9	56	n.s.

In other frog species, females show a preference for the leading call of two calling males with overlapping calls. In *Hyperolius marmoratus*, Dyson and Passmore (1988a) demonstrated a preference for the leader when females were presented with overlapping (i.e. where the onset of the second call was delayed by 40ms relative to the onset of the first call with a duration of 80ms, giving a 50% overlap) or abutting calls (i.e. the onset of the second call was delayed by 80ms relative to the onset of the first call with a duration of 80ms). The preference for the leading overlapping call prevailed even when it was reduced in amplitude by 6dB, whereas the preference for the leading abutting call was abolished when its amplitude was reduced by the same amount (Dyson and Passmore 1988b). In this species, the preference for leading calls would override the preference for low-frequency calls (Dyson and Passmore 1988a). Again, in contrast to the results in *H.*

ebraccata, females of another hylid species, *H. cinerea*, also showed a strong preference for the calls of the leading male (Table II). If one male led the other by 40ms (i.e. the calls overlapped by 78%), the preference of the females was relatively strong. The preference was, however, abolished if the sound-pressure level of the less attractive call was increased by 6dB. Although the first part of the call contains one or two short distinct pulses that are masked in the lagging call, it is unlikely that nowadays these pulses convey any particularly important information. Synthetic calls that lacked a pulsatile structure were as attractive as a typical, natural call (Gerhardt 1974) and adding a pulsatile beginning to such a synthetic call failed to enhance its attractiveness (Gerhardt 1983). The functional significance of the female bias to the first of two overlapping calls she hears is unknown. A pulsatile beginning, which is more distinctive than that in *H. cinerea*, does affect the relative attractiveness of a closely-related species, *H. gratiosa*; synthetic calls lacking pulses at the beginning were not as attractive as natural calls or synthetic calls with pulses (Gerhardt 1981). Perhaps females of *H. gratiosa* would prefer the leading call in a more intensity-independent fashion than *H. cinerea*.

Table II. The relationship between the timing of calls of two acoustically interacting males and female choice in the green treefrog, *Hyla cinerea* (call duration 180ms, 85dB SPL, 1 call/800ms). *Delay* is the delay of the lagging call (ms). *Preference* is preference for the leader (%). Note that in the second 40ms delay stimulus (bottom row of Table) the lagging call was 6dB louder.

	Number of females			
delay	tested	responding	preference	P
40	16	15	100	< 0.001
160	26	20	75	0.04
300	34	31	74	0.011
40[see note]	21	18	50	n.s.

Besides the direct effects of the relative timing of calls on female preferences, there may be other indirect benefits of non-random calling in males. Signal-to-noise ratio of a male's own call in relation to the masking background provided by the surrounding chorus is an important parameter influencing the detection of calls by females (Gerhardt and Klump 1988a). A male may enhance the signal-to-noise ratio of its calls by spatially separating them from those of neighbouring males (Schwartz and Gerhardt 1989). Spatial separation, however, requires that males can detect other calling males in their vicinity and judge their distance. Field experiments by Schwartz (1987b) with three species of hylids suggest that males may detect other males' calls more easily if they do not overlap with their own calls. Schwartz presented calling males of *P. crucifer*, *H. versicolor* and *H. ebraccata* with natural calls that either overlapped those of the target male or were non-overlapping, and then scored the number of aggressive calls given in response to the playback. Males of all three species gave more aggressive calls, which are typically used in establishing calling territories and inter-male spacing, in response to non-overlapping calls than in response to overlapping calls. Furthermore, in *H. microcephala* data on the note-duration of aggressive calls in overlapping versus non-overlapping playback situations support the notion of reduced efficiency of overlapping calls in evoking a response (see Schwartz 1987b). The hypothesis that non-overlapping calls may be more easily localisable by females than overlapping calls is not supported by the data from different species of frogs (Passmore and Telford 1981; Schwartz 1987b; Backwell and Passmore 1991).

If there are patterns in male-male interactions in the calling of frogs in a chorus that give some callers a selective advantage over others, then why have the frogs not evolved proximate mechanisms to avoid overlap in calling that would maximise a male's chance of attracting females? In some cases, e.g. in *H. ebraccata*, female choice may have driven the system into a state in which it is trapped. Given the pattern of female choice, a male *H. ebraccata* may "just make the best out of a bad job". If the male interacts with another male and gives the first call, then it is likely to be overlapped by calls of its neighbour and thus the attractiveness of his signal is reduced. Males of *H. ebraccata* counteract this strategy in part by adding secondary notes when interacting with other males. To avoid being masked, a male could wait for the other frog to start its call first. This would, however, reduce his own attractiveness because its rate of calling would be lowered. By contrast, males of the closely-related species, *H. microcephala*, which has short primary notes and can adjust note timing, are able to interdigitate both primary and secondary notes in a rather precise fashion to avoid overlap (Schwartz and Wells 1985, Schwartz, *pers. comm.*). Would males of *H. ebraccata* use such a strategy if their primary notes were shorter or their internote intervals longer?

In a communication system such as male-male vocal interaction in frogs it is possible to study simultaneously the proximate mechanisms of call timing and to model the ultimate (fitness) benefits of various strategies, e.g. using game theory to discover evolutionary stable strategies. This dual approach has the potential to provide a deeper understanding of the interactions between the proximate and ultimate factors involved in communication within assemblies of competing chorusing animals.

Acknowledgements

We thank the NATO and the organisers of the workshop on playback techniques for providing the basis for a fruitful discussion. The work on *H. cinerea* and *H. versicolor* was funded by a grant from the National Science Foundation to HCG. We thank A. Pavusa and W. Herdlicka for their help in field recordings of interacting *H. arborea*. The help of A. Köhler in preparing the Figures and measuring interaction timing is greatly appreciated. We are grateful to J.J. Schwartz and T. Friedl for comments on the manuscript.

References

Awbrey, F.T. 1978. Social interaction among chorusing pacific tree frogs, *Hyla regilla*. *Copeia*, **1978**, 208-214.

Backwell, P.R.Y. and Passmore, N.I. 1991. Sonic complexity and mate localization in the leaf-folding frog, *Afrixalus delicatus*. *Herpetologica*, **47**, 226-229.

Batschelet, E. 1981. *Circular Statistics in Biology*. Academic Press; London.

Brush, J.S. and Narins, P.M. 1989. Chorus dynamics of a neotropical amphibian assemblage: comparison of computer simulation and natural behavior. *Anim. Behav.*, **37**, 33-44.

Bucher, T.L., Ryan, M.J. and Bartolomew, G.A. 1982. Oxygen consumption during resting, calling and nest building in the frog *Physalaemus pustulosus*. *Physiol. Zool.*, **55**, 10-22.

Busnel, R.-G. and Dumortier, B. 1955. Phonoreactions de male d' *Hyla arborea* a des signaux acoustiques artificiels. *Bull. Soc. Zool. France*, **80**, 66-69.

Duellman, W.E. 1967. Social organization in the mating calls of some neotropical anurans. *Am. Midl. Nat.*, **77**, 156-163.

Dyson, M.L. and Passmore, N.I. 1988a. Two-choice phonotaxis in *Hyperolius marmoratus* (Anura: Hyperolidae): the effect of temporal variation on presented stimuli. *Anim. Behav.*, **36**, 648-652.

Dyson, M.L. and Passmore, N.I. 1988b. The combined effects of intensity and the temporal relationship of stimuli on phonotaxis in female painted reedfrogs *Hyperolius marmoratus*. *Anim. Behav.*, **63**, 1555-1556.

Forester, D.C. and Harrison, W.K. 1987. The significance of antiphonal vocalization by the spring peeper, *Pseudacris crucifer* (Amphibia, Anura). *Behaviour*, **103**, 1-15.

Foster, W.A. 1967. Chorus structure and vocal response in the pacific treefrog, *Hyla regilla*. *Herpetologica*, **23**, 100-104.

Fox, J.H. and Wilczynski, W, 1986. The augmentation of per-capita active space through chorussing in anurans: a computer model. *Soc. Neurosci. Abstr.*, **12**, 314.

Gerhardt, H.C. 1974. The significance of some spectral features in mating call recognition in the green treefrog (*Hyla cinerea*). *J. Exp. Biol.*, **61**, 229-241.

Gerhardt, H.C. 1975. Sound-pressure levels and radiation patterns of the vocalizations of some North American frogs and toads. *J. Comp. Physiol.*, **102**, 1-12.

Gerhardt, H.C. 1978. Temperature coupling in the vocal communication system of the gray treefrog *Hyla versicolor*. *Science*, **199**, 992-994.

Gerhardt, H.C. 1988. Acoustic properties used in call recognition by frogs and toads. In: *The Evolution of the Amphibian Auditory System*. (Ed. by B. Fritsch, M.J. Ryan, W. Wilczynski, T.E. Hetherington & W. Walkowiak W), pp. 455-483. Wiley, New York.

Gerhardt, H.C. and Doherty, J.A. 1988. Acoustic communication in the gray treefrog, *Hyla versicolor*: evolutionary and neurobiological implications. *J. Comp. Physiol. A*, **162**, 261-278.

Gerhardt, H.C. and Klump, G.M. 1988a. Masking of acoustic signals by the chorus background noise in the green treefrog: A limitation on mate choice. *Anim. Behav.*, **36**, 1247-1249.

Gerhardt, H.C. and Klump, G.M. 1988b. Phonotactic responses and selectivity of barking treefrogs (*Hyla gratiosa*) to chorus sounds. *J. Comp. Physiol. A*, **163**, 795-802.

Goin, C.J. 1949. The peep order of peepers: a swamp water serenade. *Quart. J. Fla. Acad. Sci.*, **11**, 59-61.

Hardy, D.F. 1959. Chorus structure in the striped chorus frog, *Pseudacris nigrita*. *Herpetologica*, **15**, 14-16.

Ibanez, R. 1991. Synchronized calling in *Smilisca sila* and *Centrolenella granulosa*. unpublished Ph.D. Dissertation, University of Connecticut.

Lemon, R.E. 1971. Vocal communication by the frog *Eleutherodactylus martiniensis*. *Can. J. Zool.*, **49**, 211-217

Lemon, R.E. and Struger, J. 1980. Acoustic entrainment to randomly generated calls by the frog, *Hyla crucifer*. *J. Acoust. Soc. Am.*, **67**, 2090-2095.

Loftus-Hills, J.J. 1974. Analysis of an Acoustic pacemaker in Strecker's chorus frog, *Pseudacris streckeri* (Anura: Hylidae). *J. Comp. Physiol.*, **90**, 75-87.

Loftus-Hills, J.J. and Littlejohn, M.J. 1971. Mating call sound intensities of anuran amphibians. *J. Acoust. Soc. Am.*, **49**, 1327-1329.

Lörcher, K. 1969. Vergleichende bioakustische Untersuchungen an der Rot- und Gelbbauchunke *Bombina bombina* (L.) und *Bombina variegata* (L.). *Oecologia*, **3**, 84-124.

Moore, S.W., Lewis, E.R., Narins, P.M. and Lopez, P.T. 1989. The call-timing algorithm of the white-lipped frog, *Leptodactylus albilabris*. *J. Comp. Physiol. A*, **164**, 309-319.

Narins. P.M. 1982. Behavioral refractory period in neotropical treefrogs. *J. Comp. Physiol.*, **148**, 337-344.

Narins. P.M. and Capranica, R.R. 1978. Communicative significance of the two-note call of the treefrog *Eleutherodactylus coqui*. *J. Comp. Physiol.*, **127**, 1-9.

Narins. P.M. and Hurley, D.D. 1982. The relationship between call intensity and function in the Puerto Rican coqui (Anura, Leptodactylidae). *Herpetologica*, **38**, 287-295.

Narins. P.M. and Zelick, R. 1988. The effects of noise on auditory processing and behavior in amphibians. In: *The Evolution of the Amphibian Auditory System*. (Ed. by B. Fritsch, M.J. Ryan, W. Wilczynski, T.E. Hetherington & W. Walkowiak W), pp. 511-536. Wiley, New York.

Oldham, R.S. and Gerhardt, H.C. 1975. Behavioral isolation of the treefrogs *Hyla cinerea* and *Hyla gratiosa*. *Copeia*, **1975**, 223-231.

Passmore, N.I. 1981. Sound levels of mating calls of some african frogs. *Herpetologica*, **37**, 166-171.

Passmore, N.I. and Telford, S.R. 1981. The effect of chorus organization on mate localization in the painted reedfrog (*Hyperolius marmoratus*). *Behav. Ecol. Sociobiol.*, **9**, 291-293.

Perkel, D.H., Gerstein, G.L. and Moore, G.P. 1967. Neuronal spike trains and stochastic point processes. II. Simultaneous spike trains. *Biophysical Journal*, **7**, 419-440.

Popp, J.W. 1989. Methods for measuring avoidance of acoustic interference. *Anim. Behav.*, **38**, 358-359.

Prestwich, K.N., Brugger, K.E. and Topping, M. 1989. Energy and communication in three species of hylid frogs: power input, power output and efficiency. *J. Exp. Biol.*, **143**, 53-80.

Rosen, M. and Lemon, R.E. 1974. The vocal behavior of spring peepers, *Hyla crucifer*. *Copeia*, **1974**, 940-950.

Schneider, H. 1967. Rufe und Rufverhalten des Laubfrosches *Hyla arborea arborea* (L.). *Z. vergl. Physiol.*, **57**, 174-189.

Schneider, H., Joermann, G. and Hödl, W. 1988. Calling and antiphonal calling in four neotropical anuran species of the family Leptodactylidae. *Zool. Jb. Physiol.*, **92**, 77-103.

Schwartz, J.J. 1987a. Spectral and temporal properties in species and call recognition in a neotropical treefrog with a complex vocal repertoire. *Anim. Behav.*, **35**, 340 -347.

Schwartz, J.J. 1987b. The function of call alternation in anuran amphibians: A test of three hypotheses. *Evolution*, **41**, 461-471.

Schwartz, J.J. 1991. Why stop calling? A study of unison bout singing in a neotropical treefrog. *Anim. Behav.*, *in press*.

Schwartz, J.J. and Gerhardt, H.C. 1989. Spatially mediated release from auditory masking in an anuran amphibian. *J. Comp. Physiol. A*, **166**, 37-41.

Schwartz, J.J. and Rand, A.S. 1991. The consequences of communication of call overlap in the tungara frog, a neotropical anuran with a frequency-modulated call. *Ethology*, *in press*.

Schwartz, J.J. and Wells, K.D. 1985. Intra and interspecific vocal behavior of the neotropical treefrog *Hyla microcephala*. *Copeia*, **1985**, 27-38.

Sokal, R.R. and Rohlf, F.J. 1969. *Biometry*. Freeman, San Francisco.

Taigen, T.L. and Wells, K.D. 1984. Energetics of vocalization by an anuran amphibian (*Hyla versicolor*). *J. Comp. Physiol. B*, **155**, 163-170.

Tuttle, M.D. and Ryan, M.J. 1982. The role of synchronized calling, ambient light, and ambient noise in anti-bat-predator behavior of a treefrog. *Behav. Ecol. Sociobiol.*, **11**, 125-131.

Wells, K.D. 1977. The social behavior of anuran amphibians. *Anim. Behav.*, **25**, 666-693.

Wells, K.D. 1988. The effects of social interactions on anuran vocal behavior. In: *The Evolution of the Amphibian Auditory System*. (Ed. by B. Fritsch, M.J. Ryan, W. Wilczynski, T.E. Hetherington & W. Walkowiak W), pp. 433-454. Wiley, New York.

Wells, K.D. and Schwartz, J.J. 1984. Vocal communication in a neotropical treefrog, *Hyla ebraccata*: Advertisement calls. *Anim. Behav.*, **32**, 405-420.

Whitney, C.L. and Krebs, J.R. 1975. Mate selection in pacific treefrogs. *Nature* **255**, 325-326.

Wickler, W. 1974. Über die Beeinflussung des Partners im Duettgesang der Schmätzerdrossel *Cossypha heuglini* Hartlaub (Aves, Turdidae). *Z. Tierpsychol.*, **36**, 128-136.

Wickler, W. and Seibt, U. 1974. Rufen und Antworten bei *Kassina senegalensis*, *Bufo regularis* und anderen Anuren. *Z. Tierpsychol.*, **34**, 524-537.

Zelick, R. and Narins, P.M. 1982. Analysis of acoustically evoked call suppression behavior in a neotropical treefrog. *Anim. Behav.*, **30**, 728-733.

Zelick, R. and Narins, P.M. 1983. Intensity discrimination and the precision of call timing in two species of neotropical treefrogs. *J. Comp. Physiol.*, **153**, 403-412.

Zelick, R. and Narins, P.M. 1985. Characterization of the advertisement call oscillator in the frog *Eleutherodactylus coqui*. *J. Comp. Physiol. A*, **156**, 223-229.

MEASURING RESPONSES OF FEMALE BIRDS TO MALE SONG

William A. Searcy

Department of Biological Sciences &
Pymatuning Laboratory of Ecology
University of Pittsburgh
Pittsburgh, Pennsylvania 15260
U.S.A.

Introduction

It has long been realised that bird song has two principal functions, one in male-male competition, usually for territories and one in attracting and stimulating females (Howard 1920). It follows that investigations of communication via song have two principal audiences to consider, conspecific adult males and conspecific adult females. In particular species, song may be directed especially at one or the other of these audiences, but across all songbirds the two audiences and functions seem to be of similar overall importance.

Nevertheless, for many years studies of song communication focussed almost entirely on the male audience. The experimental method of studying song communication used playback of recorded or synthetic songs and almost all such playbacks were done to male subjects. This bias is illustrated in Becker's (1982) thorough review of the literature on avian playback experiments published up to 1980. Becker cites 29 studies that used playback methods to demonstrate species recognition by song in passerines; 27 of these studies were done with male subjects, versus only two with females. Similarly, Becker cites 31 studies that investigated the parameters of song important to species recognition; all 31 of these used male subjects only.

The bias towards use of male subjects in playback studies did not stem from a judgment that males were the more important audience; on the contrary, many of the studies of species recognition were directed towards understanding reproductive isolation between species and here female response to male song is obviously of much greater relevance than is male response. Rather, the bias was due to methodological considerations: there existed a well-proven, general method for measuring male response to playback, while no comparable method was known for measuring female response. The method for males involved playing songs from a speaker placed on a male's territory and measuring some aspect or aspects of the male's aggressive response. I will call this

Playback and Studies of Animal Communication
Edited by P.K. McGregor, Plenum Press, New York, 1992

method *territorial playback*. Female birds were found in most cases to give little or no response to territorial playback and no general alternative method was known for testing response in females.

The problem, then, has been to find an experimental paradigm in which female birds give a reliable, graded response to playback that can be observed and quantified. Over the years, a variety of solutions to this problem have been developed. The purpose of this chapter is to review these methods, to consider their relative merits and to make recommendations on how they can be made to work. The hope is that by considering past experience in measuring female response to song, the way can be smoothed for future researchers, thus encouraging work on this important and neglected area of animal communication.

What we want in a Playback Method

Before reviewing the playback methods that have been used with female birds, I want to consider what would be the ideal attributes of such a method, in order to produce criteria for judging the relative merits of the existing methods. I will argue that these ideal attributes are *practicality*, *generality*, *sensitivity* and *interpretability*. In this section I will define each of these attributes and illustrate them with reference to the method of territorial playback to male birds.

By *practicality* I mean how easy the method is to perform; a practical method is one that can be done quickly, cheaply and with a minimum of manpower. The territorial playback method does very well on this criterion. Free-living subjects are used, so the animals do not have to be captured, housed and cared for. Often a single trial can be run in five or ten minutes and setup may be fast enough for several trials to be run in an hour. For most applications the only equipment needed is a tape recorder and speaker-amplifier, the minimum for any playback experiment. Finally, in most instances one person can perform the experiments without assistance.

Generality refers to the range of species for which the method works. One advantage of a general method is that, when beginning work on a new species, we can have confidence that such a method will produce results. A second advantage is that results from a general method can be used in cross-species comparisons. Territorial playback with males does well on this criterion also. Territorial playback has been used successfully in scores of species, with few recorded failures. One can be confident that the method will work with almost any passerine species in which males hold territories and sing.

Sensitivity means that the method is successful in detecting differences in response to different stimuli. For a method to be sensitive, the measured response must be finely graded and closely dependent on the type of stimulus presented. Again, the method of territorial playback to males does very well on this criterion. Most response measures used in territorial playbacks, such as approach or songs given in reply, are finely graded and the method has been successfully used to demonstrate discrimination between such pairs of stimuli as conspecific and heterospecific song, own and alien dialect, natural and artificially-altered song, neighbour and stranger song, etc.

By *interpretability* I mean whether a straightforward functional interpretation can be assigned to the results of the method, so that the results allow us to judge whether a stimulus is good or bad in functional terms. In other words, can we use the method to judge whether singing a particular song would be advantageous or disadvantageous? Only on this criterion is the territorial playback method lacking. Song functions in male-male communication as a signal from a territory owner to other males, warning them to keep off the defended area. Thus, a functional song is one that is good at keeping other males away. However, territorial playback actually measures the reaction of the owner, not of potential intruders and a song scores well if it evokes aggression, not avoidance. It is not clear whether owners respond most aggressively to the most effective or the least effective songs; it depends whether they can be intimidated on their own territories. The empirical evidence on this point seems mixed.

I am not arguing that the territorial playback method is unrealistic. Actually, I believe this method does mimic well the naturally-occurring situation in which an intruder enters a territory, sings and is expelled by the owner. The method provides a realistic means of testing discrimination in this situation, but it does not allow us to judge which songs or singing behaviours are selectively advantageous. There is another playback method for males, the speaker occupation experiment (Krebs et al. 1978; Yasukawa 1981), that does have a straightforward functional interpretation, but this method does so poorly on our first criterion, practicality, that it has been attempted very rarely.

Playback Methods with Females

Territorial Playbacks with Females

In this method, one simply uses the standard method of territorial playback, but observes the responses of females rather than males. The procedure works best in species in which females are active in territory defence. For example, Payne et al. (1988) used the method successfully in working with the splendid fairy-wren (*Malarus splendens*) in Australia. This is a cooperative breeding species with group territories, in which females sing and expel intruders. Payne et al. (1988) found that female fairy-wrens give the same attention and approach responses to playback as do males and that these responses could be used to show that females discriminate between songs of their own and other groups.

Females in most species are not strongly territorial, but they may still give some type of response to territorial playback. For example, female white-crowned sparrows (*Zonotrichia albicollis*) give trills, calls and flights in response to field playback (Milligan and Verner 1971; Petrinovich and Patterson 1981) and female chiffchaffs (*Phylloscopus collybita*) may approach their male and/or the speaker during playback (Salomon 1989). One problem is that in both these species, only a low proportion of females give any response. Nevertheless, Petrinovich and Patterson (1981) and Salomon (1989) used this method to demonstrate discrimination by females between own and alien dialects in white-crowned sparrows and chiffchaffs, respectively.

In some species, females answer male singing with their own vocalisations and this behaviour can be used measure response. For example, T. Dickinson and J.B. Falls (*pers. comm.*) found female eastern meadowlarks (*Sturnella magna*) respond to song playback in this way and that females were more likely to answer their mates than those of other males.

Finally, in many, perhaps most, species, females give no observable response to playback at all. For example, in my own experience, female song sparrows (*Melospiza melodia*) are usually neither seen nor heard during playback of male song.

Ratings When this method works, it has the same advantages of practicality that the male territorial playbacks have. However, the method is obviously lacking in generality. Little is known about the sensitivity of the method, but it is clear that if few of the females respond, then a large sample size will be needed to demonstrate discrimination between any stimuli, however different. In terms of interpretability, the method presents difficulties, just as the male method does. Again, it is not clear whether a male benefits from having a song that evokes aggression in a listener, whether male or female; moreover, it is not clear whether the responses given by females in species such as white-crowned sparrows and chiffchaffs are aggressive or sexual. Given the weaknesses of this method, I would not recommend using it except with species, like the splendid fairy-wren, where one knows *a priori* that females are strongly territorial.

Phonotaxis

Phonotaxis means approach to the source of sound; here I restrict the usage to cases in which the approach appears to be affiliative rather than aggressive. The method has been used primarily with captive subjects. Perhaps the best example is the work done with the zebra finch (*Taeniopygia guttata*) by Miller (1979a, b) and Clayton (1988). Miller tested female zebra finches in a flight cage 2.1m long. Contrasting stimuli were played from speakers placed at the two ends and females were scored on latency to approach and time spent in the third of the cage closest to a given speaker. This setup was used to demonstrate a preference for mate's song over songs of other males (Miller 1979a) and for father's song over other's (Miller 1979b). Clayton (1988) used similar methods to demonstrate that both male and female zebra finches prefer to approach either father's songs or songs that they were tutored with early in life rather than to approach unfamiliar songs.

With other species, this type of experiment, done with captive subjects, has had mixed success. Payne (1973) successfully used the method to demonstrate preference for conspecific over heterospecific songs in two species of parasitic indigobirds (*Vidua purpurascens* and *V. chalybeata*). Eriksson (1991) showed that captive female pied flycatchers (*Ficedula hypoleuca*), previously treated with oestradiol, prefer approaching nestboxes from which large repertoires of note types are played, compared to boxes with small repertoires. On the other hand, I was unable to show approach by female red-winged blackbirds (*Agelaius phoeniceus*) to either male song or male precopulatory calls (Searcy 1989). These last experiments were done with subjects held in a flight cage 6m long, with a speaker placed at one end; response was measured as time spent in the third of the cage closest to the speaker. Using the same apparatus and procedures, M.S. Capp (*pers. comm.*) found no tendency for female bobolinks (*Dolichonyx oryzivorus*) to approach male song of either of this species' two major song types. In the experiments with red-winged blackbirds, better results were not obtained with a two-speaker, forced-choice paradigm, nor with the visual stimulus of a male mount in addition to song (Searcy 1989, *unpublished data*).

Eriksson and Wallin (1986) successfully used phonotaxis in the field, working with two hole-nesting species, the pied flycatcher and the collared flycatcher (*Ficedula albicollis*). These authors set out nestboxes equipped with traps, with models of males placed nearby. Conspecific songs were broadcast from some of the models but not others. These experiments were not territorial playbacks, in that no males owned the nestboxes and their surroundings at the time. The result was that more females were captured at the boxes with playback than at the silent controls. Mountjoy and Lemon (1990) used similar methods to show that more female starlings (*Sturnus vulgaris*) were attracted to nestboxes with playback of starling song than to silent controls. In a second experiment they were unable to demonstrate discrimination between simple and complex song because too few females responded.

J.B. Falls, J.G. Kopachena and M. Kubisz (*pers. comm.*) attempted a similar method with a species that does not nest in holes, the white-throated sparrow (*Zonotrichia albicollis*). Falls and colleagues set out squares of mist nets in appropriate habitat early in the breeding season; songs of male white-throated sparrows were played from some of these, while others served as silent controls. No significant difference was found in the number of females captured with *vs.* without song playback and thus there was no conclusive evidence that females responded to [by approaching] song in these circumstances.

Ratings Phonotaxis with captive females presents the usual difficulties of working with captives: capturing, housing and feeding the subjects. Beyond this, however, the technique is quite simple, with little needed in the way of equipment, manpower, or start-up time. The field version does not require capture and care of subjects, but it does require substantial amounts of equipment, as each experimental nestbox must be equipped with a tape recorder and speaker-amplifier. Thus, both lab and field versions rate moderately well in terms of practicality.

Little experience exists on which to judge the generality of these methods in birds. The laboratory version has produced results with four of the six species of which I have knowledge, but undoubtedly the failures are under-represented in the literature. The field method seems unlikely to work with any but hole-nesting species; on the other hand, it seems a very good bet to work with any species that does use nestboxes.

The laboratory method, when it works, is sensitive enough to show discrimination between variations in conspecific song, rather than just discrimination between conspecific and heterospecific song. There is little indication, so far, that the field method is very sensitive.

The field method rates very high in terms of interpretability, in that it is certain that an unmated male would benefit from having a song that attracts females to his territory. The laboratory method also rates fairly high in interpretability, the only difficulty being that it is not clear that approach indicates a sexual preference. For example, female zebra finches prefer to approach their fathers' song, yet mating with fathers would very likely be disadvantageous, so approach here may indicate only a preference for associating with kin. Thus, while we can assume that approach in the laboratory context is affiliative rather than aggressive, we cannot assume that it indicates mating preferences.

Solicitation Display Assay

The single most widely used method of measuring response to playback in female birds is the solicitation display assay, in which male songs are played to captive females and the response measured is the copulation solicitation display. This method was first used by King and West (1977) with brown-headed cowbirds (*Molothrus ater*). King and West found that captive females, if first isolated from conspecifics, would respond to song of male cowbirds with copulation solicitation, a courtship display that normally precedes copulation in many birds. Using percentage of songs eliciting display as their response measure, King and West showed that female cowbirds respond more strongly to conspecific song than to heterospecific and more to songs of isolation-reared males than to songs of males with normal experience.

In 1979, Peter Marler and I were interested in measuring response to song in female song and swamp sparrows (*Melospiza melodia* and *M. georgiana*), in order to compare mechanisms of species recognition in different age and sex classes. We were aware of King and West's (1977) methods, but were doubtful that the technique could be applied to song and swamp sparrows, as experience with captives of these species suggested that females would not give solicitation display in response to song. At this point, we learned from John Wingfield that Michael Moore had found that female white-crowned sparrows give copulation solicitation displays, apparently spontaneously, if first implanted with oestradiol (Moore 1982, 1983). Marler and I quickly concluded that oestradiol treatment might prime female song and swamp sparrows to respond to song with solicitation display. Pilot experiments performed in 1979 with hand-reared females gave promising results, so we proceeded to a series of experiments with wild-caught females. In these experiments, we showed that captive females in both song and swamp sparrows respond more strongly to conspecific than to heterospecific song (Searcy and Marler 1981; Searcy et al. 1981), that female song sparrows respond more strongly to four song conspecific song types than to single song types (Searcy and Marler 1981), etc.

The solicitation display assay has now been applied successfully in at least 20 species. Attempts to use the assay have also failed for a few species, though again it is hard to say how many because negative results are seldom published. Table I summarises the results of both successful and unsuccessful applications, with emphasis on the sorts of discriminations that have been demonstrated. Below I will discuss the methodology in more detail.

All applications of the technique have used oestradiol treatment except the studies of King, West and colleagues with brown-headed cowbirds. Captive cowbirds reliably give solicitation display in response to song without hormone treatment, but even in this species oestradiol treatment produces greater response (Ratcliffe and Weisman 1987). Experiments with red-winged blackbirds indicate they do not give solicitation display without oestradiol treatment (Searcy *unpublished*). Oestradiol can be administered either via injections (see Dabelsteen and Pedersen 1988a, b) or implants. Most often, implants have been used, consisting of lengths of Silastic tubing filled with crystalline 17-ß-oestradiol and plugged at both ends with Silastic adhesive. The tubing commonly used is 1.96 mm in outer diameter. Implants can be filled with hormone using a disposable plastic pipette tip as a funnel. Care should be taken not to touch or inhale the oestradiol, which is a possible carcinogen. Implants have usually been placed under the skin but outside the body wall, in the region of either the back or the abdomen. Implantation requires making an incision 3-5 mm long and usually using a blunt probe to clear a path for insertion of the implant. A topical anaesthetic can be applied before the incision is made. If the implant is pushed away from the incision, it is not necessary to close the incision with sutures.

Steroids diffuse through Silastic at a rate proportional to the implant's surface area (Dzwik and Cook 1966), so with tubes of the same diameter, dosage is controlled by the length of tubing used. As far as I know, dosage has in all cases been ultimately derived from Moore's (1982, 1983) studies with white-crowned sparrows. Moore (1983) used 0, 1, 2 or 3 implants 14 mm long and found that the frequency of "spontaneous" solicitation display increased with increasing dosage through the range of dosages used. On the advice of John Wingfield, we have calculated dosage for other species based on a single one of Moore's implants, adjusting implant length by scaling to the metabolic rate of the species relative to white-crowned sparrows, i.e. multiplying by the ratio of body sizes raised to the 0.7 power (Searcy and Marler 1981, 1984; Searcy 1988, *in press*). Recent experiments with red-winged blackbirds indicate that when the dosage is calculated in this way it yields higher levels of display in this species than are obtained by either halving or doubling the dose (Searcy *unpublished*).

In a few of the species studied so far, females have failed to display in response to song even after oestradiol treatment (Table I). One factor that may explain some of these failures is the degree to which subjects were protected from disturbance during testing.

Table I. A summary of experiments using the solicitation display assay, noting if the display was elicited and the discrimination task was achieved.

Species	*Task*	*Response?*	*Discrimination?*	*Reference*
brown-headed cowbird - *Molothrus ater*	conspecific *vs.* heterospecific song	yes	yes	King & West 1977
	normal *vs.* isolate song	yes	yes	King & West 1977
	home *vs.* alien dialect	yes	yes	King et al. 1980
	dominant *vs.* subordinate song	yes	yes	West et al. 1981
	normal *vs.* altered dialect	yes	yes	West et al. 1979; King & West 1983a
	natural *vs.* foster father's dialect	yes	yes	King & West 1983b
	normal *vs.* altered species markers	yes	yes	Ratcliffe & Weisman 1987
song sparrow - *Melospiza melodia*	conspecific *vs.* heterospecific song	yes	yes	Searcy & Marler 1981
	normal *vs.* altered species markers	yes	yes	Searcy et al. 1981
	large *vs.* small repertoire	yes	yes	Searcy 1984
	normal *vs.* isolate song	yes	yes	Searcy et al. 1985
	isolate *vs.* deafened song	yes	yes	Searcy & Marler 1987
swamp sparrow - *Melospiza georgiana*	conspecific *vs.* heterospecific song	yes	yes	Searcy et al. 1981
	normal *vs.* altered species markers	yes	yes	Searcy et al. 1981
	large *vs.* small repertoire	yes	yes	Searcy et al. 1982
white-crowned sparrow - *Zonotrichia leucophrys*	familiar *vs.* alien dialect	yes	yes	Baker et al. 1981, 1982; Baker 1983
	familiar, alien and hybrid dialects	yes	yes	Baker et al 1987c
	conspecific *vs.* heterospecific song	yes	yes	Spitler-Nabors & Baker 1987
	normal *vs.* isolate song components	yes	yes	Spitler-Nabors & Baker 1987
white-throated sparrow - *Zonotrichia albicollis*	large *vs.* small repertoire	yes	no	Searcy & Marler 1984
	long *vs.* short songs	yes	yes	Wasserman & Cigliano 1991
field sparrow - *Spizella pusilla*	large *vs.* small repertoire	yes	no	Searcy & Marler 1984
sedge warbler - *Acrocephalus schoenobaenus*	conspecific *vs.* heterospecific songs	yes	yes	Catchpole et al. 1984
	large *vs.* small repertoire	yes	yes	Catchpoe et al. 1984
starling - *Sturnus vulgaris*	conspecific *vs.* heterospecific songs	no	no	Hindmarsh 1984
great reed warbler - *Acrocephalus arundinaceus*	short *vs.* long songs	yes	yes	Catchpole et al. 1986
	large *vs.* small repertoire	yes	yes	Catchpole 1986
yellowhammer - *Emberiza citrinella*	large *vs.* small repertoire	yes	yes	Baker et al. 1987a
	home *vs.* alien dialect	yes	yes	Baker et al. 1987a

Table I continued.

Species	*Task*	*Response?*	*Discrimination?*	*Reference*
great tit - *Parus major*	large *vs.* small repertoire	yes	yes	Baker et al. 1986
European blackbird - *Turdus merula*	conspecific *vs.* heterospecific songs	yes	yes/no	Dabelsteen & Pedersen 1988a
	normal *vs.* altered species markers	yes	yes/no	Dabelsteen & Pedersen 1988b
red-winged blackbird - *Agelaius phoeniceus*	conspecific *vs.* heterospecific songs	yes	yes	Searcy & Brenowitz 1988
	normal *vs.* altered species markers	yes	yes/no	Searcy & Brenowitz 1988
	home *vs.* alien dialect	yes	yes	Searcy 1990
	large *vs.* small repertoire	yes	yes	Searcy 1988
indigo bunting - *Passerina cyanea*	song *vs.* no song	yes	yes	Baker & Baker 1988
lazuli bunting - *Passerina amoena*	song *vs.* no song	yes	yes	Baker & Baker 1988
zebra finch - *Taeniopygia gutatta*	conspecific *vs.* heterospecific song	yes	yes	Clayton & Prove 1989
	home *vs.* alien dialect	yes	yes	Clayton & Prove 1989
	large *vs.* small repertoire	yes	yes	Clayton & Prove 1989
	foster *vs.* natural father's songs	yes	yes	Clayton 1990
Bengalese finch - *Lonchura striata*	conspecific *vs.* heterospecific song	yes	yes	Clayton & Prove 1989
	large *vs.* small repertoire	yes	yes	Clayton & Prove 1989
	foster *vs.* natural father's songs	yes	yes	Clayton 1990
canary - *Serinus canaria*	conspecific *vs.* heterospecific songs	yes	yes	Kreutzer & Vallet 1991
	own breed *vs.* other breed	yes	yes/no	Kreutzer & Vallet 1991
common grackles - *Quiscalus quiscula*	conspecific *vs.* heterospecific songs	yes	yes	Searcy *in press*
	large *vs.* small repertoire	yes	yes	Searcy *in press*
cirl bunting - *Emberiza cirlus*	song *vs.* no song	no	no	M.L. Kreutzer *pers comm*
willow warbler - *Phylloscopus trochilus*	rapid *vs.* slow song rates	no	no	B.Arvidsson & R.Neergaard *pers comm*
bobolink - *Dolichonyx oryzivorus*	conspecific *vs.* heterospecific songs	no	no	M.S. Capp *pers comm*
	alpha type *vs.* beta type	no	no	M.S. Capp *pers comm*
blue-winged warbler - *Vermivora pinus*	type I *vs.* type II songs	yes	?	D.E. Kroodsma *pers comm*
chestnut-sided warbler - *Dendoica pensylvanica*	accented *vs.* unaccented songs	yes	?	D.E. Kroodsma *pers comm*

For example, in my initial experiments with female red-winged blackbirds, none of the subjects responded to any of the playbacks. These experiments were performed in the same way as our previous experiments with song and swamp sparrows: subjects were treated with oestradiol and then held singly in sound attenuation chambers. Testing began seven days after treatment. Playback used the tape recorder's internal speaker, with the door of the subject's attenuation chamber open. The observer watched the subject from behind a blind. When this procedure did not work with red-winged blackbirds, I modified the setup by playing songs over a speaker within the attenuation chamber with the chamber door closed, while observing the subject from behind a blind and through a window in the chamber door. Under these conditions, the majority of female redwings gave displays in response to conspecific songs (Searcy 1988; Searcy and Brenowitz 1989). Evidently, female red-winged blackbirds are more nervous in captivity than are sparrows, so greater precautions must be taken to ensure that subjects are unaware of the presence of a human observer.

A second factor that may account for the failure of the method in some cases is the mode of oestradiol treatment. T. Dabelsteen (*pers. comm.*) found that female European blackbirds (*Turdus merula*) did not display in response to song when treated with implants of oestradiol in silastic tubing, whereas they did display when treated with injections of oesradiol. Injections were of 1mg of oestradiol, given subcutaneously and were repeated every 10 days (Dabelsteen and Pedersen 1988a, 1988b). Why injections worked when implants did not is unknown.

In other species, it has proved the case that song alone is not enough to elicit solicitation. Baker et al. (1987b) found that female great tits (*Parus major*), even when treated with oestradiol and tested while isolated from the observer, gave little solicitation display in response to song alone, but that they became very responsive when both songs and the visual stimulus of a live male great tit were presented. Baker and Baker (1988) found that few female lazuli buntings (*Passerina amoena*) and indigo buntings (*Passerina cyanea*) would display in response to either song alone or song plus a silent male, but most would display for the combination of song, male, plus the male's *tseep* call. Even if these other stimuli are needed to elicit display, it is still possible to test for song preferences, by presenting the other stimuli in the same manner in each test and varying only the song playback (e.g. see Baker et al. 1986, 1987a).

Finally, there are some species in which no combination of stimuli has yet been found that will elicit solicitation display from females. B. Arvidsson and R. Neergard (*pers. comm.*) found that female willow warblers (*Phylloscopus trochilus*) did not display in response to song, but they have not yet tried increasing the isolation of the females from the observer, or combining song with other stimuli such as a live male. M. Kreutzer (*pers. comm.*) found that female cirl buntings (*Emberiza cirulus*) would not solicit in response to song and he suspects that a live male and perhaps male calls will also be necessary, as in lazuli and indigo buntings. Hindmarsh (1984), working with female starlings, did not observe any displays in response to conspecific songs, with or without the visual stimulus of a male starling. Finally, M.S. Capp (*pers. comm.*) found that female bobolinks would not display for song even when rigidly isolated from the observer, nor would they display when the visual stimulus of a stuffed male bobolink was presented together with the song. It may be that female bobolinks simply do not perform solicitation; we can find no record of the display in the literature on this species.

A related problem is that, even in species in which the method "works," not all females respond. For example, in red-winged blackbirds, only about 60-70% of all subjects respond with displays to normal, conspecific songs (Searcy and Brenowtiz 1988; Searcy 1990, *unpublished*). In other species, the percentage of responding females is higher (e.g. 89% in common grackles (*Quiscalus quiscula*), Searcy (*in press*)), but it is rarely 100%. The failure of some individuals to respond means, of course, that a larger sample of subjects must be used in order to demonstrate particular discriminations.

Ratings The solicitation display assay rates only moderately well on practicality. One difficulty is that captive subjects must be used. A second is that some startup time is involved, in that one must wait at least several days after implantation before beginning testing. A third difficulty is that implantation involves an operation, albeit an extremely simple one. On the other side of the ledger, the method does not require any sophisticated skills or equipment. We have used (expensive) sound attenuation chambers to isolate subjects, especially when we are running tests on other subjects, but it is possible to get around this requirement by performing the tests in a room isolated from the area where subjects are housed, moving each subject to the room before a test and allowing some period for acclimation (Baker et al. 1981, 1986, 1987a).

The method does fairly well in terms of generality. It has been used successfully to measure female response in more bird species than any other method, perhaps in more than all other methods put together. On the other hand, there have been several failures, so one cannot have complete confidence in the method working with a new species. The method does very well in terms of sensitivity, as can been seen from the list of types of discriminations that have been successfully demonstrated (Table I).

Finally, the solicitation display assay rates very high on interpretability, in particular higher than do territorial playbacks to males. There can be no doubt that a male benefits from having a song that stimulates females to solicit copulation. The display is closely tied to the act of copulation and thus can be argued to reflect mating preferences, at least in some cases. Empirical evidence supports this interpretation in brown-headed cowbirds (West et al. 1981; Eastzer et al. 1985) and *Acrocephalus* warblers (Catchpole 1980, 1986; Catchpole et al. 1984).

Heart Rate Monitoring

A method that has been used only rarely, but which has potential for certain applications, is heart rate monitoring. A good example of this method is provided by the work of Diehl and Helb (1986) with European blackbirds (*Turdus merula*). Captive subjects were used, but these were not restrained, so the animals had to be equipped with small radio transmitters (weight 4-5g) to convey the heart rate signals to a tape recorder. Each subject had two electrodes attached subcutaneously (by sewing), one on the left side of the abdomen and one on the right shoulder. Wires ran under the skin from the electrodes to the transmitter, which was attached to the back. Diehl and Helb found their subjects responded to bird vocalisations with first an acceleration and then a deceleration of heart rate. Time to return to baseline was longer in response to conspecific than to heterospecific song.

Dooling and Searcy (1980) found a similar pattern of acceleration and deceleration in naive young swamp and song sparrows of unknown sex. These authors tested restrained birds, so that electrodes could be attached with collodion only and the wires could be lead directly to a recorder, with no intervening radio transmitter. Swamp sparrows showed longer responses to own species than other species, but song sparrows did not. Zimmer (1982) found that female chiffchaffs accelerated heart rate in response to conspecific male songs of the same dialect, but not for foreign dialects or heterospecific songs.

Ratings This method seems more difficult to carry out than most. Electrodes must be implanted under the skin of the subjects and perhaps the leads as well. One must either work with restrained captives, or use radio transmitters, which can fail, disturb the subject, or be lost. Some specialised equipment is needed for analysing the heart rate signals, in addition to the usual playback equipment.

Although few species have yet been tested with this method, it would seem likely to be general, in the sense that most species must give some response in heart rate to auditory signals. On the other hand, it is not clear that all species will give the same types of response. In three species a pattern of acceleration followed by deceleration has been found (Dooling and Searcy 1980; Diehl and Helb 1986), but in a fourth only acceleration was noted (Zimmer 1982). Little information is available on sensitivity. Response is clearly finely graded, but whether response is closely dependent on the stimulus type is unknown. It is usually necessary to average over a large number of stimulus presentations to obtain meaningful results.

The heart rate method also has weaknesses in interpretability. It seems unlikely that heart rate responses can be interpreted in terms of sexual preferences. Diehl and Helb (1986) argue that deceleration is a measure of recognition, but this argument is weakened by the fact that not all studies have demonstrated deceleration in response to conspecific song (Zimmer 1982).

Parental Behaviour

Hinde and Steel (1976) exposed captive female canaries (*Serinus canaria*) to songs and measured the number of strings these females removed from a dispenser and placed in a nest site. Females performed more nest building when played conspecific rather than heterospecific song. Using a similar method, Kroodsma (1976) showed that captive female canaries performed more nest building and laid more eggs when played large repertoires of conspecific syllable types than when played small repertoires. Spitler-Nabors and Baker (1983) found that female white-crowned sparrows, after being treated with oestradiol, performed more nest building for familiar than for alien dialects. On the other hand, D. Kroodsma (*pers. comm.*), working with swamp sparrows and T. Dabelsteen (*pers. comm.*) working with European blackbirds, both found that captive females not treated with oestradiol did not increase nestbuilding in response to playback of male song.

Ratings In terms of practicality, this method fares about as well as the solicitation display assay; as in the latter method, captive subjects must be used, oestradiol treatment is probably necessary and the experiments may take considerable time to perform. Little experience exists to judge generality, but it seems likely that the method will work without oestradiol only in species, like the canary, that readily breed in captivity. It is unknown whether oestradiol and song together elicit parental behaviour in species other than the white-crowned sparrow. From the small sample of studies performed thus far, it appears that the method is quite sensitive. Finally, the method rates high on interpretability, in that it is clearly to a male's advantage to be able to stimulate parental behaviour in conspecific females. Further, results can be reasonably interpreted in terms of mating preferences, though this perhaps requires more of a logical jump than for the solicitation display assay.

Conclusions

My ratings of the six playback methods reviewed above are summarised in Table II, using a three point scale (*poor - fair - good*) for each criterion. None of these six playback methods for female birds receives a *good* rating on all four criteria, but then neither would the standard male method, which I would rate *good* on practicality, generality and sensitivity, but *poor* on interpretability. Each of the female methods receives a *good* rating on at least one criterion. Thus, none of the methods is perfect, but all have virtues that make them suitable for some applications.

Table II. Ratings of playback methods used with female birds. The rating scale and headings are explained in the text.

Method	*Practicality*	*Generality*	*Sensitivity*	*Interpretability*
Territorial Playbacks	good	poor	poor-good	poor
Phonotaxis (lab)	fair	fair	good	fair
Phonotaxis (field)	fair	poor	poor	good
Solicitation Assay	fair	good	good	good
Heart Rate	poor	good?	?	poor
Parental Behaviour	fair	?	good?	good

Although I have rated a couple of these female methods as *good* in terms of generality, none seems likely to be as general as the male method of territorial playback. The solicitation assay is the method that has been used most extensively with females and which has succeeded in the greatest number of species, but even with this method there have been a number of stubborn failures. It seems probable, then, that no one method of measuring female response to song will work in all species. At the same time, it also seems probable that, with six methods available, at least one them will succeed with any species of interest. Therefore, in birds there is no longer any methodological reason for the bias towards studying male response to song rather than female response and, indeed, the previous imbalance in such studies has already begun to be corrected.

Acknowledgments

I would first like to thank Pete McGregor for giving me the opportunity to write this paper. I would also like to express my gratitude to all the people who shared advice on and experience with these methods, including Bjorn Arvidsson, Mike Capp, Nicky Clayton, Torben Dabelsteen, Tom Dickinson, Dag Eriksson, Bruce Falls, Andrew Hindmarsh, Andy Horn, Don Kroodsma, Michel Kreutzer, Peter Marler and D. James Mountjoy. I thank the National Science Foundation (BNS-890844) for financial support.

References

Baker, M.C. 1983. The behavioral response of female Nuttall's white-crowned sparrows to male song of natal and alien dialects. *Behav. Ecol. Sociobiol.*, **12**, 309-315.

Baker, M.C. & Baker, A.E.M. 1988. Vocal and visual stimuli enabling copulation behavior in female buntings. *Behav. Ecol. Sociobiol.*, **23**, 105-108.

Baker, M.C., Bjerke, T.K., Lampe, H. & Espmark, Y. 1986. Sexual response of female great tits to variation in size of males' song repertoires. *Amer. Natur.*, **128**, 491-498.

Baker, M.C., Bjerke, T.K., Lampe, H. & Espmark, Y. 1987a. Sexual response of female yellowhammers to differences in regional song dialects and repertoire sizes. *Anim. Behav.*, **35**, 395-401.

Baker, M.C., McGregor, P.K. & Krebs, J.R. 1987b. Sexual response of female great tits to local and distant songs. *Ornis. Scand.*, **18**, 186-188.

Baker, M.C., Spitler-Nabors, K.J. & Bradley, D.C. 1981. Early experience determines song dialect responsiveness of female sparrows. *Science*, **214**, 819-820.

Baker, M.C., Spitler-Nabors, K.J. & Bradley, D.C. 1982. The response of female white-crowned sparrows to songs from their natal dialect and an alien dialect. *Behav. Ecol. Sociobiol.*, **10**, 175-179.

Baker, M.C., Spitler-Nabors, K.J., Thompson, A.D. Jr. & Cunningham, M.A. 1987c. Reproductive behaviour of female white-crowned sparrows: effects of dialects and synthetic hybrid songs. *Anim. Behav.*, **35**, 1766-1774.

Becker, P.H. 1982. The coding of species-specific characteristics in bird sounds. In: *Acoustic Communication in Birds*. (Ed. by D.E. Kroodsma, E.H. Miller & H. Ouellet), pp. 213-252. Academic Press; New York.

Catchpole, C.K. 1980. Sexual selection and the evolution of complex songs among European warblers of the genus *Acrocephalus*. *Behaviour*, **74**, 149-166.

Catchpole, C.K. 1986. Song repertoires and reproductive success in the great reed warbler *Acrocephalus arundinaceus*. *Behav. Ecol. Sociobiol.*, **19**, 439-445.

Catchpole, C.K., Leisler, B. & Dittami, J. 1984. Differential responses to male song repertoires in female songbirds implanted with oestradiol. *Nature*, **312**, 563-564.

Catchpole, C.K., Leisler, B. & Dittami, J. 1986. Sexual differences in the responses of captive great reed warblers (*Acrocephalus arundinaceus*) to variation in song structure and repertoire size. *Ethology*, **73**, 69-77.

Clayton, N.S. 1988. Song discrimination learning in zebra finches. *Anim. Behav.*, **36**, 1016-1024.

Clayton, N.S. 1990. Subspecies recognition and song learning in zebra finches. *Anim. Behav.*, **40**, 1009-1017.

Clayton, N.S., & Prove, E. 1989. Song discrimination in female zebra finches and Bengalese finches. *Anim. Behav.*, **38**, 352-354.

Dabelsteen, T. & Pedersen, S.B. 1988a. Song parts adapted to function both at long and short ranges may communicate information about the species to female blackbirds *Turdus merula*. *Ornis. Scand.*, **19**, 195-198.

Dabelsteen, T. & Pedersen, S.B. 1988b. Do female blackbirds, *Turdus merula*, decode song in the same way as males? *Anim. Behav.*, **36**, 1858-1860.

Diehl, P., & Helb, H.-W. 1986. Radiotelemetric monitoring of heart-rate responses to song playback in blackbirds (*Turdus merula*). *Behav. Ecol. Sociobiol.*, **18**, 213-219.

Dooling, R. & Searcy, M. 1980. Early perceptual selectivity in the swamp sparrow. *Dev. Psycholbiol.*, **13**, 499-506.

Dzwik, P. J. & Cook, D. 1966. Passage of steroids through silicone rubber. *Endocr.*, **78**, 208-211.

Eastzer, D.H., King, A.P. & West, M.J. 1985. Patterns of courtship between cowbird subspecies: evidence of positive assortment. *Anim. Behav.*, **33**, 30-39.

Eriksson, D. 1991. The significance of song for species recognition and mate choice in the pied flycatcher *Ficedula hypoleuca*. *Unpubl. Ph.D. thesis*, Uppsala University, Uppsala, Sweden.

Eriksson, D. & Wallin, L. 1986. Male bird song attracts females - a field experiment. *Behav. Ecol. Sociobiol.*, **19**, 297-299.

Hindmarsh, A.M. 1984. Vocal mimicry in starlings. *Unpubl. D.Phil thesis*, University of Oxford, Oxford, UK.

Hinde, R.A. & Steel, E. 1976. The effect of male song on an oestrogen-dependent behaviour pattern in the female canary (*Serinus canarius*). *Hormones & Behaviour*, **7**, 293-304.

Howard, E. 1920. *Territory in bird life*. Collins Sons & Co.; London.

King, A.P. & West, M.J. 1977. Species identification in the North American cowbird: appropriate responses to abnormal song. *Science*, **195**, 1002-1004.

King, A.P. & West, M.J. 1983a. Dissecting cowbird song potency: assessing a song's geographic identity and relative appeal. *Z. Tierpsychol.*, **63**, 37-50.

King, A.P. & West, M.J. 1983b. Female perception of cowbird song: a closed developmental program. *Develop. Psychobiol.*, **16**, 335-342.

King, A.P., West, M.J. & Eastzer, D.H. 1980. Song structure and song development as potential contributions to reproductive isolation in cowbirds (*Molothrus ater*). *J. Comp. Physiol. Psychol.*, **94**, 1028-1039.

Krebs, J.R., Ashcroft, R. & Webber, M.I. 1978. Song repertoires and territory defence in the great tit. *Nature*, **271**, 539-542.

Kreutzer, M.L. & Vallet, E.M. 1991. Differences in the responses of captive female canaries to variation in conspecific and heterospecific songs. *Behaviour*, **117**, 106-116.

Kroodsma, D.E. 1976. Reproductive development in a female songbird: differential stimulation by quality of male song. *Science*, **192**, 574-575.

Miller, D.B. 1979a. The acoustic basis of mate recognition by female zebra finches (*Taeniopygia guttata*). *Anim. Behav.*, **27**, 376-380.

Miller, D.B. 1979b. Long-term recognition of father's song by female zebra finches. *Nature*, **280**, 389-391.

Milligan, M.M. & Verner, J. 1971. Interpopulational song dialect discrimination in the white-crowned sparrow. *Condor*, **73**, 208-213.

Moore, M.C. 1982. Hormonal response of free-living male white-crowned sparrows to experimental manipulation of female sexual behavior. *Horm. Behav.*, **16**, 323-329.

Moore, M.C. 1983. Effect of female sexual displays on the endocrine physiology and behaviour of male white-crowned sparrows, *Zonotrichia leucophrys*. *J. Zool.*, **199**, 137-148.

Mountjoy, D.J. & Lemon R.E. 1990. Song as an attractant for male and female European starlings, and the influence of song complexity on their response. *Behav. Ecol. Sociobiol.*, **28**, 97-100.

Payne, R.B. 1973. Behavior, mimetic songs and song dialects, and relationships of the parasitic indigobirds (*Vidua*) of Africa. *Ornith. Monogr.*, **11**, 1-333.

Payne, R.B., Payne, L.L. & Rowley, I. 1988. Kin and social relationships in fairy-wrens: recognition by song in a cooperative bird. *Anim. Behav.*, **36**, 1341-1351.

Petrinovich, L. & Patterson, T.L. 1981. The responses of white-crowned sparrows (*Zonotrichia leucophrys nuttalli*) to songs of different dialects and subspecies. *Z. Tierpsychol.*, **57**, 1-14.

Ratcliffe, L.M. & Weisman, R. 1987. Phrase order recognition by brown-headed cowbirds. *Anim. Behav.*, **35**, 1260-1262.

Salomon, M. 1989. Song as a possible reproductive isolating mechanism between two parapatric forms. The case of the chiffchaffs *Phylloscopus c. collybita* and *P. c. brehmii* in the western Pyrenees. *Behaviour*, **111**, 270-290.

Searcy, W.A. 1984. Song repertoire size and female preferences in song sparrows. *Behav. Ecol. Sociobiol.*, **14**, 281-286.

Searcy, W.A. 1988. Dual intersexual and intrasexual functions of song in red-winged blackbirds. *Proc. XIX Congr. Intern. Ornith.*, 1373-1381.

Searcy, W.A. 1989. Function of male courtship vocalizations in red-winged blackbirds. *Behav. Ecol. Sociobiol.*, **24**, 325-331.

Searcy, W.A. 1990. Species recognition of song by female red-winged blackbirds. *Anim. Behav.*, **40**, 1119-1127.

Searcy, W.A. *in press*. Song repertoire and mate choice in birds. *Amer. Zool.*

Searcy, W.A. & Brenowitz, E.A. 1988. Sexual differences in species recognition of avian song. *Nature*, **332**, 152-154.

Searcy, W.A. & Marler, P. 1981. A test for responsiveness to song structure and programming in female sparrows. *Science*, **213**, 926-928.

Searcy, W.A. & Marler, P. 1984. Interspecific differences in the response of female birds to song repertoires. *Z. Tierpsychol.*, **66**, 128-142.

Searcy, W.A. & Marler, P. 1987. Response of sparrows to songs of isolation-reared and deafened males: further evidence for innate auditory templates. *Devel. Pyschobiol.*, **20**, 509-519.

Searcy, W.A., Marler, P. & Peters, S.S. 1981. Species song discrimination in adult female song and swamp sparrows. *Anim. Behav.*, **29**, 997-1003.

Searcy, W.A., Marler, P. & Peters, S.S. 1985. Songs of isolation-reared sparrows function in communication, but are significantly less effective than learned songs. *Behav. Ecol. Sociobiol.*, **17**, 223-229.

Searcy, W.A., Searcy, M.H. & Marler, P. 1982. The response of swamp sparrows to acoustically distinct song types. *Behaviour*, **80**, 70-83.

Spitler-Nabors, K.J. & Baker, M.C. 1983. Reproductive behavior by a female songbird: differential stimulation by natal and alien song dialects. *Condor*, **85**, 491-494.

Spitler-Nabors, K.J. & Baker, M.C. 1987. Sexual display response of female white-crowned sparrows to normal, isolate, and modified conspecific songs. *Anim. Behav.*, **35**, 380-386.

Wasserman, F.E. and Cigliano, J.A. 1991. Song output and stimulation of the female in white-throated sparrows. *Behav. Ecol. Sociobiol.*, **29**, 55-59.

West, M.J., King, A.P. & Eastzer, D.H. 1981. Validating the female bioassay of cowbird song: relating differences in song potency to mating success. *Anim. Behav.*, **29**, 490-501.

West, M.J., King, A.P., Eastzer, D.H. & Staddon, J.E.R. 1979. A bioassay of isolate cowbird song. *J. Comp. Physiol. Psych.*, **93**, 124-133.

Yasukawa, K. 1981. Song repertoires in the red-winged blackbird (*Agelaius phoeniceus*): a test of the Beau Geste hypothesis. *Anim. Behav.*, **29**, 114-125.

Zimmer, U.E. 1982. Birds react to playback of recorded songs by heart rate alteration *Z. Tierpsychol.*, **58**, 25-30.

FIELD EXPERIMENTS ON THE PERCEPTION OF SONG TYPES BY BIRDS

Andrew G. Horn

Division of Life Sciences
Scarborough Campus
University of Toronto
Scarborough, Ontario M1C 1A4
CANADA

Introduction

The songs of passerine birds are enormously variable, but usually the many variations on the species-specific song can be divided into a set, or repertoire, of discrete, stereotyped songs known as song types (Hartshorne 1973; Dobson and Lemon 1975). Song types have been shown to have various functions in territory defence and mate attraction (reviews in Krebs and Kroodsma 1980; Catchpole 1982; Searcy and Andersson 1988). If song types are to serve any function, birds must be able to tell them apart; in fact this may be the main reason why song types are so different from one another (Kroodsma 1982). However, only recently have researchers used playbacks to ask how different songs have to be for birds to treat them as different types (Horn and Falls 1988a; Falls et al. 1988; Weary et al. 1990). At first glance, one might think that such perceptual questions are best answered in the laboratory. However, this is a question as much about how birds *evaluate and respond to* differences between songs as about whether they are *capable of sensing* the differences. Birds might treat the same structural contrast between songs as negligible in one situation and crucial in another. Provided enough is known about natural singing behaviour, this possibility can be tested through playback experiments that mimic these different situations.

The purpose of this paper is to provide a brief overview of the main playback techniques that can be used to ask how birds classify their song types. I have narrowed the focus to the responses of territorial males because most field playback methods are designed with males in mind, although some of my comments probably also apply to song perception by females (see chapter by Searcy in this volume). After outlining the methods, I show that they ask birds to make different sorts of comparisons among songs, and that in natural interactions, each of these tasks may come into play in different situations. Finally, I point out that these differences should be considered in the design of playback experiments both in the field and the lab, and that studying them can tell us a great deal about how repertoires have evolved.

Playback and Studies of Animal Communication
Edited by P.K. McGregor, Plenum Press, New York, 1992

Song Types

Before introducing the different methods, I should give at least a rough idea of what we mean by a song type (for a more extended, critical, treatment see Kroodsma 1982). The singing of most songbirds consists of many variations on the species-specific song. If one samples enough of the songs of a given individual, one usually finds that they can be classified into a few categories within which variation in temporal and frequency parameters is low relative to the differences among categories. These classes are known as song types and the number of song types each individual can sing is known as its repertoire size. Because birds learn their songs from each other, if one samples the songs from a whole population of individuals, one often finds that the song forms found in the population as a whole can also be sorted into song types.

People have several problems in classifying song types (Kroodsma 1982). Researchers may come up with very different classifications depending on how they approach these problems. In fact the main reason for doing the sorts of playback experiments reviewed here is to find out how the birds themselves deal with these problems. Later in the paper I show that different playback methods imply different solutions and also that birds may use different solutions depending on the situation. For now I will only briefly introduce them.

First, how different must different songs be to be classified as different types? This is the central question that all the studies below address and all of them approach it by varying the similarity of the playback song to a particular reference song or set of songs (an example of a reference song would be one the test bird sings). The other problems are handled in different ways by different methods. How many songs need one sample before one is sure that two groups of songs do not intergrade? If a well-sampled individual sings two distinct song forms, but the whole range of variation between the song forms can be found in the repertoires of other birds in the population, are they still song types? One might divide the variation in songs very differently, depending on how one samples songs, especially if one samples from a single individual or from the population as a whole. Ultimately, these questions can only be answered by putting them to the birds. All the studies I'll be surveying ask "how different do songs have to be to be treated as different types." However, depending on the playback method used, they may give birds different samples of songs on which to base their decision. Humans come up with different answers depending on the method used, so we should not be surprised if the birds do too.

Methods for Assaying the Perception of Song Types

Three main methods can be used to test for the sensitivity of birds to song types. Broadly, they differ according to whether a single stimulus is presented at each trial and according to the nature of the response.

Direct Comparison of Response Strengths

The most straightforward technique is to play back songs of different types and see if there is a difference in conventional measures of response strength, for example, the number of flights around the speaker and the number of songs. Significantly different responses have been found in species in which different song types are used in very different situations and therefore presumably these song types have very different meanings or functions (e.g. territorial *v.* courtship songs; Nelson and Croner 1991). However, species which use different song types in equivalent situations do not show differences in response (Nelson and Croner 1991). Also, in many species, associations between song types and different situations are so subtle (e.g. Smith et al. 1978), or imperfect (Ritchison 1988), that even if differences in response could be detected, a conventional playback may be but a crude approximation of a natural interaction.

All these difficulties conspire to make comparison of response strengths a rather limited technique for studying the perception of song types. Thus although direct comparison of response strengths has been widely used in other areas of song research, to my knowledge no-one has varied the relative similarity of songs in such an experiment to ask how different songs must be to elicit different responses.

Matched Countersinging

In many species, if a song type that is in a bird's repertoire is played back, the bird will tend to answer with a song of the same type (e.g. Krebs et al. 1981; Falls 1985). This matched countersinging (also termed song matching) provides a very direct way of asking the bird whether it classifies a playback song as similar to one of its own songs. Matched countersinging experiments also mimic natural interactions very nicely, because they can be brief and conducted over a long distance.

There are several slight variations on matched countersinging which can also be used to study the perception of song types. Some species do not match song types *per se*, but reply with a song type that shares particular features of the playback song, like its temporal cadence (e.g. McArthur 1986). In other species, singers significantly *avoid* matched countersinging (Whitney 1991), or change their singing when an experimenter matches them with playback (e.g. Todt 1975; Horn and Falls 1986), providing additional assays of song perception.

Several workers have used matched countersinging to study the perception of song types. They have varied the similarity of the playback songs to the bird's own by selecting natural variations (e.g. Falls et al. 1988) or by editing natural songs (e.g. Wolffgramm and Todt 1982; Weary et al. 1990).

An example of such an experiment involves western meadowlarks, *Sturnella neglecta*, which match playback songs (Falls 1985). In this experiment (Falls et al. 1988) males were played a song that was either the same type as one in their repertoire or of a very similar, but distinct, type (i.e. no intermediates between the two songs could be found; Horn and Falls 1988b). If they recognised the similarity between the song types, then males should answer with the similar song type (i.e. near-match), even if the playback song type was not in their repertoire. However, the frequency of near-matching was not above chance levels, showing that, at least using this matched countersinging paradigm, the birds made relatively sharp distinctions among their song types.

Song Sequences

Perhaps the commonest playback technique used on song repertoires as a whole is to compare responses to different sequences of song types. On each trial, a series of song types is played and conventional strength of response measures are recorded. This technique has been used to test for functional responses to song repertoires (reviewed in Kroodsma 1990) and different singing organisations (e.g. Simpson 1984; Kramer et al. 1985; Horn and Falls 1988a).

A variant of this technique is to test for a change in behaviour that coincides with the switch in song types. In some species this behaviour is a switch in song type or change in singing rate (e.g. Falls et al. 1990), in others it is a recovery of response after habituation to the first song type (e.g. McGregor 1986).

Two experiments have varied the relative similarity between song types presented in sequence (Horn and Falls 1988a; Falls et al. 1990). For example, Horn and Falls (1988a) played back sequences either of contrasting, or of similar, song types to western meadowlarks. Similar song types were defined in the same way as in the near-matching experiment described above (Falls et al. 1988). Sequences of contrasting song types stimulated significantly stronger responses in terms of number of song switches, number of flights, latency to closest approach and closest approach. Thus in this playback paradigm males showed a sensitivity to these similarities among song types, even though they made sharp distinctions among similar songs in the matched countersinging tests.

Overview of the Three Techniques

In summary, three main assays of the sensitivity of territorial males to song types are available: response strengths to different song types; matched countersinging; and song sequences. No doubt other techniques could be used, for example the dual speaker experiments that are used so much in work on insects and frogs (Searcy 1983; chapters by Gerhardt, and Klump and Gerhardt in this volume). However, given the dominance of the above playback designs in research on song repertoires, it is important to point out their differences.

Comparison of the Three Methods

Each of the above methods asks the birds to make different sorts of comparisons among songs. They involve different perceptual tasks that address different levels of song variation and the relevance of each depends on how birds use their songs in natural interactions.

Perceptual Tasks

In order to classify an object, you must be able to distinguish it from other objects. However, we commonly classify many objects together even though we can tell them apart, for example when we classify different shades of blue as blue or when we classify Chihuahuas and Dobermans as dogs (Harnad 1987; Medin and Barsalou 1987).

In tests of response strength and matched countersinging to different song types, a positive result indicates that the bird not only recognised that, for example, song A was a different song from song B, but also that it identified it as song type A, or more accurately as one of a class of songs for which a given response (level of responding or vocal answer) is appropriate. In tests of responses to song sequences, however, a positive result merely shows that the bird recognised that the songs were different; it does not necessarily show which particular song types they were.

This contrast I am making between the methods applies only to the *minimum* perceptual tasks they require of the birds. For example, identification (that is, classification of an isolated stimulus; Harnad 1987) may involve discrimination between a song and some internal standard (see below). Conversely, a bird might judge the difference between successive songs by identifying them first. For example, in eastern meadowlarks *Sturnella magna*, males match both song types in a sequence, whatever the relative similarity of the song types, suggesting that songs were identified at the same time as they were discriminated (Falls et al. 1990). Whatever the mechanism, the difference between the two experimental paradigms remains and it may lead to apparently conflicting results. In general, animals make finer distinctions in discrimination tasks than in identification tasks (Dooling 1989), although this depends on many factors. Field results from both sorts of task are only available for western meadowlarks, which appear to make finer distinctions in matched countersinging than in song sequence presentations (Horn and Falls 1988a; Falls et al. 1988), although there are many alternative explanations for this difference.

Standards for Comparison

Another difference between the playback methods that is closely related to the last is the standards of comparison that the birds may use to evaluate the test songs. For different song types to have different meanings, at least ones that are shared across the population, each individual must be able to associate particular song types with particular responses. This would involve comparison with internal standards that are somehow representative of song types in the population as a whole. In matched countersinging, at a minimum the bird need only compare the songs it hears with the song types in its own repertoire. When songs are presented in sequence, each song can be compared with the preceding song.

Playback experiments involving matched countersinging and sequences of song types support the idea that birds are making different sorts of comparison in each case. Several studies have shown that the likelihood of matched countersinging is at least partly a function of the similarity of the playback song to the bird's own song (Falls et al. 1982; Whitney and Miller 1983; Falls 1985; McArthur 1986), supporting the idea that they are comparing the playback song with their own. In playbacks of song sequences, some studies show habituation to each song type and recovery when the song type is switched (e.g. McGregor 1986), which suggests that switches are stimulating so long as one song type is different from the next. Further support for this idea comes from studies in which greater differences in the structure of successive song types are more effective at eliciting responses (Horn and Falls 1988a; Falls et al. 1990). Of course, provided the bird is familiar with the song types involved, sequential songs could be compared with the bird's own songs as well as with each other (e.g. Falls et al. 1990).

In effect the use of different standards means that, depending on the method one chooses, one might be addressing song variation across the population (comparison of response strengths), within one individual's total song output (matched countersinging), or within an isolated sequence of songs (song sequences). Birds might come up with different classifications in response to these playbacks because we ourselves classify song types differently according to each of these criteria. In many species, individuals may have two or more discrete classes of songs, even though other individuals in the population might sing the whole range of variants between these two versions (e.g. Slater et al. 1981; McGregor and Krebs 1982; Schroeder and Wiley 1983a, 1983b; Horn and Falls 1988b). Are these distinct song types, i.e. should songs be classified according to variation within or between individuals? The bird's answer probably depends on how the songs are used and therefore on what messages they may contain.

Different Messages

Like many other display behaviours, song types may have different meanings depending on whether they are presented alone, as part of a vocal interchange between two or more birds, or as a sequence of different song types (Smith 1986). Although these different modes of communication have usually been discussed in interspecific comparisons, there is growing evidence that many species use more than one mode in within-species communication (Derrickson 1987; Horn and Falls 1991; chapter by Dabelsteen in this volume).

Many species of passerine use different song types in different situations (examples from different taxa: Tyrannidae - Smith 1970, 1988; Mimidae - Derrickson 1987; Paridae - Gaddis 1983; Schroeder and Wiley 1983a; Parulinae - reviewed in Spector *in press*; Vireonidae - Smith et al. 1978; Emberizinae - Payne 1979; Ritchison 1988; Nelson and Croner 1991; Icterinae - Cosens and Falls 1984; Trainer 1987). When countersinging in these species has been studied, neighbouring individuals also engage in some form of matched countersinging (Payne 1979; Schroeder and Wiley 1983b; Derrickson 1987; Trainer 1988). Many of these species also vary the number of times each song type is repeated, as a message of their readiness to engage in territorial or sexual interactions (Smith 1970; 1990; Schroeder and Wiley 1983a; Derrickson 1988; Ritchison 1988; Trainer 1988). Thus song types, exchanges of song types and switches of song types are all used as displays.

Multiple functions are also apparent in the singing behaviour of birds for which there appears to be no association of song types with particular situations. Most of these species engage in some form of matched countersinging and also vary the number of times they repeat song types in different situations (e.g. Kroodsma 1977; Kroodsma and Verner 1978; Falls and d'Agincourt 1982; d'Agincourt and Falls 1984).

Thus in most species that have been studied, the messages conveyed by a single song in isolation, a song that can be matched and a sequence of song types may be very different. Often the repertoire is subdivided according to how song variants are used. For example, there may be one set of song types with separate messages and another that is used mainly in countersinging (e.g. Trainer 1988). Other species have broad categories of songs shared widely across the population, which carry different messages; finer divisions of these categories are found in the singing of individual birds, in some instances being used mainly in matched countersinging (MacNally and Lemon 1985; Schroeder and Wiley 1983b; Horn et al. *submitted*). We should not be surprised, then, if receivers treat the same difference between songs as significant in one situation and trivial in another.

Discussion

In summary, the three main ways of comparing song types in the field (response strength, matched countersinging and song sequences) differ in terms of the perceptual tasks they give the birds, the level of song variation they address and the sorts of display behaviours they mimic. These differences are important in several ways. First, they obviously must be considered in the design of field tests of how birds perceive song types. Second, they expose a number of parameters that might be important in the perception of song types and that could be teased apart using laboratory studies (see chapter by Ratcliffe and chapter by Weary in this volume). Third, differences in how birds classify song types might clarify how variation in song organisation has co-evolved with how birds perceive song.

Design of Playback Experiments

Whatever playback one uses to study song types, one must decide whether the birds should show discrimination or identification and what pool of songs the birds must select among to evaluate the stimuli. Do song types have to be discrete or merely distinguishable; do they have to be universally shared or can they be peculiar to certain individuals? Although the answers to these questions will of course vary according to a researcher's interests and study species, they should guide his or her choice of playback method.

The answers should also guide the experimenter's choice of stimuli, particularly what pool of songs the bird is expected to classify into song types. For example, two experiments might test a bird's response to a collection of song types from the same individual in one case and to song types from many different individuals in another. The first experiment would explore how birds discriminate among the songs of a given individual, the other would test how they discriminate among songs found in the population as a whole. Neither design is inherently better than the other; they are testing different levels of song variation, both of which might be relevant to different sorts of natural interactions.

Laboratory and Field Studies

Field studies of song type perception use natural responses to what we hope are relatively natural stimuli under natural noise levels, when by "natural" we mean what the birds would normally encounter in interactions with conspecifics. However, there are disadvantages to having to rely on natural behaviours. We usually do not know the previous experience of the bird, except as revealed by the songs it sings (but see McGregor and Avery 1986; chapter by Nelson in this volume). Also, it is very hard to tease apart the relative contributions of noise levels, sensory capabilities, attentiveness and display messages / meanings to a bird's responsiveness (Wiley and Richards 1982), although it might be possible (an example would be field studies of the role of degradation see McGregor et al. 1983; McGregor and Falls 1984).

A more efficient approach might be to vary these parameters systematically in the laboratory. Do birds process song types differently in identification versus discrimination tasks? What are the differences in performance when they are trained on single exemplars of each song type (perhaps drawn from their own repertoire) versus a wide range of exemplars drawn from many different individuals? How do all these processes respond to increased noise levels, doped sensory capabilities, distracted attention and different reinforcers?

Functions of Song Repertoires

For many functional questions about song types, it is often sufficient to demonstrate particular beneficial effects of song types on responses. However, functional approaches are often less concerned about whether these effects of song types are over and above those stimulated by song variation in general. If we are to understand the selective pressures on the *differentiation* between song types, we must compare the effects of song types with the effects of other forms of song variation (Kroodsma 1990).

Different perceptual mechanisms will lead to different pressures on song differentiation. The more birds perceive song types by discrimination of sequential songs, the more songs within repertoires might be expected to diverge. If birds mainly identify song types, then the songs of different repertoires might be expected to converge (Kroodsma 1982). By varying the similarity between playback songs, we could test such hypotheses directly, but only if we use a variety of playback designs.

Such studies might reveal adaptive differences in bird's strategies for perceiving song types. In particular, each perceptual strategy might be associated with different singing strategies. Within species, are rapid switches among song types directed primarily at naive or inattentive listeners that are less likely to identify each song type (Horn and Falls 1991)? Among species, populations that use song types in frequent, short-term interactions or in long-distance advertising appear to have showier singing performances, including larger repertoires, than species that use them in prolonged interactions with established territorial neighbours (Kroodsma 1977; Catchpole 1982). Perhaps new arrivals in an area might have to rely on their ability to discriminate songs presented in sequence to distinguish song types, for example to assess a male's repertoire size. Once they have settled in an area, they might be better able to identify, rather than merely discriminate, the local song types (Craig and Jenkins 1982). Conversely, showier singing performances might be specialised for situations in which receivers are more easily duped (Krebs and Dawkins 1984).

Conclusions

There are different methods for testing the response of birds to different song types. Each assumes different perceptual mechanisms, and each may send different messages to the birds. These differences among the methods should be considered in the design of such playback experiments. Future studies should involve both naturalistic observations and laboratory studies that mimic field situations accurately. This would help us distinguish among the different perceptual strategies that may have co-evolved with different sorts of song repertoires.

Acknowledgements

Above all, I thank Bruce Falls for many helpful discussions; the useful ideas in this chapter are no doubt his. The writings of W. John Smith were a strong influence, and I am grateful to Tom Dickinson for discussing these and other topics on animal communication with me. I thank all those who have discussed song types with me; Pete McGregor, Marty Leonard, Laurene Ratcliffe and Danny Weary being only my most recent victims. I could not have attended this workshop without the help of Claire Horn, Laura Horn and Marty Leonard, and of course the invitation of Pete McGregor and all the other workshop participants, to whom I am very grateful. Thanks go to the Animal Research Centre at Agriculture Canada, Ottawa, especially Dave Fraser and Brian Thompson, for discussions and the use of their facilities.

References

Catchpole, C.K. 1982. The evolution of bird sounds in relation to mating and spacing behaviour. In: *Evolution and Ecology of Acoustic Communication in Birds. Vol. I.* (Ed. by D.E. Kroodsma, E.H. Miller & H. Ouellet), pp. 297-319. Academic Press, New York.

Cosens, S.E. and Falls, J.B. 1984. Structure and use of song in the yellow-headed blackbird (*Xanthocephalus xanthocephalus*). *Z. Tierpsychol.*, **66**, 227-241.

Craig, J.L. and Jenkins, P.F. 1982. The evolution of complexity in broadcast song of passerines. *J. Theor. Biol.*, **95**, 415-422.

d'Agincourt, L.G. and Falls, J.B. 1983. Variation of repertoire use in the eastern meadowlark, *Sturnella magna*. *Can. J. Zool.*, **61**, 1086-1093.

Derrickson, K.C. 1987. Behavioral correlates of song types of the northern mockingbird (*Mimus polyglottos*). *Ethology*, **74**, 21-32.
Derrickson, K.C. 1988. Variation in repertoire presentation in northern mockingbirds. *Condor*, **90**, 592-606.
Dobson, D.W. and Lemon, R.E. 1975. Re-examination of monotony threshold hypothesis in bird song. *Nature*, **257**, 126-128.
Dooling, R.J. 1989. Perception of complex, species-specific vocalizations by birds and humans. In: *The comparative psychology of audition: perceiving complex sounds.* (Ed. by R.J. Dooling & S.H. Hulse), pp. 423-444. Lawrence Erlbaum Assoc., Hillsdale, N.J.
Falls, J.B. 1985. Song matching in western meadowlarks. *Can. J. Zool.*, **63**, 2520-2524.
Falls, J.B. and L.G. d'Agincourt. 1982. Why do meadowlarks switch song types? *Can. J. Zool.*, **59**, 2380-2385.
Falls, J.B., A.G. Horn and Dickinson, T.E. 1988. How western meadowlarks classify their songs: evidence from song matching. *Anim. Behav.*, **36**, 579-585.
Falls, J.B., Dickinson, T.E and Krebs, J.R. 1990. Contrast between successive songs affects the response of eastern meadowlarks to playback. *Anim. Behav.*, **39**, 717-728.
Falls, J.B., Krebs, J.R. and McGregor, P.K.. 1982. Song matching in the great tit (*Parus major*): the effect of similarity and familiarity. *Anim. Behav.*, **30**, 977-1009.
Gaddis, P.K. 1983. Differential usage of song types by plain, bridled, and tufted titmice. *Ornis Scand.*, **14**, 16-23.
Harnad, S. 1987. Categorical perception and representation. In:*Categorical Perception.* (Ed. by S. Harnad), pp. 535-565. Cambridge Univ. Press, Cambridge, UK.
Hartshorne, C. 1973. *Born to Sing.* Indiana Univ. Press, Bloomington, USA.
Horn, A.G. and Falls, J.B. 1986. Western meadowlarks switch song types when matched by playback. *Anim. Behav.*, **34**, 927-929.
Horn, A.G. and Falls, J.B. 1988a. Responses of western meadowlarks, *Sturnella neglecta*, to song repetition and contrast. *Anim. Behav.*, **36**, 291-293.
Horn, A.G. and Falls, J.B. 1988b. Structure of western meadowlark song repertoires. *Can. J. Zool.*, **66**, 284-288.
Horn, A.G. and Falls, J.B. 1991. Song switching in mate attraction and territory defense by western meadowlarks (*Sturnella neglecta*). *Ethology*, **87**, 262-268.
Horn, A.G., Leonard, M.L., Ratcliffe, L., Shackleton, S. and Weisman, R. *Submitted.* Frequency variation in the songs of black-capped chickadees (*Parus atricapillus*).
Kramer, H.G., Lemon, R.E. and Morris. M.J. 1985. Song switching and agonistic stimulation in the song sparrow (*Melospiza melodia*): five tests. *Anim. Behav.*, **33**, 135-149.
Krebs, J.R. 1976. Habituation and song repertoires in the great tit. *Behav. Ecol. Sociobiol.*, **1**, 215-227.
Krebs, J.R. and Dawkins, R. 1984. Animal signals: mind-reading and manipulation:*Behavioural Ecology: an Evolutionary Approach.* (Ed. by J.R. Krebs & N.B. Davies), pp. 380-402. Sinauer Press, Sunderland, USA.
Krebs, J.R. and Kroodsma, D.E. 1980. Repertoires and geographical variation in bird song. *Adv. Stud. Behav.*, **11**, 143-177.
Krebs, J.R., Ashcroft, R. and van Orsdol, K. 1981. Song matching in the great tit (*Parus major* L.). *Anim. Behav.*, **29**, 918-923.
Kroodsma, D.E. 1977. Correlates of song organization among North American wrens. *Am. Nat.*, **111**, 995-1008.
Kroodsma, D.E. 1982. Song repertoires: problems in their definition and use. In: *Evolution and Ecology of Acoustic Communication in Birds. Vol.II.* (Ed. by D.E. Kroodsma, E.H. Miller & H. Ouellet), pp. 125-146. Academic Press, New York.
Kroodsma, D.E. 1990. Using appropriate experimental designs for intended hypotheses in "song" playbacks, with examples for testing effects of song repertoire sizes. *Anim. Behav.*, **40**, 1138-1150.
Kroodsma, D.E. and Verner, J. 1978. Complex singing behaviors among *Cistothorus* wrens. *Auk*, **98**, 703-716.
MacNally, R.C. and Lemon, R.E. 1985. Repeat and serial singing modes in American restarts (*Setophaga ruticilla*): a test of functional hypotheses. *Z. Tierpsychol.*, **69**, 191-202.
McArthur, P.D. 1986. Similarity of playback songs to self song as a determinant of response strength in song sparrows (*Melospiza melodia*). *Anim. Behav.*, **34**, 199-207.
McGregor, P.K. 1986. Song types in the corn bunting *Emberiza calandra*: matching and discrimination. *J. Orn.*, **127**, 37-42.

McGregor, P.K. and Avery, M.I. 1986. The unsung songs of great tits (*Parus major*): learning neighbours' songs for discrimination. *Behav. Ecol. Sociobiol.*, **18**, 311-316.
McGregor, P.K. and Falls, J.B. 1984. The response of western meadowlarks (*Sturnella neglecta*) to the playback of undegraded and degraded songs. *Can. J. Zool.*, **62**, 2125-2128.
McGregor, P.K. and Krebs, J.R. 1982. Song types in a population of great tits (*Parus major*): their distribution, abundance, and acquisition by individuals. *Behaviour*, **79**, 126-152.
McGregor, P.K., Krebs, J.R. and Ratcliffe, L.M. 1983. The reaction of great tits (*Parus major*) to the playback of degraded and undegraded songs: the effects of familiarity with the stimulus song types. *Auk*, **100**, 898-906.
Medin, D.L. and Barsalou, L.W. 1987. Categorization processes and categorical perception. In: *Categorical Perception.* (Ed. by S. Harnad), pp. 455-490. Cambridge Univ. Press, Cambridge, UK.
Nelson, D.A. and Croner, L.J. 1991. Song categories and their functions in the field sparrow (*Spizella pusilla*). *Auk*, **108**, 42-52.
Payne, R.B. 1979. Song structure, behaviour, and sequence of song types in a population of Village Indigobirds, *Vidua chalybeata*. *Anim. Behav.*, **21**, 762-771.
Ritchison, G. 1988. Song repertoires and the singing behavior of male northern cardinals. *Wilson Bull.*, **100**, 583-603.
Schroeder, D.J. and Wiley, R.H. 1983a. Communication with repertoires of song themes in tufted titmice. *Anim. Behav.*, **31**, 1128-1138.
Schroeder, D.J. and Wiley, R.H. 1983b. Communication with shared song themes in tufted titmice. *Auk* **100**, 414-424.
Searcy, W.A. 1983. Responses to multiple song types in male song sparrows and field sparrows. *Anim. Behav.*, **31**, 948-949.
Searcy, W.A. and Andersson, M. 1986. Sexual selection and the evolution of song. *Ann. Rev. Ecol. Syst.*, **17**, 507-533.
Simpson, B.S. 1984. Tests of habituation to song repertoires by Carolina wrens. *Auk*, **101**, 244-254.
Slater, P.J.B., Ince, S.A. and Colgan, P.W. 1981. Chaffinch song types: their frequencies in the population and distribution between repertoires of different individuals. *Behaviour*, **75**, 207-218.
Smith, W.J. 1970. Song-like displays in the genus *Sayornis*. *Behaviour*, **37**, 64-84.
Smith, W.J. 1977. *The Behavior of Communicating*. Harvard Univ. Press, Cambridge, USA.
Smith, W.J. 1986. Signalling behavior: contributions of different repertoires. In: *Dolphin Cognition and Behavior: a Comparative Approach.* (Ed. by R.J. Shusterman, J.A. Thomas & F.G. Wood), pp.315-330. Lawrence Erlbaum Assoc., Hillsdale, NJ.
Smith, W.J. 1988. Patterned daytime singing of the eastern wood-pewee, *Contopus virens*. *Anim. Behav.*, **36**, 1111-1123.
Smith, W.J., Pawlukiewicz, J. and Smith, S.T. 1978. Kinds of activity correlated with singing patterns of the yellow-throated vireo. *Anim. Behav.*, **26**, 862-864.
Spector, D. *in press*. Wood warbler song systems: a review of paruline singing behaviors. *Curr. Orn.*
Stoddard, P.K., Beecher, M.D. and Willis, M.S. Response of territorial male song sparrows to song types and variations. *Behav. Ecol. Sociobiol.*, **22**, 125-130.
Todt, D. 1975. Short-term inhibition of outputs occurring in the vocal behavior of blackbirds (*Turdus merula*). *J. Comp. Physiol.*, **98**, 289-306.
Trainer, J.M. 1987. Behavioral associations of song types during aggressive interactions among male yellow-rumped caciques (*Cacicus cela*). *Condor*, **89**, 141-168.
Trainer, J.M. 1988. Singing organization during aggressive interactions among male yellow-rumped caciques. *Condor*, **90**, 681-688.
Weary, D.M., Falls, J.B. and McGregor, P.K. 1990. Song matching and the perception of song types in great tits, *Parus major*. *Behav. Ecol.*, **1**, 43-47.
Whitney, C.L. 1991. Avoidance of song matching in the wood thrush: a field experiment. *Wilson Bull.*, **103**, 96-100.
Whitney, C.L. and Miller, J. 1983. Song matching in the wood thrush (*Hylocichla mustelina*): a function of song dissimilarity. *Anim. Behav.*, **31**, 457-461.
Wiley, R.H. and Richards, D.G. 1982. Adaptations for acoustic communication in birds: sound transmission and signal detection. In: *Evolution and Ecology of Acoustic Communication in Birds. Vol.I.* (Ed. by D.E. Kroodsma, E.H. Miller & H. Ouellet), pp. 131-181. Academic Press, New York.
Wolffgramm, J. and Todt, D. 1982. Pattern and time specificity in vocal responses of blackbirds, *Turdus merula* L. *Behaviour*, **65**, 264-286.

BIRD SONG AND OPERANT EXPERIMENTS: A NEW TOOL TO INVESTIGATE SONG PERCEPTION

Daniel M. Weary

Department of Biology
McGill University
1205 Docteur Penfield Avenue
Montreal, PQ, H3A 1B1
CANADA

Introduction

Song playbacks have now been used for many years to investigate how birds respond to songs. Two methods have generally been used. Experimental songs have been played to either territorial males and the aggressive response measured, or to females treated with sex hormones and the sexual response measured (see chapter by Searcy in this volume). With either method, when birds respond differently to various songs we can conclude that they perceive a difference between them. But what can we conclude when subjects show no difference in response? Either that they do not perceive the difference between the songs, or that this difference is not relevant in distinguishing between territorial intruders or sexual partners. Recently, operant techniques have been developed which allow researchers to remove song from these biological contexts. Subjects are trained to associate certain songs with a food reward. The subjects' responses to test songs can then be used to determine their perception of the similarity between test and training songs. I review a recent operant experiment investigating the perception of song by birds and discuss the advantages of this procedure.

The Power of Playback

The playback of recorded sounds to animals has proved to be a powerful technique for those interested in how animals use sounds to communicate (see chapter by Falls in this volume). There are several reasons why workers have found this technique so useful and foremost among these may be its ease of use. Armed with only a tape-recorder, stopwatch and pencil, the aspiring bio-acoustician can set out into the woods and, if skilled and lucky, may conduct an entire experiment within the period of a few days.

Perhaps a more substantial strength of the technique is that it is biologically realistic. Careful field observations might suggest that a certain vocalisation functions, for example, in attracting mates and the playback test of this idea can mimic the natural situation very well. In two recent examples of this type of study, songs were played from certain nesting sites and not from others and the number of females visiting each type of site was monitored (Eriksson and Wallin 1986; Mountjoy and Lemon 1990). This design, like many others used with playback, may create a situation nearly identical to ones faced naturally by the animals.

A third advantage of this technique is that there are few ethical problems associated with it. In the simplest playback experiments there is little disturbance to the animal. As suggested in the previous paragraph, a playback trial can be considered as simply another repetition of something which might frequently occur to the animal. For example, playback of song to a territorial male bird might simulate an occurrence as benign as a bout of song from his neighbour, something which can occur many times a day. More commonly, subjects are captured and individually marked before being released and used in a playback experiment, but even in this case, the stress suffered by the subject seems minimal.

Problems with Song Playback

There are two substantial problems with the playback technique. One is that animals tend to habituate to the situation and the other is that it confounds the issue of how the vocalisation functions with how it is perceived. I consider the issue of habituation more minor and will discuss it first.

When animals face the same or similar situation repeatedly, their response will often tend to diminish. This very basic form of learning is known as habituation and is frequently observed in playback studies where the same subject is exposed to repetitions of the same or even different sounds. The chance of habituation introducing a systematic bias can be reduced in an experiment by randomising the order in which different stimuli are presented, but if the effect of habituation reduces the mean level of response sufficiently, any real differences might become difficult to detect. Thus, habituation limits the number of different stimuli that can be presented to the same subject. This limit might be important for many questions investigated by playback, especially given the large number of test songs sometimes necessary to avoid pseudoreplication (see consensus chapter in this volume).

A more serious problem with playback is that the tests may confuse the issue of how vocalisations function with that of how they are perceived (see also chapters by Horn and by Ratcliffe and Weisman in this volume). When a test song is played to a territorial male and the male responds strongly, we can be quite certain that the male has recognised the song as belonging to his own species. However, what can we interpret when a test evokes only a weak response? One possibility is that the male has not recognised the song as species specific. The other possibility is that the male has recognised the song but does not consider it to be a territorial threat. In other words, we cannot fully distinguish recognition from response.

Interestingly, this confound prevents us from using playback to address fully either perception or function. As an example, I will refer to my own work. Weary et al. (1986) played back manipulated songs to wild territorial male veeries (*Catharus fuscescens*), a common North American thrush. In one song we removed all the changes in frequency with time, so all notes were presented at the mean frequency of the song, but with the same amplitude changes as naturally occur. In another song we removed all the changes in amplitude with time and kept the natural changes in frequency. When we played these manipulated songs to territorial males we found a clear gradation in the territorial response. Subjects responded similarly to a normal song and to the song without variation in amplitude, but showed a much weaker response to the test song in which frequency was held constant. In fact, the response to the amplitude-constant song was statistically indistinguishable from that to the control, while the response to the frequency-constant song was significantly lower.

How should these results be interpreted? Weary et al. (1986) argued that they show that frequency changes over time are important in song recognition by veeries, while the amplitude changes are not. This reasoning might be flawed. First, consider the test song without changes in frequency. This song received a relatively weak territorial response. One reason for the diminished response might be, as Weary et al. suggested, that subjects no longer recognised the song as being one sung by a member of their own species. Another possibility is that they still recognised the song, but did not consider it to be territorially threatening. Results from a playback experiment do not allow us to distinguish between these possibilities.

Next, consider the response to the test song with no variation in amplitude. Subjects seemed to respond to this song in the same way as they respond to normal songs. Following the logic from the preceding paragraph we might conclude that this variation in amplitude does not affect how territorially threatening the song is considered. However, it is also possible that the extent to which we varied amplitude from its natural state was not perceptible to the birds. Again, results from a playback experiment do not allow us to distinguish between these possibilities.

Thus, playback does not seem well suited for testing perception independent of function. The problem of testing function independent of perception is probably less important, as it only arises in cases where there is no difference in response to stimuli (i.e. failures to reject the null hypothesis) and these should be treated cautiously by scientists in any case.

Operant Tests of Song Perception

During the past decade workers have borrowed well known operant techniques from experimental psychology and have applied them to bird song. This operant approach allows us to examine how animals perceive their vocalisations, independent of any functions these vocalisations may have, because song is removed from its natural context. The experimenter and not the animal, decides what the appropriate response to a song should be. This is usually done by training the subject to perform a certain behaviour in order to gain a reward. For example, I have trained birds to break an infra-red beam in order to have access to food. Once the animal has learnt to do this, it is then trained to perform the behaviour only after the presentation of certain stimuli. I trained birds to appreciate that beam-breaks following broadcast of certain "positive" songs were rewarded, but beam-breaks following other "negative" songs were not. In this way, subjects learnt to discriminate between songs. This type of operant procedure allows the researcher to avoid both problems associated with playback. First, there is no possibility that subjects will habituate to the stimuli. They continue to respond until they have become satiated with the reward. Secondly, the biological function of the vocalisation should not affect the operant response. Animals are trained to discriminate vocalisations just as they would any other stimuli (e.g. coloured lights). Thus, in this laboratory context the vocalisation should no longer have any biological relevance and the trained response should not be affected by the biological function of the song.

The first operant study of vocal discrimination was performed by Sinnott (1980), who trained red-winged blackbirds (*Agelaius phoeniceus*) and brown-headed cowbirds (*Molothrus ater*) to discriminate between each others' songs. Since then, a number of researchers have used operant conditioning to study the perception of vocalisations in birds (Brown et al. 1988; Cynx et al. 1990; Dooling et al. 1987a, 1987b; Park et al. 1985; Park and Dooling 1985; Shy et al. 1986; Sinnott 1987; Weary 1989, 1990, 1991; Weary and Krebs *in press*; Weary and Weisman *in press*).

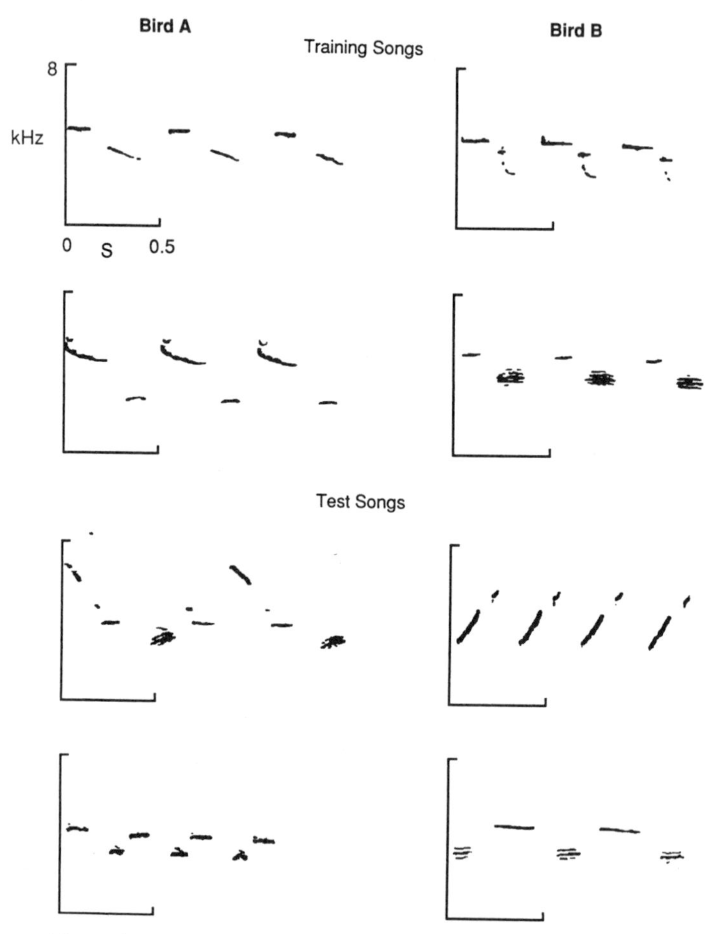

Figure 1. Sonagrams of the four songs from bird A and bird B. The two training songs for each bird are shown above the two test songs. Sonagrams were made on a Kay 6061 B using the narrow band filter.

An Example

The best way to describe this technique is by way of example. For this purpose, I will use a recent experiment on song perception in great tits (*Parus major*) carried out by John Krebs and myself (Weary and Krebs *in press*). As far as we know, it provides the first experimental evidence that birds can recognise the general voice characteristics of an individual's song repertoire. We used an operant procedure to teach male great tits to discriminate between songs from the repertoires of two individuals. Subjects were then tested with unfamiliar songs from the same two birds, which they tended to be able to assign to the correct individual.

We know that birds are able to discriminate between one another on the basis of song (e.g. Krebs 1971; Brooks and Falls 1975; Falls 1982), but do not know how they perform this discrimination. Species with repertoires tend to show poorer performance than those that sing only one song type (Falls 1982), but the fact that discrimination still occurs in these species with repertoires indicates that there are distinctive features upon which discrimination could be based. In this experiment we investigated one mechanism in particular: that all the songs in a repertoire may share a distinctive quality, as in human voice.

Experimental Songs

Great tits have an average repertoire size of three to four song types (Krebs et al. 1978; McGregor and Krebs 1982). We randomly selected two males (Bird A and Bird B), each with a repertoire of four song types (Fig. 1), from a population of wild birds. Two song types from each male were arbitrarily chosen as the training stimuli and the subjects learnt to discriminate between A's and B's repertoires using these songs. As a test for the ability of the subjects to assign novel songs to the correct singer, they were played the other two songs from each repertoire.

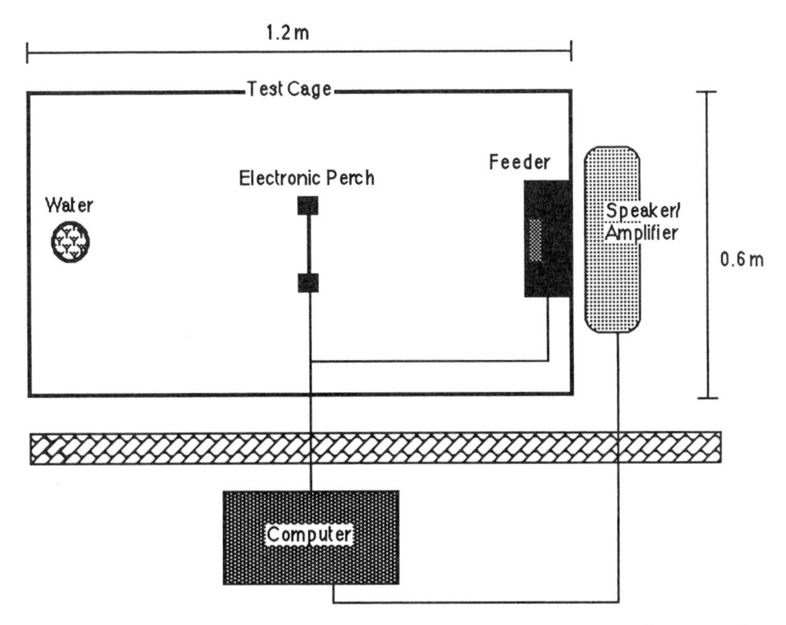

Figure 2. The test cage and some of the apparatus used to run an operant experiment on the perception of songs by great tits.

Experimental Procedure

Birds performed the experiment in the test cage illustrated in Figure 2. When the subject landed on the electronically sensitive perch in the middle of this cage a song was played. The bird could then visit the automated feeder. Visits following one type of training song were rewarded by the feeder opening. Visits following the other type of reference song were punished with a period of darkness in the test cage. A computer was used to automatically collect all data and control the experimental events.

The logic of the procedure is as follows. Certain subjects were rewarded for responses to the songs of one bird, "Bird A" for example, while their responses to the songs of Bird "B" were punished (for other subjects, the training was reversed). Once subjects learnt to discriminate correctly between these positive songs (Bird A's) and negative songs (Bird B's), they were played a series of test songs. For each they had to decide to respond or not to respond and thus classify the novel song as belonging to Bird A or Bird B. Each subject would perform several thousand trials a day, so by the end of about a week, over 80% of its visits to the feeder would be in response to the rewarded songs. At this point test songs were played, but only as probes at a low rate relative to the training songs. Responses to test songs were neither rewarded nor punished.

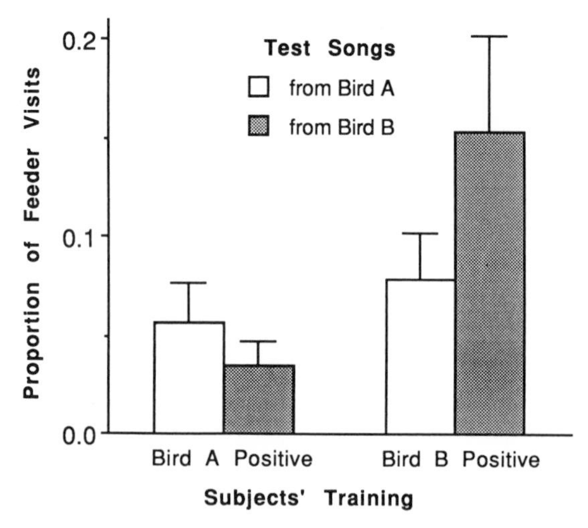

Figure 3. The proportion of visits to the feeder following playback of test songs. The responses by the three subjects trained with songs from Bird A as the positive stimuli (and Bird B as negative) are shown on the left-hand side of the figure. The responses by the two subjects trained with songs from Bird B as the positive stimuli are shown on the right. The proportion of visits to the feeder was calculated by dividing the number of visits to the feeder by the total number of times the test song was presented. The height of the column indicates the mean, while the narrow t-bar illustrates the standard error. Reprinted with permission from Animal Behaviour. Copyright 1991 Academic Press Ltd.

Results

The results from the five subjects are summarised graphically in Figure 3. Subjects were more likely to visit the feeder following playback of test songs which formed part of the positive repertoire ($F_{1,3df} = 18.36$, $P < 0.025$: repeated measure analysis of variance, the nested effect of the actual test song played was not significant $F_{1,3df} = 0.00$, NS). The two columns on the right hand show that subjects trained with B's songs as the positive stimuli, visited the feeder following playback of unfamiliar songs from this bird almost twice as frequently as when test songs from Bird A were played. Similarly, subjects trained to respond to Bird A's songs responded positively to test songs from Bird A more frequently than to those from Bird B.

Although great tits trained to respond to songs from the repertoire of Bird A (left hand side of figure) responded less, on average, than those who received the opposite training (right hand side of figure), this difference was not statistically significant ($F_{1,3df} = 1.48$, NS). While the overall level of response to test songs by subjects trained to respond to Bird A was low, subjects responded more to both types of test songs than to the negative training stimuli (i.e. the training songs from Bird B's repertoire).

Conclusion

After learning to discriminate between two songs from each of two repertoires, the subjects were correctly able to assign totally unfamiliar songs from the same two repertoires. In other words, the subjects must have learned individual voice characteristics and used these to classify the test songs. In principle they could have also learned to perform the discrimination in the training session by learning each song type individually without attending to common properties of each repertoire, but if this had been the case the correct assignment of the novel songs would not have occurred.

The experiment described above is perhaps unusual in that it is difficult to imagine how this hypothesis could be tested using playback on wild birds. Thus, for this kind of question the operant procedure may be particularly well suited. It can be usefully employed for any test of how animals perceive vocalisations and especially tests of how animals assign vocalisations to conceptual categories.

In addition to the specific techniques that I have just described, there are other operant procedures that can be used to examine song perception. Indeed, new procedures and methodologies are continuously being developed. One approach in particular deserves mention. In the "Same-Different" task (see Dooling et al. 1987) animals are presented with two sounds separated by a short time interval. If the two sounds are different and the animal responds (e.g. pecks a key) it is rewarded. Responses when the sounds are the same are punished. By training with a variety of different sounds, the subjects may learn the concept of similarity. Once subjects have been trained in this way, they can then be presented with any set of novel sounds. Differences in responses to pairs of these sounds can be used to measure the animals' perception of their similarities.

Problems with Operant Tests

Limits to Interpretation

Although operant tests clearly circumvent the problems I discussed with the more traditional playback methods, they are by no means a panacea. Perhaps the most serious limitation is that results only tell us what animals can do; they do not reveal for what purposes the animals might use these capacities. Thus, the experiment I described above demonstrates that great tits can recognise individuals using general voice characteristics, but it did not tell us that animals are doing this as a matter of course. All we can conclude is that five great tits could do so, for the songs of two specific individuals. For certain hypotheses, like the one of interest in our experiment, it is adequate to simply determine the animal's capabilities, but for other hypotheses it might be necessary to investigate the generality of this ability. When the hypothesis necessitates a large sample of subjects, the usefulness of operant procedures quickly diminishes.

Costs

The reason why operant techniques tend not to be well-suited for studies involving large samples is that they are relatively costly in both time and capital. Most set-ups require a computer, a specially designed feeder and a cage fitted with infra-red emitters, detectors and a loudspeaker. With a single set-up, the pace of experiments can be slow. It may take several weeks to train fully and to test a single experimentally naive subject. Thus, even when using a number of set-ups, designs requiring many different training conditions, or the training of many different subjects, are impractical.

Animal Welfare

There may also be some ethical responsibility to keep the number of subjects and the period of captivity for each subject to a minimum. Operant tests may be stressful for the animals. Work with wild animals requires that they be trapped and kept in captivity for a prolonged period. In almost all operant experiments the subject must perform a task in order to survive in a comfortable manner. In many procedures, the animal is maintained at a low body weight (e.g. 90% of their free feeding weight), so as to improve performance during the experiment. Although I think that it is possible to design procedures that are humane, these ethical concerns suggest the need for particularly strong justifications for large samples of subjects or long periods of experimentation on each subject.

Biological Relevance

When laboratory tests are used to investigate basic perceptual questions, like frequency difference thresholds (see Dooling 1982), the interpretation of the data poses few problems. The data allow us to draw conclusions about the animals' general perceptual capabilities. The same can not necessarily be said for tests of vocalisations. The problem is as follows. If non-human animals process their vocalisations in a different way to other sounds (see Marler and Peters 1981), as humans are thought to do for language (e.g. Mann and Liberman 1988), then we must be certain that our subjects are hearing the stimuli we present as vocalisations and not just noises. If not, the answers we come up with might not have much relevance to what animals do in nature.

The results from one set of experiments suggest that this concern might be important. Cynx et al. (1990) discovered that zebra finches (*Taeniopygia guttata*) take several thousand trials to learn to discriminate two notes that differ only in the presence of a harmonic. In contrast, subjects learnt the same task much more quickly when the syllables were embedded in an entire song (Cynx *in press*). Although one might think that the latter task would be more difficult (because the relevant information is obscured), the birds seemed to find the task easier. This difference suggests that birds might process sounds they hear as song differently to those that they do not. As these experiments involved laboratory operant procedures, they also suggest that birds can hear songs as song in this context.

My own experiments indicate that the way birds perceive song in operant laboratory tests is similar to their perception in field playback experiments. I performed two experiments, both testing how great tits use the duration of song phrases to categorise song types. One experiment used operant methods (Weary 1989) while the other used playback to wild birds (Weary et al. 1990). The results from both studies led to the same conclusion: that great tits do not rely on the duration of song phrases to categorise song types.

Conclusion

The operant procedures that I have described will probably never replace the more traditional methods of playback. Song playback to wild birds is an elegant and practical procedure, but it is not without its flaws. The principle of these, I think, is that it does not enable us to fully distinguish between the function of the vocalisation and its perception when evaluating an animal's response. When we use an operant procedure to investigate the perception of vocalisations this problem is circumvented. In this context, the vocalisation's function (e.g. mate attraction) is not associated with the animal's response (e.g. approaching a feeder or pecking a key). Operant procedures also avoid other problems associated with field playback, like habituation, but the approach also has weaknesses. Perhaps the most serious of these drawbacks, especially in the context of the issues discussed during this workshop, is that we cannot draw very general conclusions from most operant experiments. The procedure simply may be too costly to use to sample a large population. Thus, operant experiments may be best suited for showing what animals can do, but not necessarily what they routinely do. With this in mind, I argue that operant procedures form a valuable addition to the repertoire of methods employed by students of animal communication.

Acknowledgements

I thank John Krebs for encouraging me to learn the operant techniques and to apply these to birdsong. I was supported by a postdoctoral fellowship from the Natural Sciences and Engineering Council of Canada (NSERC) while writing this chapter. I am grateful to Jeff Cynx, Bob Lemon, Johanne Mongrain and Katherine Weary for their comments.

References

Brooks, R.J. and Falls, J.B. 1975. Individual recognition by song in white-throated sparrows. III. Song features used in individual recognition. *Can. J. Zool.*, **53**, 1749-1761.

Brown, S.D., Dooling, R.J. and O'Grady, K. 1988. Perceptual organization of acoustic stimuli by budgerigars (*Melopsittacus undulatus*): III. Contact calls. *J. Comp. Psychol.*, **102**, 236-247.

Cynx, J. *in press*. Varieties of perceptual experience. *Quant. Anal. Behav.*

Cynx, J., Williams, H. and Nottebohm, F. 1990. Timbre discrimination in zebra finch (*Taeniopygia guttata*) song syllables. *J. Comp. Psychol.*, **104**, 303-308.

Dooling, R.J. 1982. Auditory perception in birds. In: *Evolution and Ecology of Acoustic Communication in Birds. Vol.II.* (Ed. by D.E. Kroodsma, E.H. Miller & H. Ouellet), pp. 95-130. Academic Press, New York.

Dooling, R.J., Brown, D.D., Park, T.J., Okanoya, K. and Soli, S. D. 1987a. Perceptual organization of acoustic stimuli by budgerigars (*Melopsittacus undulatus*): I. Pure tones. *J. Comp. Psychol.*, **101**, 139-149.

Dooling, R.J., Park, T.J., Brown, D.D., Okanoya, K. and Soli, S.D. 1987b. Perceptual organization of acoustic stimuli by budgerigars (*Melopsittacus undulatus*): II. Vocal signals. *J. Comp. Psychol.*, **101**, 367-381.

Eriksson, D. and Wallin, L. 1986. Male bird song attracts females - a field experiment. *Behav. Ecol. Sociobiol.*, **19**, 297-299.

Falls, J.B. 1982. Individual recognition by sound in birds. In: *Evolution and Ecology of Acoustic Communication in Birds. Vol.II.* (Ed. by D.E. Kroodsma, E.H. Miller & H. Ouellet), pp. 237-273. Academic Press, New York.

Krebs, J.R. 1971. Territory and breeding density in the great tit, *Parus major* L. *Ecology*, **52**, 2-22.

Krebs, J.R., Ashcroft, R. and Webber, M. 1978. Song repertoires and territory defence in the great tit. *Nature*, **271**, 539-542.

Mann, V., and Liberman, A. 1988. Some differences between phonetic and auditory modes of perception. *Cognition*, **14**, 211-235.

Marler, P. and Peters, S. 1981. Birdsong and speech: Evidence for special processing. In: *Perspectives on the Study of Speech.* (Ed. by P. Eimas & J. Miller), pp. 75-112. Lawrence Erlbaum Assoc., Hillsdale, New Jersey.

McGregor, P.K. and Krebs, J.R. 1982. Song types in a population of great tits (*Parus major*): their distribution, abundance, and acquisition by individuals. *Behaviour*, **79**, 126-152.

Mountjoy, D.J. and Lemon, R.E. 1991. Song as an attractant for male and female European starlings and the influence of song complexity on their response. *Behav. Ecol. Sociobiol.*, **28**, 97-100.

Park, T.J. and Dooling, R.J. 1985. Perception of species-specific contact calls by budgerigars (*Melopsittacus undulatus*). *J. Comp. Psychol.*, **99**, 391-402.

Park, T.J., Okanoya, K. and Dooling, R.J. 1985. Operant conditioning of small birds for acoustic discrimination. *Ethology*, **3**, 5-9.

Sinnott, J.M. 1980. Species-specific coding in bird song. *J. Acoust. Soc. Am.*, **68**, 494-497.

Sinnott, J.M. 1987. Modes of perceiving and processing the information in birdsong (*Agelaius phoeniceus, Molothrus ater,* and *Homo sapiens*). *J. Comp. Psychol.*, **101**, 355-366.

Shy, E., McGregor, P.K. and Krebs, J.R. 1986. Discrimination of song types by male great tits. *Behav. Proc.*, **13**, 1-12.

Weary, D.M. 1989. Categorical perception of bird song: how do great tits (*Parus major*) perceive temporal variation in their own song? *J. Comp. Psychol.*, **103**, 320-325.

Weary, D.M. 1990. Categorization of song notes in great tits: which acoustic features are used and why? *Anim. Behav.*, **39**, 450-457.

Weary, D.M. 1991. How great tits use song-note and whole-song features to categorize their songs. *Auk*, **108**,187-189.

Weary, D.M, Falls, J.B. and McGregor, P.K. 1990. Song matching and the perception of song types in great tits. *Behav. Ecol.*, **1**, 43-47.

Weary, D.M, and Krebs, J.R. *in press.* Great tits classify songs by individual voice characteristics. *Anim. Behav.*,

Weary, D.M. and Weisman, R.G. 1991. Operant discrimination of frequency and frequency ratio in the black-capped chickadee (*Parus atricapillus*). *J. Comp. Psychol.*, **105**, 253-259.

PITCH PROCESSING STRATEGIES IN BIRDS:

A COMPARISON OF LABORATORY AND FIELD STUDIES

Laurene Ratcliffe [1] and Ron Weisman [2]
Department of Biology [1]
Department of Psychology [2]
Queen's University
Kingston, Ontario K7L 4V1
CANADA

Introduction

Over 50 studies have investigated how birds recognise conspecific song using playback (reviewed by Becker 1982; Weisman and Ratcliffe 1987). Typically, these experiments have compared the responses of territorial males (of mostly north temperate species) to broadcast of natural and altered species' songs. Song playback simulates the intrusion of a rival male and usually elicits aggressive behaviour from the subject. If the alteration of a particular song feature reduces this aggression, compared to the natural song, one may infer the feature is important in song recognition. That is, either the song lacks species-specificity, or is recognised but not considered very threatening (Weary *in press*).

Species recognition studies have benefited from several recent advances in playback design and execution, for example the use of computer-synthesised songs (e.g. Date et al. 1991) and more natural (*interactive*) stimulus presentation (Dabelsteen and Pedersen 1990, chapter by Dabelsteen in this volume). One of the most significant advances involves the use of captive birds, notably in operant conditioning procedures using both sexes (chapter by Weary in this volume) and in sexual preference assays of oestradiol-implanted females (chapter by Searcy in this volume). Experiments like these allow us to compare playback responses in different contexts. There are at least three advantages to such comparisons. First, they allow us to distinguish between a bird's *ability* to discriminate between songs and its *propensity* for responding differentially. In most field studies using song playback, failure to show a differential response is implicitly assumed to demonstrate the lack of ability to discriminate. However, for a variety of reasons, birds may respond equally aggressively to two songs which are nonetheless discriminable in another context. Only well-controlled laboratory experiments can hope to uncover the true extent of such discrimination ability. A second advantage to comparing playback studies in different contexts is that it allows us to investigate sex differences in song recognition. The bird song literature is heavily biased towards male behaviour, partly because field experiments with females are notoriously difficult to carry out. However, there are good

Playback and Studies of Animal Communication
Edited by P.K. McGregor, Plenum Press, New York, 1992

evolutionary reasons to predict that females should be even more discriminating about conspecific song than males (Searcy 1990). Finally, if we want to understand the selective forces which shape song structure, it is important to know how free-living birds rank the various potential cues to species identity contained in conspecific song (Nelson 1989). Some cues which are discriminable in the simple decision-making environment of the laboratory may not figure as prominently in the complex decision-making environment of the field (e.g. see Ratcliffe and Boag 1987).

In this chapter we compare playback studies of pitch processing in black-capped chickadees (*Parus atricapillus*) carried out in two different contexts: in captivity, where individual males were allowed to defend large indoor aviaries (Weisman and Ratcliffe 1989), and in the field, using breeding, colour-marked males. First, we briefly review what is known about pitch processing in songbirds and the reasons why some pitch features should be important for species recognition in general and black-capped chickadees in particular. We then contrast the results of the laboratory and field playback experiments. We suggest possible reasons for the lack of congruence between the two sets of findings and explore their implications for species recognition studies.

Pitch Processing in Songbirds

The songs of many Oscines are characterised by their tonality (Nowicki and Marler 1988) and songbirds are able to perceive quite small differences (of about 1%) in pitch (Dooling 1980). Table I summarises experiments on pitch perception which have been carried out since Becker's (1982) review of the field. Features of song like absolute pitch range and the presence or absence of pitch change frequently emerge as important cues for song recognition (Becker 1982; Nelson 1988; Weary 1990). Recently, researchers have also begun to ask whether songbirds attend to the pitch *relationships* among notes. That is, do birds in any way perceive what we would call the musical pattern of their songs? The significance of this question lies not in establishing whether bird song is music, but whether there might be fundamental similarities in the way different vertebrate brains represent the patterning of acoustic information. As it turns out, theories of human music perception have proven useful in answering this question (see reviews by Weisman and Ratcliffe 1991; Weary *in press*).

As shown in Table I, pitch processing can be described by four concepts. *Absolute pitch* perception, *sensu strictu*, refers to the ability to identify the frequency of a single note without reference to an external standard, sometimes called perfect pitch in humans (Ward and Burns 1982). The term can also refer to the ability to distinguish between notes of different frequency in a broader sense, or to identify a note as belonging to a particular frequency range. *Relative pitch* perception refers to the general ability to perceive frequency relationships between adjacent notes and can be of two types. Listeners may identify *pitch contour*, directional change in frequency upwards or downwards (Deutsch 1982). Listeners may also identify *pitch ratio*, the exact magnitude of the directional change. Pitch ratio is calculated by dividing the frequency of one note by that of the adjacent note (Burns and Ward 1982). Humans are skilled at pitch ratio perception, as shown by the ease of recognising a melody when it is transposed to a new key, but we are much less adept at absolute pitch perception (Miyazaki 1989).

Given the melodic nature of bird song, there is reason to suspect that birds should be capable of relative pitch perception. Curiously, until recently there was little empirical evidence of relative pitch perception in any vertebrate apart from humans. For example, experiments on songbirds using operant procedures with rising and falling sequences of tones have shown that absolute pitch predominates over relative pitch in transfer to novel sequences (Hulse and Cynx 1985, 1986; Page et al. 1989). Over the last few years, we and a number colleagues have investigated relative pitch perception in natural songs, using two lines of approach. First, we have measured the variability in pitch ratios during song production. If birds can produce constant pitch intervals with precision, this implies an ability to perceive such constancy during song development. Selection may have also favoured the use of such constant features in species recognition. Therefore, as a second step, we have used playback to compare birds' responses to songs with normal *versus* altered (increased or decreased) species-specific pitch ratios. A reduced response to songs with altered ratios implies an ability to perceive relative pitch, at least at the level of pitch contour. Three species (white-throated sparrow *Zonotrichia albicollis*; veery *Catharus fuscescens*; and Carolina chickadee *P. carolinesis*) have been investigated in the field and two of these show excellent relative pitch production and perception (Hurly et al. 1990, 1991; Weary et al. 1991) (Table I). Carolina chickadees, on the other hand, do not provide any convincing ability of relative pitch production (Lohr et al. 1991) nor do they attend to relative pitch during playback trials (Lohr *in prep.*). A fourth species, the black-capped chickadee, provided the first evidence of relative pitch production (Weisman et al. 1990) and the first laboratory evidence of relative pitch perception (Weisman and Ratcliffe 1989). However, as described below, field confirmation of these effects has been more difficult to obtain.

Table I. A summary of experimental evidence for three types of pitch perception in songbirds in relation to experimental context. *AP* = absolute pitch, *PC* = pitch contour, and *PR* = pitch ratio.

Species	Context	AP	PC	PR	Reference
starling *Sturnus vulgaris*	Operant	yes	yes	no	Hulse & Cynx 1985, 1986; Page et al. 1989
cowbird *Molothrus ater*	Operant	yes	yes	no	Hulse & Cynx 1985
mockingbird *Mimus polyglottos*	Operant	yes	yes	no	Hulse & Cynx 1985
blackbird *Turdus merula*	Field	-	-	-	Dabelsteen & Pedersen 1985
field sparrow *Spizella pusilla*	Field	yes	-	-	Nelson 1988
white-throated sparrow *Zonotrichia albicollis*	Field	yes	yes	yes	Hurly et al. 1990 Hurly et al. *submitted*
veery *Catharus fuscescens*	Field	yes	yes	yes	Weary et al. 1986, 1991
great tit *Parus major*	Operant+Field	yes	-	-	Weary 1990
black-capped chickadee *Parus atricapillus*	Lab territories Operant	yes yes	yes yes	yes yes	Weisman & Ratcliffe 1989 Weary & Weisman 1991
Carolina chickadee *Parus carolinensis*	Field	yes	no	no	Lohr et al. *in prep.*

Pitch Processing in Chickadees

Structure of the Fee Bee Song

Black-capped chickadees have a complex repertoire of calls (Ficken et al. 1978) but the territorial song is simple and highly invariant, consisting of two whistled notes, *fee bee* (Fig. 1). We have analysed fee bee songs from more than 150 individuals (Weisman et al. 1990). Frequency descends gradually within the first note, from Fee_{start} to Fee_{end} (mean $\pm$ se: 204 $\pm$ 72 Hz), followed by a major pitch change between Fee_{end} and Bee (426 $\pm$ 68 Hz). Here we term the first frequency change the *glissando* and the major pitch change the *frequency interval*. For both the glissando and the interval, the change in frequency over time is better predicted by a ratio function (i.e. Fee_{start} / Fee_{end} and Fee_{end} / Bee, respectively) than a difference function. Chickadees typically sing multiple fee bees at a single absolute frequency with great precision (coefficients of variation within birds, within bouts are less than 2%), before shifting to a new frequency (Ratcliffe and Weisman 1985; Hill and Lein 1987). Whereas the absolute pitch of song can vary substantially within and among birds, both the glissando and the frequency interval vary by less than 2%. Even when individuals from different geographic areas shift their songs over a wide range of frequencies (Ratcliffe and Weisman 1985; Hill and Lein 1987; Horn et al. *submitted*), these relative parameters remain remarkably constant. This invariance prompted Hulse and Cynx (1986) to suggest black-capped chickadees might be capable of relative pitch recognition.

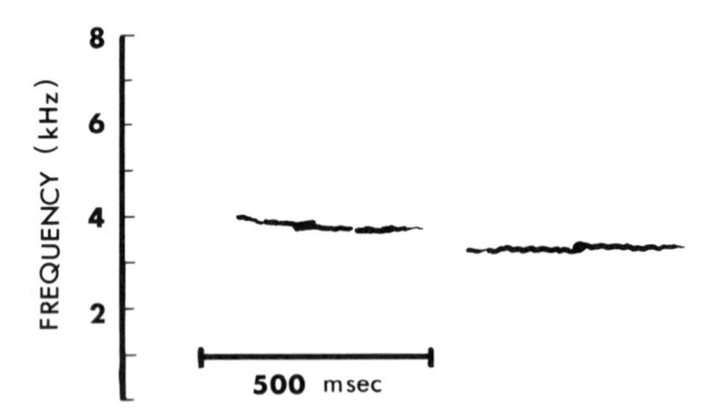

Figure 1. Narrow band (45Hz) sonagram of a typical black-capped chickadee *fee bee* song.

Laboratory Experiments

We carried out two sets of playback trials on captive birds to investigate the responses of black-capped chickadees to songs of different absolute pitch and pitch ratio. (By pitch ratio, we refer to the major pitch change between fee and bee described above.) These experiments are described in Weisman and Ratcliffe (1989). Subjects were wild, adult males which had been captured at least two months before testing. Each male was allowed to "defend" an indoor aviary (dimensions 1.3 x 2.1 x 2.7m) for one week during testing. Previous work (Ratcliffe and Weisman 1986) had shown that after a brief period of social isolation, adult males respond to playback of conspecific song with the same sorts of songs and calls typically heard in the field. The test songs were digitised versions of two natural fee bees recorded from local males three years previously. The trials were videotaped, from which we extracted the number and kind of vocalisations elicited by each playback stimulus.

In the first experiment, we varied the absolute pitch of song while maintaining the original pitch ratio between fee and bee, i.e. the ratio in the original recordings. Variation in absolute pitch influenced what we term *social* calls (*tseets* and *chatters*), but not *gargle* calls or fee bee songs, which are used in more aggressive contexts (Fig. 2a). Birds gave the most calls to fee bees at the pitch of the original recording; songs beginning appreciably higher or lower elicited significantly less response. Thus, in the laboratory, black-capped chickadees detect absolute pitch differences in conspecific song, but this is evident from their social responses rather than aggressive behaviour.

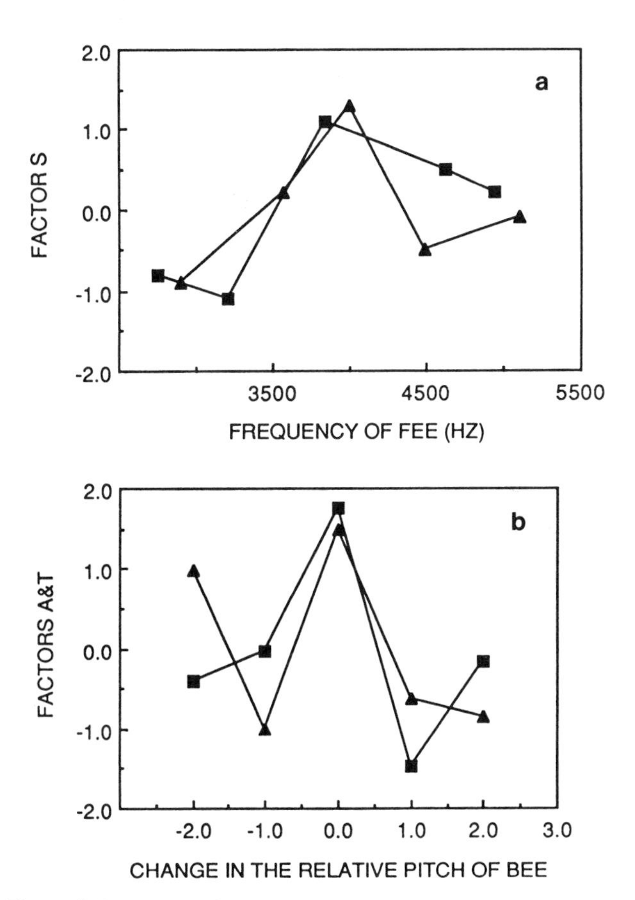

Figure 2. Responses of laboratory black-capped chickadees to altered songs. **a.**) Mean social responses (Factor S=*tseets* and *chatter* calls) as a function of the absolute frequency of fee bee song. **b.**) Mean territorial and agonistic responses (Factor T = *fee bee* songs; Factor A = *gargle* calls) as a function of the relative pitch of *bee*. Adapted from Anim. Behav. **38**, 689-690 by permission of Academic Press.

In the second experiment, we held the absolute pitch of fee constant, while varying the relative pitch of bee to produce songs with normal, increased, or decreased pitch ratios (Fig. 2b). In this case, songs with the original (normal) ratio evoked significantly more *gargles* and fee bees than songs with altered ratios. In contrast to the first experiment, there was no effect of altered ratios on social calls like *tseets* and *chatters*. From these results, we concluded that black-capped chickadees perceive relative pitch change in conspecific song. Since manipulating the pitch ratio influenced aggressive responses like songs and *gargles* in the laboratory, it seemed reasonable to expect songs with abnormal pitch ratios would elicit less aggressive responses from free-living birds.

Field Experiments

Methods We attempted to repeat the experiments on absolute and relative pitch perception described above using breeding males from a colour-marked population at Queen's University Biological Station in southern Ontario, Canada. The population has been followed continuously (including winters) since 1987. Black-capped chickadees are single-brooded, primary cavity nesters. Males often retain their mates and territories between years (Odum 1941; Glase 1973; *pers. obs.*) To reduce variability in response due to stage of breeding, the first experiment (absolute pitch) was carried out over three weeks in April; the second experiment (relative pitch), during three weeks in May. All the subjects in our study were defending territories.

Playback songs were synthesised from digitised versions of natural songs recorded at least one year previously from local birds no longer present in the study area. To ensure that song frequency effects would not be due to the peculiarities of particular tapes or individuals, two songs from different males were used for each type of playback stimulus. We created two versions of each stimulus by joining the fee of one individual with the bee of another, using SoundEdit software on a Macintosh computer. Stimulus songs varied in absolute pitch and pitch ratio according to the experimental design (described below) but otherwise were similar to local songs in amplitude, duration and inter-note interval (Weisman et al. 1990).

Playback tests were performed mostly between 0600 and 1100 hours. A few trials took place between 1500-1800 hours. A speaker/amplifier (Sony SRS-35) was placed at a height of 2m in a tree located in the centre of a subject's territory. All territories had been mapped previously by observation of singing posts and boundary disputes. The observer stood 10-15m distant and controlled the playback recorder (Sony Walkman Professional WMDC6). To minimise possible bias, the observer used coded playback tapes and was naive regarding the experimental predictions.

A trial was begun once the subject was located within its territory. Test songs were played at a rate of 10 songs per minute, which is a typical singing rate of males involved in a territorial encounter (Ratcliffe, *unpublished*). During the 1 minute of playback and for 5 minutes afterwards, the observer dictated observations of all vocalisations and movements into a second cassette recorder. The following response variables were extracted from the tapes for the playback period (*During*), the period following playback (*Post*) and both periods combined (*All*): number of fee bee songs, calls, flights, distance from the speaker and time spent within 10m of the speaker. We also measured the latencies (s) from the start of playback to first vocalisation and first approach within 10m. These responses are similar to those recorded by Hill and Lein (1989) and Ratcliffe (1990).

Each male received one playback trial in each experiment. Order of presentation of the playback songs was randomised in blocks of six (absolute pitch experiment) or eight birds (relative pitch experiment). Males to be tested on a given day were selected randomly and adjacent neighbours were not tested on the same day.

Absolute Pitch Results We presented songs at three absolute pitches, *LOW*, *MEDIUM* and *HIGH*, corresponding to the 10th percentile, mean and 90th percentile of the distribution of fee bee frequencies analysed by Weisman et al. (1990) (Table II). Birds showed no significant differences in any response to these songs, for the periods *Post* and *All* ($F_{2,21df} = 2$, $P > 0.1$). *During* playback, only time spent within 10 m of the speaker varied significantly with pitch ($F_{2,21df} = 6.42$, $P < 0.01$). Pair-wise (Fisher's LSD) comparisons ($P < 0.05$) among the means revealed that subjects spent less time close to the speaker during playback of LOW songs compared to either *Medium* or *High* songs (Table II).

Table II. The territorial response of 24 breeding male black-capped chickadees *During* playback of three absolute pitch modifications of fee bee song: LOW (3339, 3349Hz), *Medium* (3617, 3638Hz) and *High* (3876, 3911Hz). Values are means ± 1se. Sample sizes are 8 for each modification and ANOVA degrees of freedom are 2,21 except for Min. Distance (n = 6,7,8; 2,18df), Time Close (n = 8,7,7; 2,19df) and Latency to Approach (n = 7,8,8; 2,20df). Units are m for distance and s for time measures. F-ratios and P values are for 1-way ANOVA.

Response measure	LOW	MEDIUM	HIGH	F	P
Number of Songs	9.5±1.6	11.6±1.7	8.9±2.8	0.46	<0.5
Number of Calls	0.3±0.3	0.4±0.3	1.9±1.2	1.52	>0.2
Number of Flights	0.9±0.3	1.6±0.6	2.0±1.1	0.64	>0.5
Min. Distance	9.0±1.8	7.3±0.6	5.0±1.3	2.47	>0.1
Time Close	8.8±4.0	32.9±3.6	27.1±7.1	6.42	<0.01
Latency to Vocalise	23.8±5.7	25.0±3.8	26.3±7.3	0.05	>0.9
Latency to Approach	60.0±8.7	47.5±11.8	40.0±5.7	1.19	>0.3

Relative Pitch Results The laboratory trials tested relative pitch perception across a very narrow range of absolute pitches (3800-4000 Hz). In the field trials, we used a broader range of absolute frequencies. Stimulus songs were of four types (Table III). These were categorised by absolute pitch (*HIGH* vs. *MEDIUM*) and pitch ratio (*NORMAL* vs. *ALTERED*). The *High Altered* songs had pitch ratios increased from normal (the species-specific mean of 1.13). The *Medium Altered* songs had decreased ratios. Within the *High* and *Medium* stimulus groups, we took care to match the frequencies of the fee notes of song versions with different pitch ratios.

Table III. Frequencies and pitch ratios of playback songs (see text for details) used in field tests of relative pitch. Time intervals were matched to original songs.

Playback song	Version	Fee_{end} (Hz)	Bee (Hz)	Pitch Ratio
High Normal	1	3921	3431	1.14
	2	3880	3533	1.10
High Altered	1	3921	3043	1.29
	2	3880	3084	1.26
Medium Altered	1	3410	3431	0.99
	2	3390	3431	0.99
Medium Normal	1	3410	3022	1.13
	2	3390	3063	1.11

Analysis of *All* mean responses only revealed a significant effect of playback on number of flights. Two-factor ANOVA showed an effect of pitch ratio ($F_{1,20df} = 4.76$, $P < 0.05$), no significant effect of absolute pitch ($P > 0.3$) and a significant interaction between absolute pitch and ratio ($F_{1,20df} = 13.41$, $P = 0.001$). Further analysis of flight responses broken down into the *During* and *Post* periods showed that the ratio effect occurred during the *Post* period, whereas significant interactions occurred during both periods (Table IV). Pair-wise (Fisher's LSD) comparisons ($P < 0.05$) showed that in the *Post* period, *Medium Normal* songs elicited significantly more flights than *Medium Altered, High Normal,* or *High Altered* songs (Fig. 3a). Thus, any effects of pitch ratio were confined to *Medium* songs.

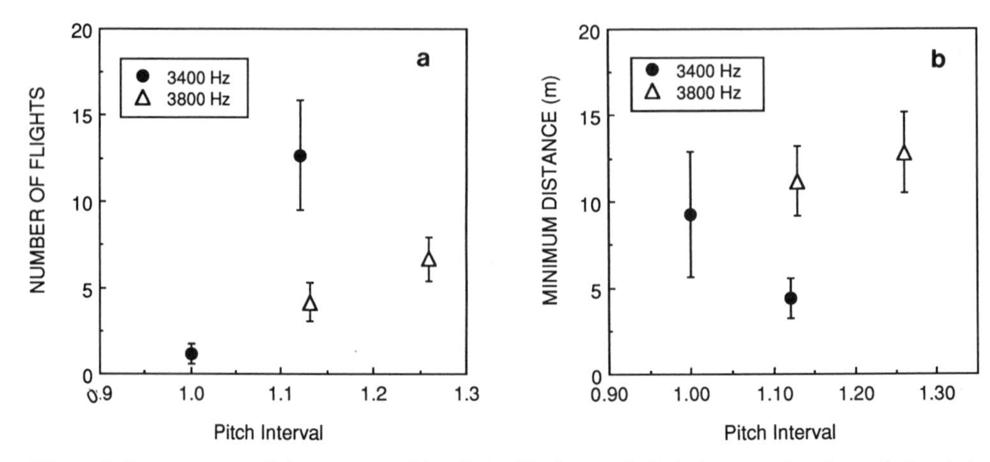

Figure 3. Two measures of the response of free-living black-capped chickadees as a function of the pitch interval of fee bee song. **a.**) Mean number of flights *Post* playback. **b.**) Closest approach *During* playback.

Effects of absolute pitch were detected in two other variables, minimum distance and time spent within 10 m of the speaker, *During* playback ($F_{1,20df} = 4.8$, $P < 0.05$) (Table IV). Both types of *Medium* song tended to elicit more close approaches than either form of *High* song (Fig. 3b), however only *Medium Normal* was consistently more potent than either *High* song (pair-wise comparisons, Fisher's LSD, $P < 0.05$).

Taken together, the field experiments suggest the *High-* and *Low*-pitched fee bee songs are somewhat less potent than *Medium*-pitched songs, in terms of eliciting close approaches, but not markedly so. The pitch ratio experiments show that free-living black-capped chickadees discriminate against songs in which the pitch change has been reduced to 1.0 (i.e. the fee and bee are at the same frequency). This is evidence of one type of relative pitch perception - pitch contour. However, in contrast to the laboratory experiments, there is no evidence of weaker aggressive responses to songs with increased intervals. Moreover, there is a strong interaction between absolute pitch and the pitch ratio.

Table IV. Two-way ANOVA on territorial responses of 24 male black-capped chickadees to playback of fee bee song with modified absolute pitch and pitch ratio. D = *During* playback, P = *Post* playback, A = *All* playback (see text). Degrees of freedom are 1,20 (except for Minimum Distance D = 1,17 and P = 1,18; and Time < 10m D = 1,10). * = $P < 0.05$. ** = $P < 0.01$.

Response	Period	Absolute Pitch	Pitch Ratio	Interaction
Number of Songs	D	0.42	0.26	0.001
	P	1.40	0.35	1.13
Number of Calls	D	0.19	0.75	0.19
	P	2.07	1.23	0.14
Number of Flights	D	2.34	0.72	6.50*
	P	0.59	5.28*	12.77**
Minimum Distance	D	4.89*	1.94	0.46
	P	1.03	2.25	4.98*
Time < 10m	D	5.88*	0.41	0.15
	P	0.78	0.34	2.08
Latency to Vocalize	A	0.10	1.39	0.19
Latency to Approach	A	0.09	0.05	0.81

Discussion

The playback experiments in the laboratory clearly demonstrated that adult male black-capped chickadees attend to differences in the absolute pitch of conspecific songs and discriminate against fee bees with increased and decreased pitch ratios, compared to normal songs. Yet the field trials provided only weak evidence for absolute pitch discrimination and failed to show any effect of increased pitch ratios (though songs with decreased ratios were significantly less potent in eliciting an aggressive response). In a sense, these field data are not much of an advance over an earlier laboratory study (Ratcliffe and Weisman 1986) which showed that fee bee songs lacking a major downward pitch change are not as potent as normal songs.

Perhaps it is not surprising we failed to see large effects of absolute pitch on the strength of response in the field. In the laboratory, only *social* calls like *tseets* and *chatters* varied with absolute frequency. These sorts of responses occur in very low numbers during a typical field trial compared to songs. Moreover, recent song matching studies (Horn et al. *submitted*) show that wild black-capped chickadees do detect differences in absolute pitch. Had we assessed the quality (i.e. the absolute frequency of response songs), rather than the quantity of vocal response elicited by playback, we might have found more subtle effects.

The difference in pitch ratio results between the field and the laboratory are more difficult to reconcile. One explanation might be that we simply ran poor field trials, with a design that was inadequate to detect the birds' known discrimination abilities. For example, we may have failed to control adequately for individual variation in responsiveness. This is essentially a technical problem, of the sort inherent in all field playback studies (see the consensus chapter in this volume). Although we feel this is unlikely, we were able to use a repeated measures design with 10-13 subjects in the laboratory experiments, whereas the field trials used independent subjects, with a maximum of only eight birds per playback stimulus. However, Ratcliffe (1990) was able to document neighbour/stranger discrimination (probably a more difficult perceptual feat) in this same population, using an independent subjects design with 8-10 birds. Moreover, Shackleton et al. (*submitted*) have also found no discrimination between songs with normal and increased pitch ratios, using a much larger sample of 24 wild subjects in a repeated measures design. Interestingly, this latter study implicates the glissando as important in distance estimation of neighbours. Finally, the sample was adequate to detect differential response to normal songs compared to those with a decreased pitch ratio of 1.0.

A second possibility concerns the interaction of absolute and relative pitch cues. Recently Weary and Weisman (1991) have shown that in operant experiments, black-capped chickadees' responses vary with both the absolute frequency of fee *and* bee and the pitch ratio between them. Also, there is an interaction between these effects, such that responses are strongest to songs with intermediate frequencies of both notes and the correct ratio. This is what the field trials showed. In the laboratory work of Weisman and Ratcliffe (1989), the songs with increased ratios, which elicited weaker responses, also had low bee frequencies, for which we did not control. Thus the apparent discrepancy between the two studies may be the result of a small difference in the choice of songs presented.

Finally, it is important to consider that in the field, black-capped chickadees may not rank cues about exact pitch ratio very highly in making decisions to respond to territorial intruders, even though they can discriminate such differences. Other cues, like tonality, the pitch range and some sort of downwards pitch change may be more salient. If this is truly the case, the challenge remains to explain the selective pressures maintaining pitch ratio stereotypy. We plan to investigate this question next by looking at the effects of song structure on mate choice by females.

Implications for Future Species Recognition Studies

We believe these studies have a number of implications for studies of species recognition by song in birds. There has been a common tendency in playback work to choose song features for experimentation based on their absolute properties as measured from sonagrams. Clearly we should be more interested in choosing features as they are perceived by the birds themselves, admittedly a more difficult task. Our work also points to the importance of investigating relational cues among song features.

The choice of playback context, e.g. operant procedures, field tests with territorial males, etc. influences what sorts of questions can be answered and should be given due consideration when planning studies. Our laboratory trials, for example, elicited only a small portion of the kinds of responses typically seen in the field. Ideally, researchers should attempt to cross-check their conclusions using different contexts. Of course, not all species are equally amenable to testing under different conditions. Black-capped chickadees are quite difficult to test in the field, because they have large territories and sing only sporadically after dawn. Until we perfected a technique for *priming* subjects (Shackleton et al. *submitted*), success rates for playback trials varied from 30-60%.

Finally, it is important to remember that failure to show differential response, especially in the field, does not necessarily imply lack of discrimination ability. Researchers would do well to be cautious about drawing such conclusions unless they are convinced they have explored birds' true cognitive powers.

Acknowledgements

We thank Ingrid Johnsrude for preparing the playback songs, Lesley James for field assistance and Queen's University Biology Station for logistical support. Financial support was provided by an operating grant from the Natural Sciences and Engineering Research Council of Canada to LR.

References

Becker, P.H. 1982. The coding of species-specific characteristics in bird sounds. In: *Acoustic Communication in Birds*. (Ed. by D.E. Kroodsma, E.H. Miller & H. Ouellet), pp. 213-252. Academic Press; New York.

Burns, E.M. and Ward, W.D. 1982. Intervals, scales, and tuning. In: *The Psychology of Music*. (Ed. by D. Deutsch), pp. 241-265. Academic Press; New York.

Dabelsteen, T. and Pedersen, S.B. 1985. Correspondence between messages in the full song of the blackbird *Turdus merula* and meanings to territorial males, as inferred from responses to computerized modifications of natural song. *Z. Tierpsychol.*, **69**, 149-165.

Dabelsteen, T. and Pedersen, S.B. 1990. Song and information about aggressive responses of blackbirds, *Turdus merula*: evidence from interactive playback experiments with territory owners. *Anim. Behav.*, **40**, 1158-1168.

Date, E.M., Lemon, R.E., Weary, D.M. and Richter, A.K. 1991. Species identity by birdsong: discrete or additive information? *Anim. Behav.*, **41**, 111- 120.

Deutsch, D. 1982. The processing of pitch combinations. In: *The Psychology of Music*. (Ed. by D. Deutsch), pp. 271-312. Academic Press; New York.

Dooling, R.J. 1980. Behavior and psychophysics of hearing in birds. In: *Comparative Studies of Hearing in Vertebrates*. (Ed. by A.N. Popper and R.R. Fay), pp. 261-288. Springer-Verlag; Berlin.

Ficken, M.S., Ficken, R.W. and Witkin, S.R. 1978. The vocal repertoire of the black-capped chickadee. *Auk*, **95**, 34-48.

Glase, J.C. 1973. Ecology of social organization in the Black-capped Chickadee. *Living Bird*, **12**, 235-267.
Hill, B.G. and Lein, M.R. 1987. Function of frequency-shifted songs of Black-capped Chickadees. *Condor*, **89**, 914-915.
Hill, B.G. and Lein, M.R. 1989. Natural and simulated encounters between sympatric Black-capped Chickadees and Mountain Chickadees. *Auk*, **106**, 645-652.
Horn, A.G., Leonard, M.L., Ratcliffe, L.M., Shackleton, S. A. and Weisman, R.G. *submitted*. Frequency variation in the songs of Black-capped Chickadees (*Parus atricapillus*).
Hulse, S.H. and Cynx, J. 1985. Relative pitch perception is constrained by absolute pitch in songbirds (*Mimus*, *Molothrus*, and *Sturnus*). *J. Comp. Psychol.*, **99**, 176-196.
Hulse, S.H. and Cynx, J. 1986. Interval and contour in serial pitch perception by a songbird *Sturnus vulgaris*. *J. Comp. Psychol.*, **100**, 215-228.
Hurly, T.A., Ratcliffe, L., Weary, D. and Weisman, R. *submitted*. White-throated sparrows can perceive pitch change using the frequency ratio independent of the frequency difference.
Hurly, T.A., Ratcliffe, L. and Weisman, R. 1990. Relative pitch recognition in white-throated sparrows, *Zonotrichia albicollis*. *Anim. Behav.*, **40**, 176- 181.
Hurly, T.A., Weisman, R.G., Ratcliffe, L. and Johnsrude, I.S. 1991. Absolute and relative pitch production in the song of the white-throated sparrow (*Zonotrichia albicollis*). *Bioacoustics*, **3**, 81-91.
Lohr, B., Nowicki, S. and Weisman, R. 1991. Pitch production in Carolina chickadee songs. *Condor*, **93**, 197-199.
Miyazaki, M. 1989. Absolute pitch identification: effects of timbre and pitch region. *Music Perception*, **7**, 1-14.
Nelson, D.A. 1988. Feature weighting in species song recognition by the field sparrow, *Spizella pusilla*. *Behaviour*, **106**, 158-182.
Nelson, D.A. 1989. The importance of invariant and distinctive features in species recognition of bird song. *Condor*, **91**, 120-130.
Nowicki, S. and Marler, P. 1988. How do birds sing? *Music Perception*, **5**, 391- 426.
Odum, E.P. 1941. Annual cycle of the Black-capped Chickadee. *Auk*, **58**, 314- 333.
Page, S.C., Hulse, S.H. and Cynx, J. 1989. Relative pitch perception in the European starling (*Sturnus vulgaris*); further evidence for an elusive phenomenon. *J. Exp. Psych.: Anim. Behav. Proc.*, **15**, 137-146.
Ratcliffe, L.M. 1990. Neighbour/stranger discrimination of whistled songs in black-capped chickadees. *Proc. XX Int. Orn. Congress*, (Abstract #423).
Ratcliffe, L.M. and Boag, P.T. 1987. Effects of colour bands on male competition and sexual attractiveness in zebra finches (*Poephila guttata*). *Can. J. Zool.*, **65**, 333-338.
Ratcliffe, L.M. and Weisman, R.G. 1985. Frequency shift in the song of the Black-capped Chickadee. *Condor*, **87**, 555-556.
Ratcliffe, L.M. and Weisman, R.G. 1986. Song sequence discrimination in the Black-capped Chickadee (*Parus atricapillus*). *J. Comp. Psychol.*, **100**, 361-367.
Searcy, W.A. 1990. Species recognition of song by female red-winged blackbirds. *Anim. Behav.*, **40**, 1119-1127.
Shackleton, S.A., Ratcliffe, L. and Weary, D.M. *submitted*. Relative frequency parameters and song recognition in black-capped chickadees.
Ward, W.D. and Burns, E.M. 1982. Absolute pitch. In: *The Psychology of Music*. (Ed. by D. Deutsch), pp. 431-449. Academic Press; New York.
Weary, D.M. 1990. Categorization of song notes by great tits, which acoustic features are used and why? *Anim. Behav.*, **39**, 450-457.
Weary, D.M. *in press*. How birds use relative and absolute pitch to recognize their songs. In: *Quantitative Analyses of Behavior: Vol.14*.
Weary, D.M., Lemon, R.E. and Date, E.M. 1986. Acoustic features used in song discrimination by the Veery. *Ethology*, **72**, 199-203.
Weary, D.M. and Weisman, R.G. 1991. Operant discrimination of frequency and frequency ratio in the black-capped chickadee (*Parus atricapillus*). *J. Comp. Psychol.*, *in press*.
Weary, D.M., Weisman, R.G., Lemon, R.E., Chin, T. and Mongrain, J. 1991. Use of the relative frequency of notes by veeries in song recognition and production. *Auk*, *in press*.
Weisman, R.G. and Ratcliffe, L.M. 1987. How birds identify species information in song: a pattern recognition approach. *Learning and Motivation* **18**, 80- 98.
Weisman, R.G. and Ratcliffe, L.M. 1989. Absolute and relative pitch processing in black-capped chickadees, *Parus atricapillus*. *Anim. Behav.*, **38**, 685- 692.

Weisman, R.G. and Ratcliffe, L.M. 1991. The perception of pitch constancy in bird songs. In: *Cognitive Aspects of Stimulus Control*. (Ed. by W.K. Honig & J.G. Fetterman), pp. 243-261. Lawrence Erlbaum Assoc.; Hillsdale, New Jersey.

Weisman, R.G., Ratcliffe, L.M., Johnsrude, I.S. and Hurly, T.A. 1990. Absolute and relative pitch production in the song of the Black-capped Chickadee. *Condor*, **92**, 118-124.

Participants

Clive Catchpole
Department of Biology, Royal Holloway & Bedford New College, University of London, Egham, Surrey, TW20 0EX, UK
Torben Dabelsteen
Institute of Population Biology, University of Copenhagen, Universitetsparken 15, DK-2100 Copenhagen Ø, Denmark
Bruce Falls
Department of Zoology, University of Toronto, Toronto, Ontario M5S 1A1, Canada
Leonida Fusani
Dipartimento de Biologia Animale e Genetica, Universita di Firenze, via Romana 17, Firenze 50125, Italy
Carl Gerhardt
Division of Biological Sciences, University of Missouri, Columbia, Missouri 65211, USA
Francis Gilbert
Behaviour and Ecology Research Group, Department of Life Science, University of Nottingham, University Park, Nottingham, NG7 2RD, UK
Andy Horn
Division of Life Sciences, Scarborough Campus, University of Toronto, Scarborough, Ontario M1C 1A4, Canada
Georg Klump
Institut für Zoologie, Technische Universität München, Lichtenbergstrasse 4, W-8046 Garching, FRG
Don Kroodsma
Department of Zoology, University of Massachusetts, Amherst, MA 01003, USA
Marcel Lambrechts,
CNRS-CEFE, Route de Mende, BP-5051, 30433 Montpellier Cedex, France
Karen McComb
Large Animal Research Group, Department of Zoology, University of Cambridge, Downing Street, Cambridge CB2 3EJ, UK
Peter McGregor
Behaviour and Ecology Research Group, Department of Life Science, University of Nottingham, University Park, Nottingham, NG7 2RD, UK
Doug Nelson
Animal Communication Laboratory, Department of Zoology, University of California, Davis, California 95616, USA
Irene Pepperberg
Departments of Ecology & Evolutionary Biology and Psychology, University of Arizona Tucson, Arizona AZ 85721, USA

Laurene Ratcliffe
Department of Biology, Queen's University, Kingston, Ontario K7L 4V1, Canada
Bill Searcy
Department of Biological Sciences, University of Pittsburgh, Pittsburgh, Pennsylvania 15260, USA
Danny Weary
Department of Biology, McGill University, 1205 Docteur Penfield Avenue, Montreal, PQ, H3A 1B1, Canada

Index

Absolute pitch, 212
Action-based learning, 121

Bayesian
  approach, 59, 69-70
    v. classical, 70-74
  further reading, 75
  statistics, 71-74
Broad hypotheses, 7, 16

Call (*see also* Vocal interaction)
  activation, 166 (*see also* triggering)
  -cycle period, 166
  entrainment, 161
  suppression, 166
  triggering, 167-168
  timing, *see* Vocal interaction, timing
Choruses
  background noise in, 67
  call timing (*see also* Vocal interaction)
    alternation, 153-154
    mechanisms, 164-166
    overlapping, 154, 171
    predation risk, 154
  female choice, 169-172
  insects, 19
  interference in, 153
  masking, 168
  simulation, 168-169
Communication
  definitions, 11-12
  form, meaning and context, 47
  as social interaction, 47
Constraints
  in choruses, 153
Constraints (continued)
  on validity, 6
Copulation solicitation display, *see* Oestradiol
Cross-correlation
  and call timing, 161

Degradation
  and distance estimation, 26, 84
  and execution errors, 8
Digital sound emitter, 102-103
Discrimination (*see also* Operant conditioning)
  individual, *see* individual discrimination
  neighbour-stranger, *see* Neighbour -stranger discrimination
  parent-offspring
    mammals, 21-22
  sex bias, 211-212
  species
    birds, 23, 221
    fish, 19-20
    frogs, 20
    insects, 17-18
    operant conditioning, 203
Distortion (*see also* Degradation)
  by equipment, 63-67
  by field environment, 66-67
  harmonic, 63-66

Execution errors in playback, 7-8
Experimental philosophy, 16, 35-36, 43-44
External validity, 6, 43-44
  relationship to internal validity, 6

Female choice (*see also* Oestradiol)
amplitude, 170-171
and call overlapping, 169-172
in birds, 23
in fish, 19-20
in frogs, 20
methods of assessment, 186
generality, 176, 186
heart rate monitoring, 184-186
interpretability, 176-177, 186
parental behaviour, 185-186
phonotaxis, 178-179, 186
practicality, 176, 186
sensitivity, 176, 186
solicitation display, *see* Oestradiol
territorial playback, 177-178, 186
Function
song, *see* Song function
song repertoires, 197-198
Further topics for playback investigation, 27

Generality
geographical effects, 69
limits, 69
of playback results, 6-7
Group calling (*see also* Chorus)
assessing odds of success, 114-116
effects of
familiarity, 116-117
individual identity, 116-117
number of intruders, 113-114
number of defenders, 114
other vertebrates, 117
social mammals, 112-113

Habituation
and input level, 50
and playback, 202
as a playback method, 15
Hypotheses, *see* Broad *and* Narrow

Individual discrimination (*see also* Operant conditioning)
in mammals, 21-22
Infra-sound, 22
Input
contextual applicability, 50-51
Input (continued)
contextual applicability (continued)
and interaction, 51-52
optimal
amount, 50
level, 49
quality, 48-49
referentiality, 50-51
and interaction, 51-52
Interaction intensity, 52, (*see also* Vocal interaction)
Interactive playback, 15-16, 52-53, 97, 154, 164
equipment, 102-103, 154 (*see also* Digital sound emitter)
limitations, 107-108
prospects, 107
Internal
standard, 195
validity, 43-44
relationship to external validity, 6

Jamming avoidance, 153

Level of response to playback, 16
Limitations
interactive playback, 107-108
operant conditioning, 207-208
interpretation, 208-209
pitch perception experiments, 220

Matched countersinging, 24-25
singing performance, 142-143
and song type perception, 193
Mate choice (*see also* Oestradiol)
costs in choruses, 68
Measures of response to playback
approaches, 79
embarrassment of riches problem, 79
in general, 12, 16, 79
many-measures, 82-83
suggested guidelines, 83
multi-variate composite measure, 84-93
suggested guidelines, 92
principal components analysis, *see* Principal components analysis
single-measure, 80-81

Measures of response to playback (continued)
single-measure (continued)
suggested guidelines, 81

Narrow hypotheses, 7, 16,
Near-matching, 193
Neighbour-stranger discrimination
birds, 24, 51, 53
fish, 19
Neighbour-stranger effects, 126-127

Oestradiol
implant technique, 40-42, 179-184
interpretation, 148, 184, 186
and singing performance, 146-147
Operant conditioning
costs, 208
equipment, 205
and individual characteristics, 203-207
interpretation, 207-209
paradigm, 205-206
and playback, 15, 201
and perception, 203-210
reinforcement, 203
species discrimination, 203
subject welfare, 208

Perception of song types
and function, 202-203
methods
matched countersinging, 193
operant systems, 203-210 (*see also* Operant conditioning)
response strength, 192-193
sequences, 193-194
Perceptual
significance of self song, 131
tasks, 194
Phase angle, 157-158
Philosophy
and experimental rationale, 16, 35-36, 43-44
of statistical analysis, 69-72 (*see also* Bayesian)
Pitch
contour, 212
ratio, 212
Pitch perception
limits to interpretation, 220
and music perception, 212
in songbirds, 212-213
Playback
controversy, 1 (*see also* Pseudo-replication debate)
debate, *see* Pseudoreplication debate
definition, 1, 12
design, *see* Playback methods
early history, 13-14
errors of execution, 7-8
and ethics, 202, 208
execution of, 7-8
fidelity, 63-67
field *v.* laboratory, 13, 15, 66-67, 197, 201- 203, 219-221
generalising from experiments, 6-7, 16
habituation, *see* Habituation
interactive, *see* Interactive playback
measures of response, *see* Measures of response to playback
methods
with birds, 15-16
habituation, 15
interactive, *see* Interactive playback
operant conditioning, 15
sequential presentation, 15
social group contests, 113-116
two-speaker design, 15, 67-
natural input, 48, *see* Input
rationale, 47
response to, *see* Response to playback
role in integrated approach, 35-36, 44
and serendipity, 27
uses, 13-14 (*see also* Uses of playback)
welfare, 208
Principal components
analysis
component extraction, 87
data suitability, 86, 94-95
interpretation, 87-89
rationale, 84
stages, 85
suggested guidelines, 92-93

Principal components (continued)
scores
computing, 89
Pseudoreplication
and ANOVA designs, 3-4
in bioacoustics, 2
debate, 2, 35
definition, 2
in ecology, 2
and external validity, 6
remedies for
question clarity, 4-5
number of exemplars, 4-5
and sample size, 2-5
a specific example, 3-4
and synthetic calls, 5-6

Quality and singing performance, *see* Singing performance

Receiver, 11-12
Refractory period, 166
Relative pitch, 212
Repertoire (*see also* Song types)
composition, 135
and individual distinctiveness, 204-207
size
comparative study, 38-39
and oestradiol implants, 40-42, 181-182
syllables, 38
Resource holding potential, 111
Response to playback
individual differences, 53-54, 140
level, *see* Level of response to playback
measures of, 12 (*see also* Measures of response to playback)
socially appropriate, 51

S/N ratio, *see* Signal to noise ratio
Sensitisation, 50
Signal, 11-12
sender, 11-12
to noise ratio
and execution errors, 8
Singing performance
effect of playback, 138-140
individual differences, 136-137
Singing performance (continued)
and female choice, 145-148
as information exchange, 142 (*see also* Matched counter-singing)
measures, 137
perception of differences
in the field, 141
operant procedures, 141-142
and quality, 135-136
Social modelling theory, 48
Song function, 25, 175
song repertoires, 197-198
Song learning
plastic song, 127-129
and recognition, 131
selective attrition, 124
neighbour effects, 125-127
Song ontogeny, 25-26
Song sharing
acquisition, 121
similarity constraints, 130-131
Song type
context effects, 197
definition, 192
different meanings, 196-197
and experimental design, 197
function, *see* Function *and* Song function
perception, *see* Perception of song types
in relation to function, 202-203
switching, 194
Sound pressure level
and execution errors, 8
SPL, *see* Sound pressure level
Standard, *see* Internal standard
Synthetic calls
generation, 60, 215-216
level of parameter change, 60
and pseudoreplication, 5-6

Temporal gap stimuli, 163-164
Timing, *see* Vocal interaction, timing
Two-speaker playback design, 15, 17-18, 67-69, 178

Uses of playback
in amphibia, 20-21
in birds, 23-26

Uses of playback (continued)
  in birds (continued)
    calls, 26
    female response, 23
    male response, 24-25
    ontogeny, 25-26
    sound perception, 26
  in insects, 17-19
  in fish, 19-20
  in mammals, 21-22

Varying signal properties, 60-63
  inadvertent generation
    frequency effects, 61
    time effects 61-63

Vocal interaction
  between groups, 111-112
  between male frogs, 154-155
  blackbird song, 97-99
  matched singing, 105-107
  quantification, 155-156
  timing
    mechanisms, 164-166
    refractory period, 166
    relationship to phase angle, 158
    speed of sound constraints, 99-101, 162-163
    spontaneous patterns, 157-158

Welfare of operant subjects, 208